AF247602

ORDINARY DIFFERENTIAL EQUATIONS

HOLDEN-DAY SERIES IN MATHEMATICS

Earl A. Coddington and Andrew M. Gleason, Editors

ORDINARY DIFFERENTIAL EQUATIONS

Otto Plaat

University of San Francisco

Holden-Day, Inc.

San Francisco, Cambridge, London, Amsterdam

Copyright © 1971 by Holden-Day, Inc.
500 Sansome Street, San Francisco, California 94111

All rights reserved.

No part of this book may be reproduced in any form, by mimeograph or
any other means, without permission in writing from the publisher.

Library of Congress Catalog Card Number: 70-156869
ISBN: 0-8162-6844-4

Printed in the United States of America

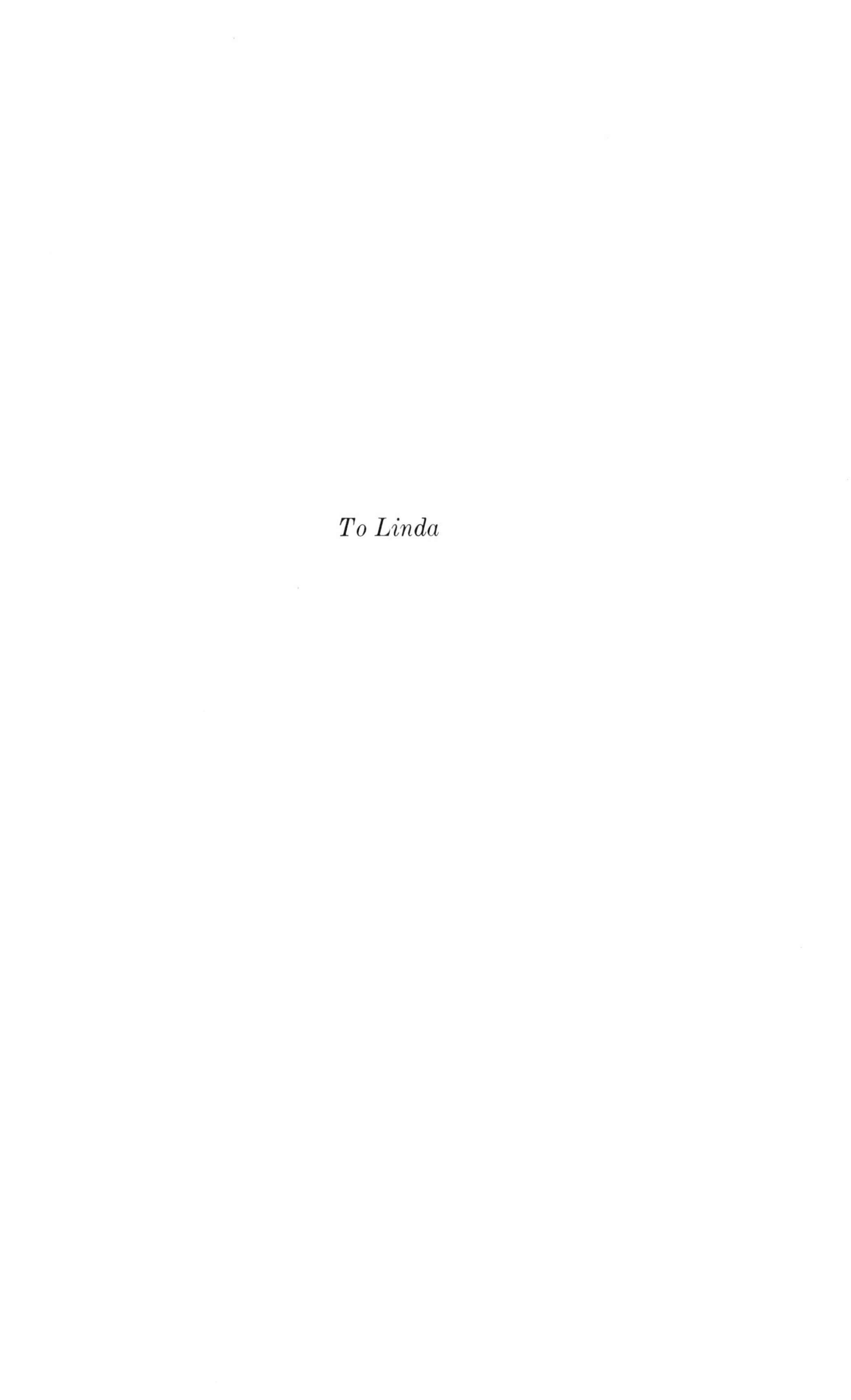

To Linda

PREFACE

This book is designed to introduce students to the central ideas and methods of the theory of ordinary differential equations, both linear and non-linear. It departs in certain respects from more traditional introductions to the subject, which tend to emphasize its formal and analytic details at the expense of its basic geometric character. It seems to me of the utmost importance that the student should acquire a firm grasp of the geometric nature of differential equations and the problems associated with them, for without it he will see no coherence in the diversity of special methods in which the subject abounds, nor will he acquire the geometric habit of thought which is the most permanently valuable tool of all who work with differential equations. This book represents what may be called the qualitative point of view of differential equations, a point of view which not only dominates contemporary research, but is ideally suited to make the subject accessible and meaningful to the beginner. To enable the student to think about differential equations, rather than merely to manipulate them, has been my most basic objective.

In the selection of subject matter I have departed from custom primarily by including a thorough introduction to plane autonomous systems (Chapter 6). I consider this topic the most important in the book. It is one of the starting points of the modern theory of differential equations, it is the most beautiful and perfectly developed part of that theory, and its intuitive appeal gives it a unique value for initiating students into the ideas of non-linear differential equations in general. The inclusion of this topic also allowed me to do justice to exact systems and their integrals, a subject which is deprived of its real meaning and its applications when, as is customary, it is dealt with in the context of first-order equations.

Except for one exercise, there is no mention of power series in the book. Power series are extremely specialized numerical tools, and their discussion in an introductory textbook seemed to me both mathematically and pedagogically out of place.

The reader should be familiar with elementary calculus. The small amount of matrix theory needed for the discussion of systems of differential equations is presented in a separate chapter. The last chapter, which is devoted to the proofs of existence and uniqueness theorems, presupposes a little more mathematical maturity, and I have occasionally omitted it when teaching classes comprised largely of engineers.

The book can be used for a variety of courses, the chapters being largely independent. To cover the whole book would require two semesters for most students. A first course of one semester's duration would probably emphasize the first three chapters. A course for students who have been introduced to differential equations in the calculus sequence could begin with a review of Chapter 1 and then proceed to Chapter 4, or, if the students have the requisite matrix theory, to Chapter 5. I have indicated at the beginning of several chapters both the prerequisites for the chapter and the portions of the chapter needed later on. By a suitable selection from the earlier chapters it is possible to design a one-semester course emphasizing any of the later chapters. Chapter 5 is not, in a strictly technical sense, prerequisite to Chapter 6, but pedagogical considerations suggest that Sections 1, 2, 3, and 5 of Chapter 5 should precede Chapter 6. Chapter 7 is independent of the others and may be taken up at any time.

I wish to thank Professors Henry A. Antosiewicz and Earl A. Coddington for their reading of the manuscript and their advice and criticism. I am grateful to Mr. Edward F. Riley, Vice President of Holden-Day, for his friendly assistance in the production of the book. My thanks also go to Mrs. Eleanor Thornhill for her excellent typing of the manuscript.

Otto Plaat

TABLE OF CONTENTS

1

INTRODUCTION

This chapter is designed to give the reader a preliminary survey of the various types of ordinary differential equations and systems, and to introduce him to some basic ideas and terminology. The classification presented in Section 2 will enable him to place the material of subsequent chapters in the context of the subject as a whole, and he will want to refer to it repeatedly. Section 3 contains a precise definition of a solution of a differential equation; this section also contains the only formal proof of the chapter. Section 6 contains statements of various basic theorems. The proofs of these theorems are given in Chapter 7.

1. Physical Examples

The theory of differential equations, from its inception at the time of Newton and Leibniz to the present day, has been strongly influenced by its manifold contacts with science and technology. From these contacts it has drawn ceaseless inspiration, transmuting physical problems into mathematical problems and sometimes into entire mathematical theories. In order to show the physical sources of the idea of a differential equation, we consider a few simple examples of physical models and the equations which describe them. The significance of these examples lies not so much in the technique by which the equations are derived as in the plain fact that the equations do describe the models. The reader who is uncomfortable with physical applications may prefer to omit this section on first reading and return to it in connection with Section 5.

EXAMPLE 1.   A vessel contains 10 liters of salt solution, diluted by an inflow of 0.1 liter per second (l/sec) of pure water. The volume of solution in the vessel is kept constant by an outflow valve through which the solution runs off at the rate of 0.1 l/sec. The solution is kept perfectly mixed, so that its concentration (mass of salt per unit of volume) is uniform throughout the vessel. Denote the mass of salt (in grams) in the vessel by x, and the time (in seconds) by t. The rate of change of x with respect to t, dx/dt, is equal to the rate at which salt enters the vessel, which is zero, minus the rate at which it leaves. Because the outflow rate is 0.1 l/sec, and the concentration of salt in the vessel—and hence in the outflow—is $x/10$ grams per liter (g/l), the salt leaves the vessel at the rate of $(x/10)(0.1)$ g/sec. Thus

$$\frac{dx}{dt} = -0.01x. \tag{1.1}$$

Equation (1.1) is an ordinary differential equation. It involves a differentiated variable, x, and a variable of differentiation, t.

We alter the example slightly by considering the case that the inflow does not consist of pure water, but of a salt solution whose concentration is 10 g/l. The rate at which salt enters the vessel is now $(10 \text{ g/l})(0.1 \text{ l/sec}) = 1$ g/sec. Equating dx/dt to rate of entry minus rate of exit yields

$$\frac{dx}{dt} = 1 - 0.01x. \tag{1.2}$$

Let us see how (1.2) reflects the most obvious features of the model it describes. We note that if $x = 100$, then $dx/dt = 0$. The significance of this is clear. If $x = 100$, then the concentration in the vessel is $100/10 = 10$ g/l, which is the concentration in the inflow, so that the inflow does not alter the mass of salt in the vessel. If $x > 100$, then $dx/dt < 0$, and if $x < 100$, then $dx/dt > 0$, reflecting the fact that the inflow in the first case reduces, and in the second case, increases, the mass of salt in the vessel.

EXAMPLE 2.   Two vessels (Fig. 1) with respective volumes of 4 and 2 liters are filled with salt solutions. The solutions flow from one vessel to the other at the rate of 0.1 l/sec via connecting tubes of negligible length and volume. Within each vessel, the concentration is uniform. Let x and y denote the masses of salt in the larger and smaller vessel, respectively. Salt enters the larger vessel at the rate at which it leaves the smaller one, which is $(y/2)(0.1)$, and it enters the smaller vessel at the rate at which it leaves the larger one, namely $(x/4)(0.1)$. Equating rate of change of mass to rate of entry minus rate of exit for each of the two vessels, we obtain the equations

$$\frac{dx}{dt} = -\frac{1}{40}\,x + \frac{1}{20}\,y$$
$$\frac{dy}{dt} = \frac{1}{40}\,x - \frac{1}{20}\,y. \tag{1.3}$$

These equations form a system of two simultaneous ordinary differential equations. Concerning this model, we note that there are infinitely many values of x and of y for which both dx/dt and dy/dt vanish, namely those which satisfy $x = 2y$. These are, of course, the values which make the concentrations in the two vessels equal. The fact that $\frac{dx}{dt} = -\frac{dy}{dt}$ also has a simple physical meaning.

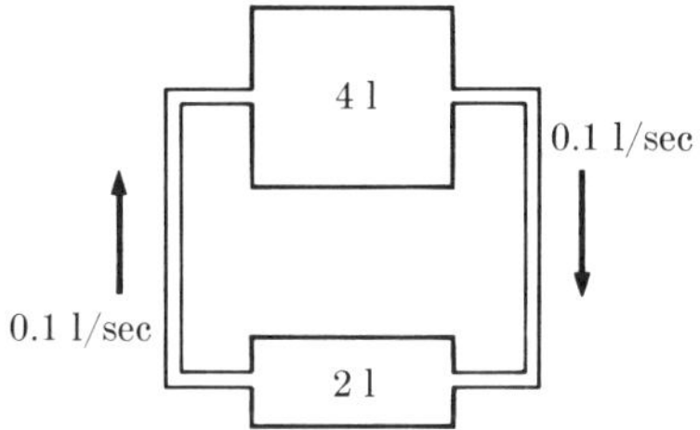
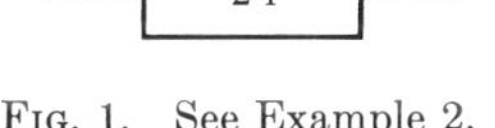

FIG. 1.   See Example 2.

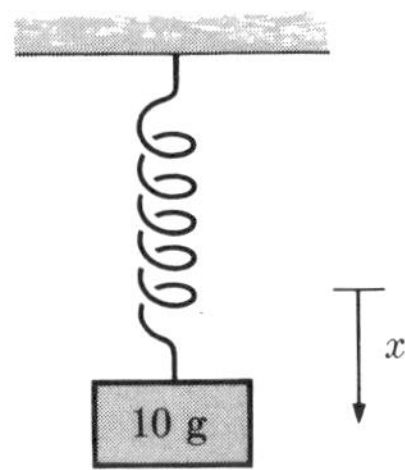

FIG. 2.   See Example 3.

EXAMPLE 3.   We turn now to a mechanical example. A mass of 10 grams is suspended from a spring whose upper end is attached to a fixed support (Fig. 2). We assume that the mass is constrained to move in a vertical line passing through the point of support. The forces acting on the mass are its weight (the force of gravity), which is $mg = (10)(981)$ dynes,* and the tension in the spring. We assume that the spring obeys Hooke's law, according to which the tension is proportional to the distance the spring is stretched. Let x be the distance of the mass below the position it occupies when the spring is relaxed, i.e., neither stretched nor compressed. The tension in the spring is then equal to kx, being negative (compression!) when x is negative. The constant of proportionality k is the elastic constant of the spring; we take it to be 1000. The total force on the mass is thus weight —

---

* The letter g conventionally denotes the acceleration of a freely falling body at the surface of the earth. If distance is measured in centimeters and time in seconds, then $g = 981$ cm/sec². One dyne is the force which gives an acceleration of 1 cm/sec² to a mass of 1 gram.

tension $= mg - kx = 9810 - 1000x$ dynes. Because $m = 10$, Newton's second law, $F = m\, d^2x/dt^2$, yields

$$\frac{d^2x}{dt^2} = 981 - 100x \tag{1.4}$$

as the differential equation which governs the motion of the mass. This equation is distinguished from the preceding ones by the occurrence of a second derivative. It is possible to avoid the second derivative at the cost of replacing Eq. (1.4) by two equations. We simply introduce the variable v for the velocity, so that $dx/dt = v$. Now (1.4) may be replaced by the system of two simultaneous differential equations

$$\frac{dx}{dt} = v$$
$$\frac{dv}{dt} = 981 - 100x, \tag{1.5}$$

in which only derivatives of the first order occur. This trick can always be used to eliminate higher-order derivatives—sometimes a desirable objective.

In the preceding examples, the time variable occurred only under the differentiation sign. To put it another way, not the time itself, but only its differential, played a role in the equations. We now consider an example in which the time does play a role. We suppose that the point of support of the spring is made to execute vertical sinusoidal oscillations about its normal position with an amplitude of 5 cm and a frequency of 1/sec. Its displacement above the normal position is, therefore, $5 \sin (2\pi t + \phi)$ at time t, where ϕ is a constant whose value depends on our choice of the time origin. Because we are free to measure time from any instant we choose, we make a choice for which $\phi = 0$. (Note that no choice of any kind was necessary in the preceding examples.) We measure the position of the mass with respect to the same fixed point in space as before. At time t, the spring is stretched a distance $x + 5 \sin 2\pi t$, so that its tension is $1000(x + 5 \sin 2\pi t)$. The motion of the mass is thus governed by the differential equation

$$\frac{d^2x}{dt^2} = 981 - 100x - 500 \sin 2\pi t, \tag{1.6}$$

in which the time occurs "explicitly."

2. Classification of Differential Equations and Systems

The equations, and systems of equations, derived above have a common feature: they contain derivatives with respect to *one* variable. This is what makes them *ordinary* differential equations. An equation which contains derivatives with respect to two or more variables is called a *partial* differential equation. The derivatives are then written with the ∂ symbol. An example is $\partial z/\partial t = \partial^2 z/\partial x^2$. In this book, we consider only ordinary differential equations. The variable of differentiation will always be t, and may be thought of as denoting time.

An equation of the form

$$\frac{dx}{dt} = f(t,x) \tag{2.1}$$

is called a differential equation of the *first order*. The qualification "first order" refers to the fact that only the first derivative occurs in the equation. It is not excluded that the function f is constant with respect to t or x or both. Indeed, $dx/dt = 0$ is of the form (2.1), the function f being zero for all values of t and x. Equations (1.1) and (1.2) are first-order equations. It is possible, although we shall not do so, to define a first-order equation more broadly as an equation of the form

$$F\left(t,x, \frac{dx}{dt}\right) = 0.$$

When this is done, an equation like (2.1) is called a "normal" or "explicit" differential equation. Examples of the more general type of equation are

$$\sin\left(\frac{dx}{dt}\right) - x = 0$$

$$x\frac{dx}{dt} + x = 0$$

$$\frac{dx}{dt} + x = 0.$$

Such an equation is not usually equivalent to a normal equation, i.e., it does not determine dx/dt uniquely as a function of t and x. Of the above equations, only the last is equivalent to a normal equation. We shall deal only with normal differential equations.

An equation of the form

$$\frac{d^2x}{dt^2} = f\left(t,x, \frac{dx}{dt}\right) \tag{2.2}$$

is called a differential equation of the *second order*. By a differential
equation *per se*, we mean any equation of the form

$$\frac{d^n x}{dt^n} = f\left(t, x, \frac{dx}{dt}, \ldots, \frac{d^{(n-1)}x}{dt^{(n-1)}}\right), \tag{2.3}$$

the positive integer n (the order of the highest derivative) being called
the *order* of the equation. Equations (1.4) and (1.6) are second-order
equations.

A *system of first-order differential equations* is a system of simultaneous
equations of the form

$$\frac{dx_1}{dt} = f_1(t, x_1, x_2, \ldots, x_n)$$

$$\frac{dx_2}{dt} = f_2(t, x_1, x_2, \ldots, x_n) \tag{2.4}$$

$$\cdots \cdots \cdots \cdots \cdots \cdots \cdots$$

$$\frac{dx_n}{dt} = f_n(t, x_1, x_2, \ldots, x_n).$$

Note that the number of equations is equal to the number of variables,
not counting t. Equations (1.3) and (1.5) are examples with $n = 2$.
Any first-order equation is, of course, an example with $n = 1$.

Systems of second-order equations occur frequently in the applications. Such a system is of the form

$$\frac{d^2 x_1}{dt^2} = f_1\left(t, x_1, x_2, \ldots, x_n, \frac{dx_1}{dt}, \frac{dx_2}{dt}, \ldots, \frac{dx_n}{dt}\right)$$

$$\frac{d^2 x_2}{dt^2} = f_2\left(t, x_1, x_2, \ldots, x_n, \frac{dx_1}{dt}, \frac{dx_2}{dt}, \ldots, \frac{dx_n}{dt}\right) \tag{2.5}$$

$$\cdots \cdots \cdots \cdots \cdots \cdots \cdots \cdots \cdots \cdots \cdots \cdots$$

$$\frac{d^2 x_n}{dt^2} = f_n\left(t, x_1, x_2, \ldots, x_n, \frac{dx_1}{dt}, \frac{dx_2}{dt}, \ldots, \frac{dx_n}{dt}\right).$$

We attain complete generality with a system of n simultaneous equations in n variables, in which the equations may have any order
whatever, and not necessarily the same order. We require, of course,
that the system be normal, i.e., that the derivative of highest order
of each variable occur in only one equation, and on the left side of
that equation. This requirement may also be formulated by saying
that the system expresses the highest derivative of each variable as
a function of the time, the variables, and the lower derivatives of the
variables.

We define the *order of a system* to be the sum of the orders of the
equations of the system. A unifying simplification arises from the fact

that an arbitrary system of the nth order can be written as a system of n first-order equations. We already illustrated the procedure in converting Eq. (1.4) into the system (1.5). An arbitrary equation of order n (2.3) is transformed to a system of first-order equations by the introduction of the variables $x_2 = dx/dt$, $x_3 = d^2x/dt^2$, . . . , $x_n = d^{(n-1)}x/dt^{(n-1)}$. Writing x_1 for x, the equation thus becomes

$$\frac{dx_1}{dt} = x_2$$

$$\frac{dx_2}{dt} = x_3 \tag{2.6}$$

$$\cdot \ \ \cdot \ \ \cdot \ \ \cdot \ \ \cdot \ \ \cdot \ \ \cdot \ \ \cdot \ \ \cdot \ \ \cdot \ \ \cdot \ \ \cdot$$

$$\frac{dx_n}{dt} = f(t,x_1,x_2, \ . \ . \ . \ ,x_n),$$

which is a system of n first-order equations. The same method of introducing new variables applies to an arbitrary system of order n, as will be apparent from the example of two simultaneous second-order equations:

$$\frac{d^2x}{dt^2} = f\left(t,x,y,\frac{dx}{dt},\frac{dy}{dt}\right)$$

$$\frac{d^2y}{dt^2} = g\left(t,x,y,\frac{dx}{dt},\frac{dy}{dt}\right).$$

Let $x_1 = x$, $x_2 = y$, $x_3 = dx/dt$, $x_4 = dy/dt$. We then obtain a system of four first-order equations:

$$\frac{dx_1}{dt} = x_3$$

$$\frac{dx_2}{dt} = x_4$$

$$\frac{dx_3}{dt} = f(t,x_1,x_2,x_3,x_4)$$

$$\frac{dx_4}{dt} = g(t,x_1,x_2,x_3,x_4).$$

This method merely amounts to writing a system without the explicit use of higher-order derivatives.

It is thus apparent that *all differential equations and systems can be put into the form* (2.4). When there is no danger of confusion, we shall use the terms "equation" and "system" interchangeably. By regarding x_1, x_2, . . . , x_n as the components of a vector, the system (2.4) may indeed be thought of as a single *vector* differential equation.

The system (2.4) is *autonomous* if the functions $f_1, f_2, \ldots, f_n$ are independent of t, so that the system has the form

$$\frac{dx_i}{dt} = f_i(x_1, x_2, \ldots, x_n), \qquad i = 1, \ldots, n. \tag{2.7}$$

An arbitrary system of the nth order is autonomous if the equivalent system of first-order equations is autonomous. Thus a single nth-order equation is autonomous if it has the form

$$\frac{d^n x}{dt^n} = f\left(x, \frac{dx}{dt}, \ldots, \frac{d^{(n-1)}x}{dt^{(n-1)}}\right).$$

A non-autonomous system is said to be *time-dependent* or to depend "explicitly" on the time, i.e., on the variable of differentiation. Equations (1.1) to (1.5) are autonomous; Eq. (1.6) is non-autonomous. We remarked in connection with (1.6) that the occurrence of t was connected with the necessity of choosing a time origin. Autonomous systems are characterized by their invariance with respect to all translations $t' = t + c$ of the time axis.

Perhaps the most important distinction is that between linear and non-linear systems. We shall say that a function $f(t, x_1, x_2, \ldots, x_n)$ is *linear* in $x_1, \ldots, x_n$ if $f(t, x_1, x_2, \ldots, x_n) = a_1(t)x_1 + a_2(t)x_2 + \cdots + a_n(t)x_n + b(t)$. It is not excluded that the functions $a_i(t)$ or $b(t)$ are constant. If $b(t) = 0$, we say that f is a *homogeneous linear* function of the x_i. (It should be pointed out that only homogeneous linear functions are called "linear" in linear algebra.)

The system (2.4) is *linear* if all the functions f_i are linear in the x's. A linear system of first-order equations thus has the form

$$\frac{dx_i}{dt} = \sum_{j=1}^{n} a_{ij}(t)x_j + b_i(t), \qquad i = 1, \ldots, n. \tag{2.8}$$

If all the functions b_i are zero, the system is *homogeneous*, and otherwise *non-homogeneous*. An arbitrary system is linear (homogeneous) if the equivalent system of first-order equations is linear (homogeneous). Equations (1.1) to (1.6) are all linear; Eqs. (1.1) and (1.3) are, in addition, homogeneous. A *non-linear* system is simply one which is not linear. An example of a non-linear equation is $dx/dt = x^2$. Note that the equation $dx/dt = t^2$ is linear. If in Example 3 the mass is constrained to move in a vertical line which does not pass through the point of support of the spring, the differential equation that governs its motion is non-linear, owing to the fact that the stretch of the spring is a non-linear function of the displacement of the mass.

EXERCISES

Find the order of each of the following equations or systems. State whether it is autonomous (A) or non-autonomous (NA), linear (L) or non-linear (NL). If it is linear, state whether it is homogeneous (H) or non-homogeneous (NH).

1. $\dfrac{dx}{dt} = e^t.$

2. $\dfrac{d^3x}{dt^3} = x + y, \quad \dfrac{dy}{dt} = x - \dfrac{d^2x}{dt^2}.$

3. $\dfrac{dx}{dt} = e^t y, \quad \dfrac{dy}{dt} = x + y.$

4. $\dfrac{d^2x}{dt^2} = x\,\dfrac{dx}{dt}.$

5. $\dfrac{dx}{dt} = e^x.$

6. $\dfrac{dx}{dt} = y, \quad \dfrac{dy}{dt} = xy + \sin t.$

7. $\dfrac{d^2x}{dt^2} = (\sin t)x + \dfrac{dy}{dt}, \quad \dfrac{d^2y}{dt^2} = x + (\cos t)y + 1.$

8. $\dfrac{dx}{dt} = \sin xy, \quad \dfrac{dy}{dt} = \cos (x + y).$

3. Solutions

Second in importance only to the idea of a differential equation itself is the idea of a solution of such an equation. In order to make this idea precise, we introduce the notion of an interval.

Definition. An *interval* is a set I of real numbers with the properties:

(1) I contains at least two numbers
(2) If t_1 and t_2 belong to I, and if $t_1 < t < t_2$, then t belongs to I

We denote the set of all real numbers by R. Note that R is an interval. For every other interval I, there exists a number a, or two numbers a and b, $a < b$, such that I consists of one of the following sets of numbers t:

$$
\begin{array}{llll}
\text{(i)} & t < a & \text{(v)} & a < t < b \\
\text{(ii)} & t \leq a & \text{(vi)} & a \leq t < b \\
\text{(iii)} & t > a & \text{(vii)} & a < t \leq b \\
\text{(iv)} & t \geq a & \text{(viii)} & a \leq t \leq b.
\end{array}
$$

An interval may, alternatively, be defined as a set of real numbers of one of these nine types. Every interval except R has at least one endpoint. A *closed* interval is one whose endpoints, if any, belong to it. The closed intervals are therefore R and the intervals of types (ii), (iv) and (viii). An *open* interval is one whose endpoints, if any, do not belong to it. R is both open and closed. The other open intervals are those of types (i), (iii) and (v). An interval I is *bounded* if it has two endpoints, or equivalently, if there is a number M such that $|t_1 - t_2| \leqq M$ for all numbers t_1 and t_2 in I. Clearly the bounded intervals are those of types (v) to (viii); we may take $M = b - a$.

Definition.   A *solution* of the equation

$$\frac{dx}{dt} = f(t,x)$$

is a function u defined on an interval I, such that

$$u'(t) = f(t,u(t))$$

for all numbers t in I. More generally, u is a solution of

$$\frac{d^n x}{dt^n} = f\left(t, x, \frac{dx}{dt}, \cdots, \frac{d^{(n-1)}x}{dt^{(n-1)}}\right)$$

if u is defined on an interval I, and

$$u^{(n)}(t) = f(t, u(t), u'(t), \ldots, u^{(n-1)}(t))$$

for all t in I.

In short, a solution of a differential equation is a function u defined on an interval such that the substitution of $u(t)$ for x reduces the equation to an identity for all t in the interval. For example, the function u defined for t in R by $u(t) = 100 - 100e^{-0.01t}$ is a solution of (1.2), as is also the constant function $u(t) \equiv 100$.* The function u defined on the interval $t < 1$ by $u(t) = 1/(1 - t)$ is a solution of $dx/dt = x^2$; another solution is that defined on the interval $t > 1$ by the same formula. The function defined for all $t \neq 1$ by this formula is *not* a solution, although it satisfies the differential equation, because its domain of definition is not an interval.

It is clear that a solution of a first-order equation is necessarily a differentiable function, and therefore continuous, and that a solution of an nth-order equation is n times differentiable.

---

* If c is a number, we write "$u(t) \equiv c$" as an abbreviation for "$u(t) = c$ for all t."

Definition.   A *solution* of a system of first-order equations

$$\frac{dx_i}{dt} = f_i(t,x_1,x_2, \ldots ,x_n), \qquad i = 1, \ldots , n,$$

is an n-tuple of functions $u_1, u_2, \ldots , u_n$ defined on a common interval I such that

$$u_i'(t) = f_i(t,u_1(t),u_2(t), \ldots ,u_n(t)), \qquad i = 1, \ldots , n,$$

for all t in I. Similarly, a solution of an arbitrary system of n equations in the variables $x_1, x_2, \ldots , x_n$ is an n-tuple of functions $u_1, u_2, \ldots , u_n$ defined on a common interval I such that the substitution of $u_i(t)$ for x_i, $i = 1, \ldots , n$, reduces the equations to identities for all t in I.

We now consider the connection between the solutions of an nth-order equation

$$\frac{d^n x}{dt^n} = f\left(t,x, \frac{dx}{dt}, \ldots , \frac{d^{(n-1)}x}{dt^{(n-1)}}\right) \tag{3.1}$$

and the solutions of the equivalent system of n first-order equations

$$\frac{dx_i}{dt} = x_{i+1}, \qquad i = 1, \ldots , n - 1$$
$$\frac{dx_n}{dt} = f(t,x_1,x_2, \ldots ,x_n), \tag{3.2}$$

where, as usual, we have written x_1 in place of x.

Theorem 1.1. u *is a solution of* (3.1) *if and only if the n-tuple of functions* $u_1 = u, u_2 = u', \ldots , u_n = u^{(n-1)}$ *is a solution of* (3.2).

Proof.   Let u be a solution of (3.1). Because u is n times differentiable, we may define $u_1 = u, u_2 = u', \ldots , u_n = u^{(n-1)}$, and each of these functions is again differentiable. In fact, $u_1' = u_2, u_2' = u_3, \ldots , u_n' = u^{(n)}$. These relationships together with the hypothesis imply

$$u_i'(t) = u_{i+1}(t), \qquad i = 1, \ldots , n - 1$$
$$u_n'(t) = f(t,u_1(t),u_2(t), \ldots , u_n(t)),$$

for all t in the interval of definition of u.

Conversely, let $u_1, u_2, \ldots, u_n$ be a solution of (3.2) and let $u = u_1$. Then u is differentiable and $u' = u_2$. But u_2 is also differentiable, and therefore u is twice differentiable and $u'' = u_3$. Continuing, we see that u is n times differentiable and that $u^{(n)} = u_n'$. Therefore,

$$u^{(n)}(t) = f(t, u_1(t), u_2(t), \ldots, u_n(t)) = f(t, u(t), u'(t), \ldots, u^{(n-1)}(t))$$

for all t in the interval of definition of $u_1, u_2, \ldots, u_n$. $\square$

This simple theorem tells us that if a function is a solution of (3.1), it together with its first $n - 1$ derivatives forms a solution of (3.2), and conversely every solution of (3.2) consists of a solution of (3.1) together with its first $n - 1$ derivatives. This theorem may be extended in an obvious way to an arbitrary system and the equivalent system of first-order equations.

To illustrate the theorem, we note that (1.4) has the solution $x = 9.81 + \sin 10t$, and this function paired with $v = 10 \cos 10t$ is a solution of (1.5).

EXERCISES

1.   Verify that the differential equation has the given function or functions as solutions.

(a) $\quad \dfrac{dx}{dt} = \sin^2 2x + \cos^2 x; \qquad u(t) \equiv \dfrac{\pi}{2}, \quad v(t) \equiv \dfrac{3\pi}{2}$

(b) $\quad \dfrac{dx}{dt} = e^{\sin x} + \cos x; \qquad u(t) \equiv \pi$

(c) $\quad \dfrac{d^2x}{dt^2} = x; \qquad u(t) = e^t, \quad v(t) \equiv 0$

(d) $\quad \dfrac{d^2x}{dt^2} = -x; \qquad u(t) = \sin t + \cos t, \quad v(t) \equiv 0$

(e) $\quad \dfrac{dx}{dt} = (x - t)^{2/3}; \qquad u(t) = 1 + t$

(f) $\quad \dfrac{d^2x}{dt^2} = -x - 2\dfrac{dx}{dt}; \qquad u(t) = te^{-t}$

(g) $\quad \dfrac{dx}{dt} = 2tx^2; \qquad u(t) = \dfrac{1}{1 - t^2} \ \text{ for } -1 < t < 1, \quad v(t) \equiv 0$

(h) $\quad \dfrac{d^3x}{dt^3} = 2\dfrac{d^2x}{dt^2} - \dfrac{dx}{dt}; \qquad u(t) = (1 + t)e^t, \quad v(t) \equiv 10$

2. Verify that the differential system has the given function pair as a solution.

(a) $\dfrac{dx_1}{dt} = 2x_2, \quad \dfrac{dx_2}{dt} = -2x_1; \quad u_1(t) = \sin 2t, \quad u_2(t) = \cos 2t$

(b) $\dfrac{dx_1}{dt} = x_1 x_2 - 2x_1, \quad \dfrac{dx_2}{dt} = x_2 - x_1 x_2; \quad u_1(t) \equiv 1, \quad u_2(t) \equiv 2$

(c) $\dfrac{dx_1}{dt} = x_2, \quad \dfrac{dx_2}{dt} = x_1; \quad u_1(t) = e^{-t}, \quad u_2(t) = -e^{-t}$

3. Show that very solution of $dx/dt = x^2 + 1$ is an increasing function.

4. Show that no solution of $d^2x/dt^2 = e^t + dx/dt$ has a relative maximum.

5. We say that u is a *constant* function if $u(t) \equiv c$ for some number c. Find all constant solutions, if any, of

(a) $\dfrac{dx}{dt} = t(x - 1)$ (b) $\dfrac{d^2x}{dt^2} = x\dfrac{dx}{dt}$

(c) $\dfrac{dx}{dt} = x^2 - 1$ (d) $\dfrac{dx}{dt} = e^x$

(e) $\dfrac{dx}{dt} = x^2 + x - 2$

4.   Equations with Parameters

We propose, for purposes of illustration, to generalize the formulation of Example 1. Instead of specifying the volume, the flow rate, and the concentration of the inflow, let us merely denote them by v, r, and c, respectively, with $v \neq 0$. Then the mass of salt in the vessel is governed by the differential equation

$$\frac{dx}{dt} = rc - \frac{r}{v}x. \tag{4.1}$$

If $v = 10$, $r = 0.1$, and $c = 0$, we again obtain Eq. (1.1), and if $v = 10$, $r = 0.1$, and $c = 10$, we obtain Eq. (1.2). Strictly speaking, therefore, (4.1) is not *one* differential equation, but a family of infinitely many such equations, any one of which is obtained by replacing v, r, and c by numbers. The variables v, r, and c are called *parameters* of the family of differential equations (4.1). A parameter is a variable appearing in a system of differential equations which never occurs under the differentiation sign: it is not differentiated, and no variable is differentiated with respect to it. (Of course, letters like π, e, and so forth, when used in their conventional sense of denoting specific numbers, are not variables, and therefore not parameters.) Instead of speaking of a *family* of differential equations or systems depending on one or more parameters, we simply speak of a differential equation or system *with parameters*. The general form of a system of first-order equations with parameters μ_1, μ_2, . . . , μ_k is

$$\frac{dx_i}{dt} = f_i(t, x_1, x_2, \ldots , x_n, \mu_1, \mu_2, \ldots , \mu_k), \qquad i = 1, \ldots , n.$$

EXERCISES

1.   Let the inflow rate in Example 1 be r_1, the outflow rate r_2, and the salt concentration in the inflow c. Find the system of differential equations which governs x and v, the mass of salt in the vessel, and the volume of solution in the vessel, in terms of the parameters r_1, r_2, and c.

2.   Two vessels with respective volumes of v_1 and v_2 liters are filled with salt solutions. Pure water runs into the first vessel at the rate of r liters per second. The solution passes out of the first vessel and into the second at the rate r, and the solution is drained from the second vessel at the same rate, so that the volume of solution in each vessel remains constant. Find the differential equations for x_1 and x_2, the masses of salt in the two vessels.

3.   A cylindrical oil drum floats with its axis vertical on the surface of a calm lake. The mass of the drum is m and its radius is r. Let x be the distance of the bottom of the drum below the water surface. Find the differential equation for x, assuming that the only forces acting on the drum are its weight and the buoyant force of the water. (Take the weight per unit volume of water to be 1.)

4.   Suppose that the water surface in Exercise 3 is rising and falling owing to the passage of waves. Let the height of the water surface above its normal level be $\sin \omega t$ at time t, and let x be the distance of the bottom of the drum below the normal water level. Find the differential equation for x.

5.   A cylinder of mass m and radius r floats with its axis vertical in water contained in an upright cylindrical tank of radius R. Let x be the height of the water surface above the position it would occupy were the float removed from the water. Find the differential equation for x.

6. A mass m is suspended from a spring of elastic constant k whose upper end is attached to a fixed support. Let the mass be constrained to move in a vertical line which is at a distance s from the point of support of the spring, and suppose that the spring is relaxed when the mass is in the horizontal plane of the point of support. Let x be the distance of the mass below this plane. Find the differential equation for x.

5. The Geometric Interpretation of a Differential Equation

A first-order differential equation

$$\frac{dx}{dt} = f(t,x) \tag{5.1}$$

may look at first sight like a problem: find x as a function of t so that the equation is satisfied. In this naive view, x is an "unknown"; the whole content of the equation consists in the problem of solving it, i.e., of finding some or all of its solutions. One thing that contributes to this view is the symbol dx/dt, which appears to make no sense unless we regard x as a (generally unknown) function of t. Nevertheless, this view is quite erroneous, and indeed is an impediment to understanding the real problems posed by a differential equation. It was to combat it that we took as our point of departure the use of differential equations for describing physical systems. To think of differential equations as problems is simultaneously to reduce most physical laws to the status of problems also, and this is clearly absurd. Furthermore, the idea of "solving" a differential equation, or "finding" its solutions, is by no means free of ambiguities, for the solutions of differential equations are not usually expressible by means of formulas, and it is therefore not at all clear what is meant by "finding" a solution.

To understand what a differential equation is we must focus on the equation itself, not on its solutions. Let us suppose for simplicity that the function f in Eq. (5.1) is defined for all values of t and x; to any point (t,x) in the (t,x)-plane f assigns the number $f(t,x)$. What is the significance of this number? Our examples suggest that, in some way, it is the rate of change of x with respect to t at this point. We interpret this geometrically by passing a line, or line segment, of slope $f(t,x)$ through the point (t,x). The length of the segment is immaterial; only its slope matters. Thus every point in the plane is assigned a line segment or, what amounts to the same thing, a slope. (Of course, no matter how short the segment is, it will necessarily pass through infinitely many points other than that to which it is assigned. Note that no segment is parallel to the x-axis, for such a segment has no slope.) The assignment of lines (or slopes) to points of the (t,x)-plane will be called a *direction field*. The content of the differential equation (5.1) is that it defines a direction field. Figure 3 shows a portion of the direction field of Eq. (1.2), $dx/dt = 1 - 0.01x$. (Negative values of x have no significance in terms of the physical model from which the equation was derived; they are included in the figure only for their mathematical interest.) The autonomous character

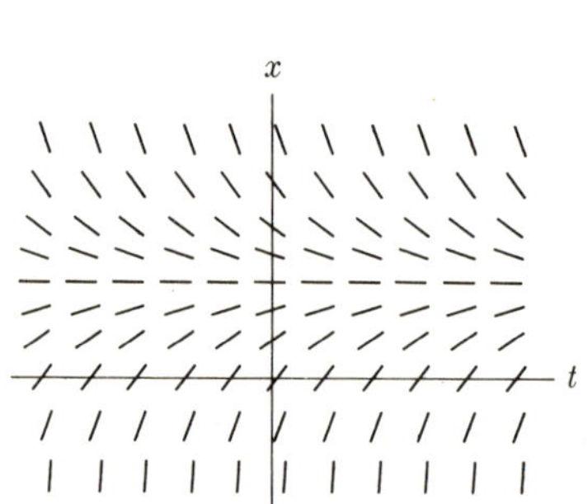

FIG. 3.   Direction field of $dx/dt - 1 - 0.01x$.

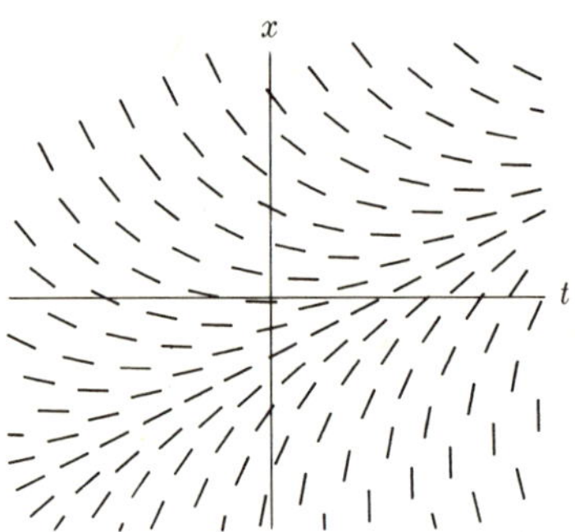

FIG. 4.   Direction field of $dx/dt = t - 2x$.

of the equation is reflected in the fact that points lying on a line parallel to the t-axis are assigned the same slope: the slope at the point (t,x) is $1 - 0.01x$, and is therefore independent of t.

Figure 4 shows the direction field of the equation

$$\frac{dx}{dt} = t - 2x. \tag{5.2}$$

In both examples, the function f was defined in the entire (t,x)-plane. Examples of differential equations defined only in a portion of the plane are $dx/dt = \sqrt{tx}$, which is defined only for $tx \geqq 0$, i.e., in the first and third quadrants inclusive of their boundaries, and the equation $\dfrac{dx}{dt} = \dfrac{1}{\sqrt{1 - t^2 - x^2}}$, which is defined only for $t^2 + x^2 < 1$, i.e., in the interior of the circle with radius 1 and center at the origin.

The curves $f(t,x) = $ constant are called the *isoclines* of the differential equation $dx/dt = f(t,x)$. Frequently the direction field is most easily sketched if a number of isoclines are drawn first. Because the direction field is constant on each isocline, it then only remains to draw line segments of slope c through a sampling of points of each isocline $f(t,x) = c$. The isoclines of an autonomous equation are the lines parallel to the t-axis. The isoclines of (5.2) are the lines $t - 2x = c$.

Now let us suppose that we can find a curve in the (t,x)-plane which is tangent at each of its points to the line assigned to that point by the differential equation $dx/dt = f(t,x)$. (We shall imagine such a curve as the path traced by a point which moves always in the direction of the direction field; this ensures that the curve is connected—i.e., it consists of one piece.) Such a curve has, by definition, a tangent at each of its points. These tangents, furthermore, are never parallel to the x-axis. The curve is therefore the graph $x = u(t)$ of a function u defined on some interval of the t-axis and having a derivative at each point of this interval. The slope of the curve at any point $(t,u(t))$ on it is, by definition, the slope $f(t,u(t))$ assigned to that point. On the other hand, this slope is equal to $u'(t)$, the value of the derivative of u at t. Thus $u'(t) = f(t,u(t))$ for every t in the interval of definition of u. It follows that u is a solution of the differential equation.

Conversely, if we have a function u which is a solution of $dx/dt = f(t,x)$, it is obvious that its graph $x = u(t)$ is tangent at each of its points to the line assigned to that point by the differential equation.

The graph of a solution will be called a *solution curve*. Figures 5 and 6 show several solution curves of Eqs. (1.2) and (5.2).

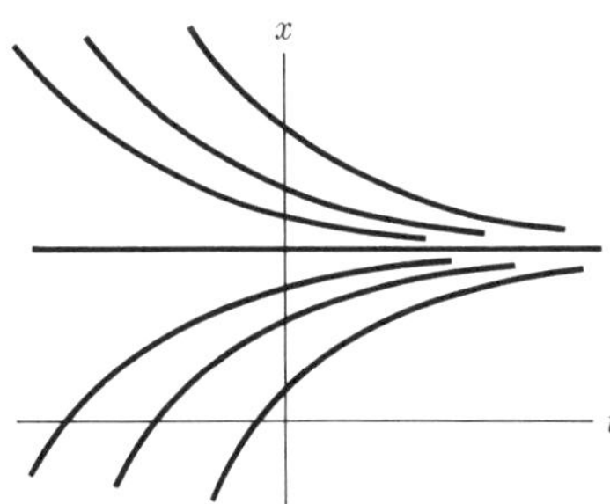

Fig. 5. Solution curves of
$dx/dt = 1 - 0.01x$.

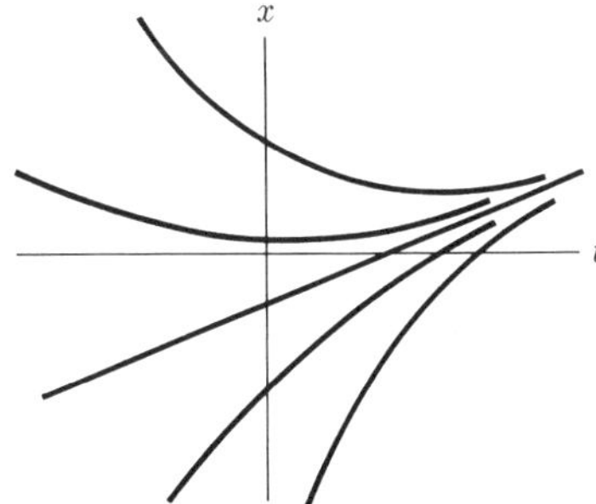

Fig. 6. Solution curves of
$dx/dt = t - 2x$.

We summarize this discussion by formulating the definition of a differential equation and its solutions in geometrical terms:

A first-order differential equation is a direction field in a region of the (t,x)-plane. A solution is a function defined on an interval of the t-axis whose graph is tangent to the field.

Now, a word in passing to the reader who is concerned about the symbol dx/dt in the differential equation. The equation of the line assigned by the differential equation $dx/dt = f(t,x)$ to a point (t_0,x_0) is $(x - x_0)/(t - t_0) = f(t_0,x_0)$. If we set $x - x_0 = dx$, $t - t_0 = dt$, we recover the differential equation, except for the zero subscripts, which may now be dropped. This shows that the symbols dx and dt in the differential equation are rectangular coordinates relative to the point (t,x) as origin. This is in agreement with the way differentials are defined in calculus: if u is a function differentiable at t, $dx = u'(t)\, dt$ is the equation of the line tangent to the graph of u at t, dt and dx being rectangular coordinates with origin at $(t,u(t))$.

The geometrical interpretation of a first-order differential equation generalizes easily to systems of such equations and thereby gives an intrinsic meaning to all differential equations and systems. Consider a system of two first-order equations:

$$\frac{dx}{dt} = f(t,x,y)$$

$$\frac{dy}{dt} = g(t,x,y). \tag{5.3}$$

In order to obtain a geometric interpretation, we must go to three-dimensional space with a rectangular (t,x,y)-coordinate system. If f and g are defined at a point (t_0,x_0,y_0), we assign to this point the direction numbers 1, $f(t_0,x_0,y_0)$, $g(t_0,x_0,y_0)$ or, equivalently, the line with parametric equation $t = t_0 + s$, $x = x_0 + f(t_0,x_0,y_0)s$, $y = y_0 + g(t_0,x_0,y_0)s$. We recall that a solution of (5.3) is a pair of functions u, v defined on an interval I such that $u'(t) = f(t,u(t),v(t))$, $v'(t) = g(t,u(t),v(t))$ for t in I. The graph of this solution is the curve $x = u(t)$, $y = v(t)$ in the (t,x,y)-space. The direction numbers of its tangent line at t_0 are 1, $u'(t_0)$, $v'(t_0)$, and these are precisely the direction numbers assigned by the system (5.3). Conversely, if u and v are functions defined and differentiable on an interval I such that the tangent line at any point on the curve $x = u(t)$, $y = v(t)$ is the line assigned to this point by the system (5.3), then u, v is a solution of (5.3). It is therefore natural to interpret (5.3) as defining a lineal, or direction, field in the (t,x,y)-space. A solution is a pair of functions u, v defined on an interval I of the t-axis whose graph is, at each of its points, tangent to the field.

There is no difficulty in extending this interpretation to systems of higher order, provided we are familiar with the elements of n-dimensional analytic geometry. It cannot be denied, of course, that space of dimension greater than 3 is artificial, and the intuitive visual value of the geometric interpretation is considerably diminished.

The fundamental problem of the theory of differential equations is to describe the solution curves of a given differential equation. The reader is familiar with a similar problem from calculus and analytic geometry: given a function u, describe the curve $x = u(t)$. In fact, this elementary problem is in a sense not a different problem at all. In both cases we want to know the shape of the graph of a given function. The difference between the two problems is merely one of difficulty. In calculus the function u is defined by a formula, i.e., it is given as a combination of certain standard functions—powers, exponentials, trigonometric functions, and so forth. Since the graphs of these standard functions are known, it is usually not very difficult to deduce the shape of the graph of u. In differential equations, however, the function u is not defined by a formula, but by the condition that it satisfies a given differential equation and that its graph passes through a given point (t_0, x_0). In general, therefore, the problem of describing the graph is a more difficult one.

In exceptional cases the solutions of a differential equation can be expressed by formulas. In these cases the problem of determining the shape of the solution curves reduces to a problem of calculus. To find formulas for the solutions of a differential equation is called "solving" the equation. It must be remembered that solving an equation is only a means to an end, and that in most cases this end must be attained by other means.

The reader will recall that after studying the elementary functions of calculus he learned how to find power series expansions for them. A power series enables one to construct a table of values for a function in a neighborhood of the point about which the function is expanded. The problem of constructing a table of values arises also in connection with differential equations. It may happen, for example, that a numerical table is needed for some particular solution for purposes of accurate short-term prediction, as in astronomy. A procedure which can be used to construct a table of values for a solution of an arbitrary differential equation is described in Section 7 of the next chapter. This procedure is a simplified version of the methods which are routinely used on computers, and for which machine codes are available in every computer library. Series expansions are specialized tools of limited interest, and are not discussed in this book.

EXERCISES

1. Sketch the direction field for the following equations.

(a) $\dfrac{dx}{dt} = -x$ (b) $\dfrac{dx}{dt} = x$

(c) $\dfrac{dx}{dt} = t - x$ (d) $\dfrac{dx}{dt} = t$

(e) $\dfrac{dx}{dt} = |x|$ (f) $\dfrac{dx}{dt} = \dfrac{1}{\sqrt{1 - t^2 - x^2}}$

2. Strange as it may seem, it is possible for two different solution curves of a first-order differential equation to pass through the same point. For example, both $x \equiv 0$ and $x = t^3$ are solution curves of the equation $dx/dt = 3x^{2/3}$, and both pass through the origin.

Let u and v be solutions of a first-order differential equation, both defined on an interval $a < t < b$, and suppose that $u(t_0) = v(t_0)$ for some t_0.

(a) Prove that the curves $x = u(t)$ and $x = v(t)$ are tangent at $t = t_0$.

(b) Define the functions w and z on $a < t < b$ by $w(t) = u(t)$ for $a < t \leq t_0$, $w(t) = v(t)$ for $t_0 \leq t < b$, and $z(t) = v(t)$ for $a < t \leq t_0$, $z(t) = u(t)$ for $t_0 \leq t < b$. Prove that w and z are also solutions of the differential equation.

3. Explain in geometrical terms why the equation $dx/dt = \sqrt{-|x - t|}$ has no solution, noting that the equation is defined only on the line $x = t$.

6. Basic Theorems

As Exercises 2 and 3 of the preceding section show, it is possible that a differential equation has no solution at all, and it is also possible that two different solution curves of the same first-order differential equation intersect each other. We want to be able to exclude such cases. We want to be sure that a solution curve passes through every point of the region in which the differential equation is defined, and we also want to be sure that only *one* solution curve passes through each point. The first of these demands probably strikes the reader as more reasonable than the second. There are numerous purely mathematical reasons for the latter, but the simplest is scientific rather than mathematical. When we say that a physical system is "governed" by a first-order differential equation $dx/dt = f(t,x)$, we mean that if at time t_0 the system is in the state x_0, its future course is completely determined by the differential equation—it is, so to speak, guided by the direction field. Thus the differential equation can govern a physical system only if to a given point (t_0,x_0), there corresponds exactly one solution u which satisfies $u(t_0) = x_0$. When this is the case, we say that the equation has the "uniqueness" property. Fortunately, it is possible to guarantee that an equation has the uniqueness property if the function f has certain properties which can be directly verified.

Now we must make a definition. We say that a subset D of the plane is *open* if every point of D is the center of a rectangle which is contained in D. More precisely, D is open if for every point (t_0,x_0) in D, there are positive numbers a and b such that any point (t,x) which satisfies $|t - t_0| \leqq a$, $|x - x_0| \leqq b$ also belongs to D. The plane itself is open; other examples are (1) the set of points (t,x) satisfying $t^2 + x^2 < 1$; (2) the set of points (t,x) satisfying $x < 0$; (3) the plane with a line deleted. Examples of sets which are not open are (1) a line; (2) the set $x \geqq 0$; (3) the set $0 < t < 1$, $0 \leqq x < 1$.

After these preliminaries, we are ready to state the fundamental existence and uniqueness theorem for first-order equations.

Theorem 6.1. *Let the function f in the differential equation $dx/dt = f(t,x)$ be defined and continuous on an open set D of the (t,x)-plane. Suppose further that f has a partial derivative with respect to x at every point of D and that $\partial f/\partial x$ is continuous on D. Let (t_0,x_0) be a point of D.*

Then the equation has a solution u defined on an interval about t_0 and such that $u(t_0) = x_0$. Furthermore, if v is a solution defined on the same interval as u, and if $v(t_0) = x_0$, then $v = u$.

The proof of this theorem is given in Chapter 7.

The given point (t_0,x_0) is frequently called an "initial condition," and the problem of finding a solution u for which $u(t_0) = x_0$ is called an "initial-value problem." We sometimes state this problem "$x = x_0$ when $t = t_0$" instead of "$u(t_0) = x_0$." Theorem 6.1 asserts that under certain circumstances an initial-value problem has a unique solution. This is not quite a correct description of the theorem, nor were we quite precise in introducing the concept of uniqueness. A little reflection will show that what we called the uniqueness property is an impossibility. If there is a solution whose graph passes through a given point, we can always find another solution whose graph passes through the same point by the simple trick of taking the interval of definition of the second solution to be smaller than the interval of definition of the original solution, but otherwise taking the second solution to be identical with the first. This, to be sure, is a hair-splitting criticism, but it raises an important point which deserves discussion.

Let u be a solution of $dx/dt = f(t,x)$ defined on an interval I. Let v be a solution defined on an interval I' which contains I but does not coincide with it. If $v(t) = u(t)$ for t in I, we say that v is a *continuation* of u. The graph of u is then simply a piece of the graph of v. If u has no continuation, we say that u is a *maximal* solution. Assuming, as we shall henceforth, that f satisfies the hypotheses of Theorem 6.1, a solution u defined on an interval of the form $[a,b]$ cannot be maximal. For, letting $t_1 = b$ and $x_1 = u(t_1)$, we need only apply the theorem to the point (t_1,x_1) and thus obtain a solution v defined on an interval $[a_1,b_1]$, where $a_1 < t_1 < b_1$, such that $v(t_1) = x_1$. Because $u(t_1) = v(t_1)$, a simple argument based on the uniqueness part of the theorem shows that $u(t) = v(t)$ for $a_2 \leqq t \leqq b$, where a_2 is the larger of the two numbers a and a_1. Now we define the function w on the interval $[a,b_1]$ by $w(t) = u(t)$ for $a \leqq t \leqq b$ and by $w(t) = v(t)$ for $b \leqq t \leqq b_1$. Then w is a continuation of u "to the right." The same procedure applied to the point $(a,u(a))$ in place of $(b,u(b))$ shows that u has a continuation to the left. Thus if the interval of definition of a solution has either a right or a left endpoint which belongs to the interval, the solution is not maximal: the interval of definition of a maximal solution is always an *open* interval.

The following theorem, a simple extension of Theorem 6.1, has equal claim to the title of "fundamental existence and uniqueness theorem." It is proved in Chapter 7.

Theorem 6.2.   *Let f and $\partial f/\partial t$ be defined and continuous on an open set D of the (t,x)-plane, and let (t_0,x_0) be a point of D.*

Then the equation $dx/dt = f(t,x)$ has a maximal solution u such that $u(t_0) = x_0$. If v is any other solution and $v(t_0) = x_0$, then u is a continuation of v.

In the sequel, we shall sometimes suppress the adjective "maximal"; the reader will be able to decide from the context whether or not it is implied.

Suppose that u is a maximal solution. What can prevent it from being defined for all t? One thing that can obviously do so is the failure of f to be defined for all (t,x). But even if the set D is the whole plane, some or all maximal solutions may be defined on intervals with endpoints. Consider the equation $dx/dt = x^2$. Here $f(t,x) = x^2$, $\partial f(t,x)/\partial x = 2x$, and both functions are continuous everywhere. Nevertheless, the solution $u(t) = -1/t$, defined for $t < 0$, is maximal. Any continuation would have to be defined and continuous at $t = 0$ while agreeing with u for $t < 0$, which is clearly impossible. This example shows that the interval of definition of a maximal solution u may have an endpoint because $u(t) \to \infty$ or $u(t) \to -\infty$ as t approaches this point. It turns out that this is the only possibility.

Theorem 6.3.   *Let f and $\partial f/\partial x$ be continuous in the whole plane. Let u be a maximal solution of $dx/dt = f(t,x)$. If the interval of definition of u has an endpoint α, then $|u(t)| \to \infty$ as $t \to \alpha$.*

The proof is given in Chapter 7.

The preceding theorems can all be generalized so as to apply to systems of first-order differential equations

$$\frac{dx_i}{dt} = f_i(t,x_1,x_2, \ldots ,x_n), \qquad i = 1, \ldots , n.$$

Because the f_i are functions of $n + 1$ variables, we must replace the open subset D of the plane by an open subset D of $(n + 1)$-dimensional Euclidean space, which we now proceed to define.

Let m be a positive integer. The set of all m-tuples $(x_1,x_2, \ldots ,x_m)$ of real numbers will be denoted by R^m. Thus $R^1 = R$, the set of all real numbers, and R^2 is the plane, the set of all pairs of real numbers. We shall refer to an m-tuple as a "point" of R^m. The number x_i is called the ith coordinate of the point $(x_1,x_2, \ldots ,x_m)$.

Let D be a set of points of R^m. We say that D is open if for every point $(x_1^0,x_2^0, \ldots ,x_m^0)$ in D, there are positive numbers $a_1, a_2, \ldots , a_m$ such that any point $(x_1,x_2, \ldots ,x_m)$ which satisfies $|x_i - x_i^0| \leqq a_i$, $i = 1, 2, \ldots , m$, also belongs to D. R^m itself is obviously open. The set $x_1^2 + x_2^2 + \cdots + x_m^2 < 1$ is open. An example of a set which is not open is $x_1 + x_2 + \cdots + x_m = 0$.

Theorem 6.4.   *Let the functions $f_1, f_2, \ldots , f_n$ in the system*

$$\frac{dx_i}{dt} = f_i(t,x_1,x_2, \ldots ,x_n), \qquad i = 1, \ldots , n, \tag{6.1}$$

be defined and continuous on an open set D of R^{n+1}. Suppose further that each f_i has partial derivatives with respect to $x_1, x_2, \ldots , x_n$ at every point of D, and that these partial derivatives $\partial f_i/\partial x_j$ $(i,j = 1, \ldots ,n)$ are continuous on D. Let $(t_0,x_1^0,x_2^0, \ldots ,x_n^0)$ be a point of D.

Then there exists an interval $[a,b]$, $a < t_0 < b$, and functions u_1, u_2, . . . , u_n defined on $[a,b]$, such that u_1, u_2, . . . , u_n is a solution of (6.1) and $u_i(t_0) = x_i^0$, $i = 1, . . . , n$.

Furthermore, if v_1, v_2, . . . , v_n is a solution defined on $[a,b]$ and $v_i(t_0) = x_i^0$, then $v_i = u_i$, $i = 1, . . . , n$.

Clearly this theorem reduces to Theorem 6.1 when $n = 1$. There are similar straightforward generalizations of Theorems 6.2 and 6.3.

EXERCISES

1. Apply the fundamental existence and uniqueness theorems to show that each of the following differential equations has just one (maximal) solution satisfying the given initial condition.

$$\text{(a)} \quad \frac{dx}{dt} = \frac{1}{x} \quad (x \neq 0); \qquad u(0) = 1$$

$$\text{(b)} \quad \frac{dx}{dt} = t^2 + x^2; \qquad u(0) = 0$$

$$\text{(c)} \quad \frac{dx}{dt} = \sqrt{1 - t^2 - x^2}; \qquad u(0) = 0$$

2. Use Theorem 6.4 to show that the systems (1.3) and (1.5) have unique solutions for every initial condition (t_0, x_0, y_0) and (t_0, x_0, v_0), respectively.

3. The equation $dx/dt = \sqrt{-|x - t|}$ (see Exercise 3, page 20) has no solution. In what way does it violate the hypotheses of Theorem 6.1?

4. The equation $dx/dt = 3x^{2/3}$ (see Exercise 2, page 20) has more than one solution satisfying the initial condition $u(0) = 0$. How are the hypotheses of Theorem 6.1 violated?

5. Give an example of a function $f(t,x)$ which is not continuous, but which has a continuous partial derivative with respect to x.

MISCELLANEOUS EXERCISES FOR CHAPTER 1

1. Given two differential equations $dx/dt = f(t,x)$ and $dx/dt = g(t,x)$, construct a differential equation with a parameter μ such that for $\mu = 0$ the first equation results and for $\mu = 1$ the second. (Assume that f and g have the same domain of definition.)

2.   Calculate the constant solutions of $dx/dt = x^2 - 2\mu x + \mu$ in terms of the parameter μ. For which values of μ has the equation no constant solution?

3.   We say that a solution $u_1, u_2, \ldots, u_n$ of a system is a *constant* solution if all the functions $u_1, u_2, \ldots, u_n$ are constant. Calculate the constant solutions of

$$\text{(a)} \quad \frac{dx}{dt} = x(1 - y), \quad \frac{dy}{dt} = y(1 + x)$$

$$\text{(b)} \quad \frac{dx}{dt} = x + y, \quad \frac{dy}{dt} = x - y$$

4.   Find the constant solutions of the following systems in terms of the parameter μ:

$$\text{(a)} \quad \frac{dx}{dt} = \mu x - y, \quad \frac{dy}{dt} = x - \mu^2 y$$

$$\text{(b)} \quad \frac{dx}{dt} = \mu x - y + \mu, \quad \frac{dy}{dt} = x - \mu^2 y$$

5.   Let $u_1, u_2, \ldots, u_n$ be a solution of a linear homogeneous system with constant coefficients

$$\frac{dx_i}{dt} = \sum_{j=1}^{n} a_{ij} x_j, \qquad i = 1, \ldots, n.$$

Let $v_i = u_i', \quad i = 1, \ldots, n$. Show that each v_i is differentiable, and that $v_1, v_2, \ldots, v_n$ is a solution of the system.

6.   Compute the constant solutions of

$$\text{(a)} \quad \frac{d^2 x}{dt^2} = \frac{dx}{dt} \qquad\qquad \text{(b)} \quad \frac{d^2 x}{dt^2} = x$$

$$\text{(c)} \quad \frac{d^2 x}{dt^2} = \sin x + \frac{dx}{dt}$$

7.   Let $f(x,y)$ be continuous for all x, y. Show that if the differential equation $d^2x/dt^2 = f(x, dx/dt)$ has no constant solution, then it has no periodic solution. [A function $u(t)$ is periodic if there is a positive number p such that $u(t + p) = u(t)$ for all t. For example, $\sin t$ is periodic, because $\sin(t + 2\pi) = \sin t$ for all t.]

2

FIRST-ORDER EQUATIONS

In this chapter we consider differential equations of the first order, i.e., equations of the form $dx/dt = f(t,x)$. The reader who merely wishes to acquire the technical prerequisites for later chapters may confine himself to reading Sections 1 and 3, although Sections 6 to 8 are also recommended as further preparation.

1. Separation of Variables; Description of the Method

If the function f in the differential equation $dx/dt = f(t,x)$ is the product of a function of t alone and a function of x alone, so that the equation has the form

$$\frac{dx}{dt} = g(t)h(x),$$

it is possible in certain cases to find formulas for the solutions. The technique involved is known as "separation of variables," and an equation of the form $dx/dt = g(t)h(x)$ is called an equation with "variables separable." Note that autonomous first-order equations are of this type.

We shall first give a deliberately oversimplified description of the technique, filling in the gaps as we do the examples which follow. Following the examples, we outline the technique in more detail. In the next section, we shall examine the technique theoretically.

The first step in the procedure is to write the equation in the form

$$\frac{dx}{h(x)} = g(t)\ dt,$$

so that the variables are "separated," with only x's on the left and t's on the right. We take indefinite integrals of both sides, getting

$$\int \frac{dx}{h(x)} = \int g(t)\,dt + c,$$

where c is an arbitrary constant. It is not necessary to introduce an arbitrary constant on the left, because this constant can be combined with the one on the right. The last equation may be written

$$H(x) = G(t) + c,$$

where H and G are indefinite integrals of $1/h$ and g, respectively. We now solve this equation for x. For each c (possibly restricted to a certain range), the resulting function will yield one or more solutions of the differential equation.

EXAMPLE 1.
$$\frac{dx}{dt} = 2tx^2. \tag{1.1}$$

We separate the variables,

$$\frac{dx}{x^2} = 2t\,dt,$$

and take indefinite integrals of both sides:

$$-\frac{1}{x} = t^2 + c.$$

Solving for x yields

$$x = \frac{-1}{t^2 + c}. \tag{1.2}$$

We verify that for each c, the function $\phi(t) = \dfrac{-1}{t^2 + c}$ satisfies (1.1).

Differentiation yields $\phi'(t) = \dfrac{2t}{(t^2 + c)^2}$, and since also $2t\phi^2(t) = \dfrac{2t}{(t^2 + c)^2}$, we have $\phi'(t) = 2t\phi^2(t)$ and the equation is satisfied. We now separate these functions into solutions, i.e., into functions defined on intervals. For every positive c, formula (1.2) yields a function defined for all t. All of these solutions are negative; their graphs are the inverted bell-shaped curves in Fig. 7. For $c = 0$, we obtain two solutions, defined respectively for $t < 0$ and $t > 0$; both are negative. For each negative value of c, we obtain three solutions. One of these is positive and is defined for $-\sqrt{-c} < t < \sqrt{-c}$; the other two are negative and are defined for $t < -\sqrt{-c}$ and $t > \sqrt{-c}$, respectively. Note that all the solutions obtained are maximal.

Can we be sure that there are not still other (maximal) solutions? This question is answered by the existence and uniqueness theorem (Theorem 6.2, page 22). Because $f(t,x)$, which is equal to $2tx^2$, and $\partial f(t,x)/\partial x$, which is equal to $4tx$, are both continuous in the whole plane, one and only one solution curve passes through any given point (t_0,x_0). If it should turn out that there is a point which lies on none of the

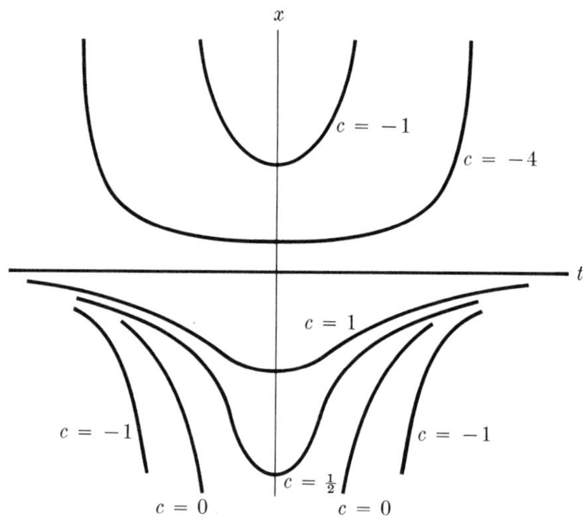

Fig. 7.   Solution curves of $dx/dt = 2tx^2$.

solution curves which we have found, then clearly the family of solutions which we have found is incomplete. Conversely, if for every point (t_0,x_0) we can supply from our family a solution u which satisfies $u(t_0) = x_0$, then our family is complete. The first assertion follows from the *existence* part of the theorem: for every point (t_0,x_0), there exists a solution u for which $u(t_0) = x_0$. The second assertion follows from the *uniqueness* part: there cannot be more than one solution curve through any point; if our solutions already cover the whole plane, there are no others.

The question is thus reduced to this: given a point (t_0,x_0), is there a number c so that $x_0 = \dfrac{-1}{t_0^2 + c}$? Solving the equation, we have that $c = -t_0^2 - 1/x_0$ provided $x_0 \neq 0$. What if $x_0 = 0$? We see that none of our solutions ever vanishes; no point on the t-axis lies on any of our solution curves, and our family is therefore incomplete. Now inspection of the differential equation shows that $x \equiv 0$ is itself a solution. We lost it when we divided the equation by x^2. The method as we described it is likely to miss constant solutions because it involves division by $h(x)$, a procedure justifiable only for values of x for which $h(x) \neq 0$.

$\square$

EXAMPLE 2.
$$\frac{dx}{dt} = x. \tag{1.3}$$

This equation has the constant solution $x \equiv 0$. For $x \neq 0$, the equation may be written

$$\frac{dx}{x} = dt$$

and integration yields

$$\log |x| = t + c.$$

(We always use the notation $\log x$ for the *natural* logarithm of x.) To solve for x, we take the exponential of both sides:

$$|x| = e^{t+c} = e^c e^t = k e^t,$$

where k is an arbitrary positive constant. Thus

$$x = \pm k e^t.$$

This formula is equivalent to

$$x = c e^t, \tag{1.4}$$

where c is an arbitrary non-zero constant.

Differentiation yields $dx/dt = ce^t = x$, so that (1.3) is satisfied. We also note that $c = 0$ produces the constant solution, so that (1.4) yields solutions for all values of c. All of these solutions are defined for all t. Finally, we observe that $x_0 = ce^{t_0}$ has the solution $c = x_0 e^{-t_0}$, so that the curves (1.4) cover the plane. Because (1.3) has the uniqueness property, we have found all solutions. The solution curves are shown in Fig. 8. $\square$

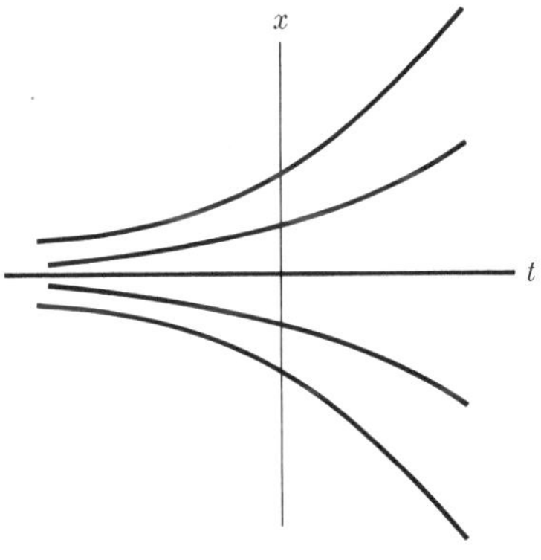

FIG. 8. Solution curves of $dx/dt = x$.

EXAMPLE 3.
$$\frac{dx}{dt} = \cos^2 x. \tag{1.5}$$

The solutions of $\cos^2 x = 0$ yield infinitely many constant solutions of the differential equation, given by $x = (n + \frac{1}{2})\pi$, where n is an arbitrary integer. If $\cos^2 x \neq 0$, we may write

$$\frac{dx}{\cos^2 x} = dt.$$

Because $1/\cos^2 x = \sec^2 x$ and $\int \sec^2 x \, dx = \tan x$, we have, after integrating,

$$\tan x = t + c. \tag{1.6}$$

In solving this equation for x, we must not be hasty and simply write $x = \arctan (t + c)$. If we did so, we would be assuming that x lies between $-\pi/2$ and $\pi/2$; all the solutions of (1.5) thus obtained naturally lie in this strip. We must, instead, allow x to assume values in each of the infinitely many intervals between consecutive zeros of $\cos^2 x$. Suppose, then, that $-\pi/2 + n\pi < x < \pi/2 + n\pi$ for some integer n, so that $-\pi/2 < x - n\pi < \pi/2$. Because $\tan x = \tan (x - n\pi)$, we can write (1.6) as

$$\tan (x - n\pi) = t + c$$

and taking the inverse tangent of both sides yields

$$x = n\pi + \arctan (t + c). \tag{1.7}$$

In this formula, c is arbitrary, and n is an arbitrary integer. We leave it to the reader to check that the functions defined by this formula are indeed solutions, and that together with the constant solutions they cover the plane. Because (1.5) has the uniqueness property, we are thus assured that all solutions have been found. Note that each of them is defined for all t. $\square$

Example 4. $\dfrac{dx}{dt} = 2t \sec x,$ $-\pi/2 < x < \pi/2.$ (1.8)

Because the secant has no zeros, there are no constant solutions. We separate the variables and integrate, obtaining

$$\sin x = t^2 + c. \tag{1.9}$$

In the preceding examples, the constant of integration was always completely arbitrary. Here, however, it would be senseless to allow c to assume the value 2, for example. If we did, we would be unable to solve the equation for x, no matter how we restricted t. The right side would never be less than 2, while $\sin x$ cannot exceed 1. We therefore allow only those values of c for which Eq. (1.9) is satisfied by at least one pair of values (t,x). This requirement together with the restriction $-\pi/2 < x < \pi/2$ implies $c < 1$. Having chosen such a c, we restrict t so that $-1 < t^2 + c < 1$. At last, we are able to solve and obtain the formula

$$x = \arcsin (t^2 + c). \tag{1.10}$$

It yields one solution of (1.8) if $-1 < c < 1$, and two solutions if $c \leqq -1$. $\square$

The foregoing examples illustrate all the crucial steps in the method of separation of variables. These steps may be systematically arranged as follows:

1. *Solve the equation $h(x) = 0$ in order to obtain the constant solutions.*
2. *Restrict x to an interval on which $h(x)$ does not vanish, separate variables and integrate. (This step must be carried through for every admissible interval.)*
3. *Restrict the constant of integration c so that the equation $H(x) = G(t) + c$ has at least one solution (t,x). Solve the equation for x, remembering that the form of the solution will in general depend on the interval to which x was restricted in step 2.*
4. *Separate the values of t admitted by step 3 into intervals.*

Assuming that the differential equation has the uniqueness property, which is guaranteed if g is continuous and h has a continuous derivative, this procedure will yield all the solutions of the equation. In fact, it yields all the solutions not merely in the sense of providing a solution for every initial condition, but it automatically provides the maximal solution satisfying the given initial condition. As a check, the answers should always be tested against the differential equation and against arbitrary initial conditions. It is also instructive to sketch a number of the solutions, and to compare them with a rough sketch of the direction field. This will reveal the geometrical significance of the formulas, and is often helpful in uncovering gross errors.

EXERCISES

1. Using the formulas derived in the text, compute the solutions of Eq. (1.5) satisfying the following initial conditions:

(a)	$u(0) = \pi$	(b)	$u(1) = 0$
(c)	$u(0) = -\pi/4$	(d)	$u(0) = \pi/2$

2. Compute formulas for the solutions of the following differential equations. Compute the solution satisfying the given initial condition and specify its interval of definition.

$$\text{(a)} \quad \frac{dx}{dt} = x^2; \qquad u(0) = 1$$

$$\text{(b)} \quad \frac{dx}{dt} = 1 - x; \qquad u(0) = 0$$

$$\text{(c)} \quad \frac{dx}{dt} = -tx; \qquad u(0) = 0$$

$$\text{(d)} \quad \frac{dx}{dt} = t \tan x \quad (-\pi/2 < x < \pi/2); \qquad u(0) = \pi/6$$

$$\text{(e)} \quad \frac{dx}{dt} = x \cos t; \qquad u(\pi) = 1$$

$$\text{(f)} \quad \frac{dx}{dt} = e^{t-x}; \qquad u(1) = 0$$

$$\text{(g)} \quad \frac{dx}{dt} = x^2 - x; \qquad u(0) = 2$$

$$\text{(h)} \quad \frac{dx}{dt} = x^3; \qquad u(1) = -\tfrac{1}{3}$$

$$\text{(i)} \quad \frac{dx}{dt} = 2t \sin x; \qquad u(0) = \pi/2$$

$$\text{(j)} \quad \frac{dx}{dt} = e^{-x}; \qquad u(0) = 0$$

3. Show that the substitution $y = at + bx$ transforms the equation $dx/dt = F(at + bx)$ into $dy/dt = a + bF(y)$, in which the variables are separable.

4. Solve the following equations by the method of problem 3.

$$\text{(a)} \quad \frac{dx}{dt} = t + x \qquad\qquad \text{(b)} \quad \frac{dx}{dt} = t - 2x$$

$$\text{(c)} \quad \frac{dx}{dt} = \sin^2 (t - x) \qquad\qquad \text{(d)} \quad \frac{dx}{dt} = (t + x)^2$$

2.   Separation of Variables; Theory

This section is devoted to some theoretical aspects of the method of separation of variables, and may be omitted without loss of continuity. Our primary object is to prove the general validity of the method. We begin by recalling some definitions and theorems from calculus.

An *indefinite integral* of a function $f(t)$ is by definition a function $u(t)$, defined on an interval, such that $u'(t) = f(t)$ for all t in the interval.* An indefinite integral of f is, therefore, simply a solution of $dx/dt = f(t)$; the symbol $\int f(t)\, dt$ denotes neither more nor less than such a solution.

If f is continuous on an interval I, then it has an indefinite integral on I. To deduce this fact from the general existence theorem for differential equations would be a roundabout argument. It is actually included in the fundamental theorem of calculus:

Let f be continuous on an interval I, and let a be a number in I. Let u be defined by

$$u(t) = \int_a^t f(s)\, ds$$

for t in I. Then u has a derivative, and $u' = f$.

The existence of an indefinite integral of f is, therefore, a consequence of the existence and properties of the definite integral. We remark that the fundamental theorem of calculus has the corollary that if $F' = f$, then $\int_a^b f(t)\, dt = F(b) - F(a)$. This corollary is also called the fundamental theorem of calculus.

Separation of variables involves, besides two indefinite integrations, the solution of the equation $H(x) = G(t) + c$. We shall prove that a solution defined on a t-interval always exists. Here we shall use the theorem that if a function H has a derivative which is positive (or negative) throughout an interval, then there is a function W, called the *inverse* of H, such that $W(H(x)) = x$. [It is by taking W of both sides of the equation $H(x) = G(t) + c$ that we get the solution $x = W(G(t) + c)$.] We shall also use the fact that W has a derivative given by $W'(y) = 1/H'(W(y))$.

Theorem 2.1.   *Consider the equation*

$$\frac{dx}{dt} = g(t)h(x) \tag{2.1}$$

for t in an open interval I_t on which g is continuous, and x in an open interval I_x on which h is continuous and does not vanish. Let G and H

---

* The requirement that an indefinite integral be defined on an interval is not always stressed in calculus presentations. It is, nevertheless, always understood. See Exercise 4, below.

be indefinite integrals of g and $1/h$ on I_t and I_x, respectively. Let c be a number for which the equation

$$H(x) = G(t) + c \tag{2.2}$$

is satisfied by at least one pair of numbers (t,x).

If (t_0,x_0) satisfies (2.2), then (2.2) defines x implicitly as a function u of t on some interval about t_0. Furthermore, u is a solution of (2.1) and $u(t_0) = x_0$.

Proof. Because $H'(x) = 1/h(x)$, H' is continuous and never zero. Thus H' is of one sign, and therefore H has an inverse W. Our task now is to show that there is a t-interval for which $W(G(t) + c)$ is defined. Because I_x is an open interval and H is increasing (or decreasing), the range of H is also an open interval. Thus, if x_0 is a number in I_x, there is a number $\epsilon > 0$ such that the interval $H(x_0) - \epsilon < y < H(x_0) + \epsilon$ lies in the range of H and, therefore, in the domain of W. If $G(t_0) + c = H(x_0)$, the continuity of the function $G(t) + c$ implies the existence of a $\delta > 0$ such that if $|t - t_0| < \delta$, then $|G(t) + c - H(x_0)| < \epsilon$. Hence, if $|t - t_0| < \delta$, then $H(x_0) - \epsilon < G(t) + c < H(x_0) + \epsilon$, i.e., $G(t) + c$ lies in the domain of W. The existence of a suitable t-interval is thereby proved. With t in such an interval, we take W of both sides of (2.2) and obtain the solution $x = W(G(t) + c)$. Denoting this function by u, we have

$$u'(t) = W'(G(t) + c)G'(t) = \frac{1}{H'[W(G(t) + c)]} G'(t)$$
$$= h[W(G(t) + c)]g(t) = h(u(t))g(t).$$

Thus u is a solution of (2.1). Also, $u(t_0) = W(G(t_0) + c) = W(H(x_0)) = x_0$. $\square$

This theorem assures us that the procedure of separating variables, integrating, and solving for x always leads to solutions of the differential equation. Note that we did not need to assume that h has a continuous derivative, or even that it has a derivative at all, and that no appeal was made to the fundamental existence and uniqueness theorem. The following corollary is thus not only a guarantee that by means of the steps outlined in Section 1, we can obtain solutions satisfying every initial condition; it is incidentally a special existence theorem for separable equations.

Corollary 2.2.   *If g and h are continuous on open intervals I_t and I_x, if t_0 is in I_t and x_0 is in I_x, then the differential equation $dx/dt = g(t)h(x)$ has a solution u for which $u(t_0) = x_0$ and which can be obtained by the method of separation of variables.*

Proof. If $h(x_0) = 0$, set $u(t) = x_0$ for t in I_t. If $h(x_0) \neq 0$, then the continuity of h implies the existence of an open interval I'_x about x_0 on which h does not vanish. Take this interval to be the interval I_x of Theorem 2.1, and let $c = H(x_0) - G(t_0)$. Then the function u of Theorem 2.1 has the desired properties. $\quad\square$

We mentioned in the preceding section that whenever the differential equation has the uniqueness property, the solutions produced by separation of variables are maximal. The converse is also true: if all the solutions produced by the method are maximal, then the differential equation has the uniqueness property. We omit the proof of these statements.* However, we shall illustrate by means of the equation

$$\frac{dx}{dt} = x \tag{2.3}$$

how separation of variables can be used to prove uniqueness. As we showed in the preceding section, separation of variables produces the solutions

$$x = ce^t, \tag{2.4}$$

each defined for all t, and hence maximal. Given (t_0, x_0), exactly one of the solutions (2.4) satisfies $x = x_0$ when $t = t_0$, viz., the solution for which $c = x_0 e^{-t_0}$. To prove that (2.3) has the uniqueness property, we must, therefore, show that there are no maximal solutions other than (2.4).

Let u be a maximal solution of (2.3); denote its interval of definition by I. If u is identically zero, it is included in (2.4). Next, suppose u does not vanish at all. Because

$$u'(t) = u(t),$$

we have

$$\frac{u'(t)}{u(t)} = 1$$

for t in I, and integration yields

$$u(t) = ce^t$$

for some c and for all t in I. Because u is maximal, I must obviously be equal to R, i.e., u is one of the solutions (2.4). The only remaining case in that in which u assumes both zero and non-zero values. We shall show that this is impossible. If $u(t) \neq 0$ for some t, the continuity

---

* A proof may be found in the article *Separation of Variables*, American Mathematical Monthly, October 1968, page 844.

of u implies that there is an interval I' such that $u(t) \neq 0$ for t in I'. It follows, as before, that

$$u(t) = ce^t$$

for some $c \neq 0$ and for all t in I'. By our assumption, we can choose I' so that it has an endpoint t_1 with $u(t_1) = 0$. But then

$$\lim_{t \to t_1} ce^t = 0,$$

which is absurd. Hence, the first two cases exhaust the possibilities, and uniqueness is established.

Two equations without the uniqueness property are given in Exercise 5. The reader is urged to work through this exercise to see how non-uniqueness is connected with the failure of separation of variables to produce maximal solutions.

Remarks on Formulas.   The indefinite integrations involved in the separation of variables technique, and the inversion of H, may be impossible to effect by means of elementary functions. In other words, it may be impossible to express H or G by means of formulas, and even when formulas for them exist, there may be no formula for W. In such a case, the method fails to yield a formula for the solutions. This is not a shortcoming of the method, however, but an indication that no formula for the solutions exists. To clarify this point, let us agree for the moment that a function is representable by a formula if it is a finite combination of the functions studied in calculus: powers, exponentials, logarithms, trigonometric functions, and so on. The impossibility of representing the solutions of a differential equation by a formula is thus an intrinsic property of the solutions, just as it is an intrinsic property of the solutions of $x^2 - 2 = 0$ that they cannot be expressed as ratios of integers. An example of a differential equation whose solutions cannot be expressed by a formula in this sense is $dx/dt = e^{t^2}$; no finite combination of calculus functions has a derivative equal to e^{t^2}. Of course, we can write $x = \int e^{t^2}\, dt + c$, and, in that sense, we have a formula for the solutions, just as the solutions of $x^2 - 2 = 0$ are given by $x = \pm\sqrt{2}$, the radical sign playing a role analogous to the integral sign. But, even if we allow integral signs to enter our formulas, equations which may be formally solved remain exceptional. The only substantial class of first-order equations whose solutions are given by formulas (with integral signs allowed) are the linear equations, which are discussed in the next section. Even so simple a non-linear equation as $dx/dt = x^2 + t$ cannot be formally solved.

Efforts to discover formal methods of solution occupy a prominent place in the early history of the theory of differential equations. The method of separation of variables, and the formula for the solutions of linear equations, were discovered by Leibniz and Johann Bernoulli toward the end of the 17th century. By the beginning of the 19th century, this vein was essentially exhausted, and Liouville gave the first proofs of the impossibility of solving certain equations by formulas. Investigations of this kind belong to algebra rather than to the theory of differential equations.

EXERCISES

1. Consider the differential equation $dx/dt = f(t)$, where f is defined and continuous on an interval I. Let t_0 be in I and let x_0 be an arbitrary number. Use the fundamental theorem of calculus and its corollary to prove that the equation has exactly one solution u defined on I satisfying $u(t_0) = x_0$, and that this solution is given by

$$u(t) = x_0 + \int_{t_0}^{t} f(s)\ dx.$$

2. Compute functions u satisfying the following equations, and state in each case the values that t may assume.

(a) $\quad u(t) = 1 + \int_0^t u(s)\ ds$

(b) $\quad u(t) = 1 + \int_0^t u^2(s)\ ds$

(c) $\quad u(t) = t + \int_0^t u^2(s)\ ds$

(d) $\quad u(t) = \int_0^t s e^{u(s)}\ ds$

3. If f is defined and continuous on an interval I, and a is a number in I, then $\int_a^t f(s)\ ds$ is an indefinite integral of f. Is every indefinite integral of f included among these definite integrals (produced by different choices of a)? Discuss.

4. Let $F(x) = \log(-x)$ for $x < 0$ and $F(x) = \log(x) + 1$ for $x > 0$. Then $F'(x) = 1/x$ for all $x \neq 0$, but F is not of the form $\log|x| + c$.

Is there something wrong with the formula $\displaystyle\int \frac{dx}{x} = \log|x| + c$?

Discuss.

5. (a) Use separation of variables to solve the equation $dx/dt = \sqrt{|x|}$. Sketch the solution curves. ($\sqrt{}$ denotes, as always, the non-negative square root.) Show that the constant solution is the only maximal solution produced by separation of variables and that every

other solution obtained has infinitely many distinct maximal continuations.

(b) Use separation of variables to solve the equation $dx/dt = \frac{3}{2} x^{\frac{1}{3}}$. Sketch the solution curves. Show that infinitely many distinct maximal solutions pass through every point on the t-axis.

6. Use separation of variables to solve the equation $dx/dt = |x|$. Sketch the solution curves. Prove that this equation has the uniqueness property, imitating the proof used in the text for $dx/dt = x$.

7. Prove: If g and h are continuous (on some intervals) and h is never zero, then the equation $dx/dt = g(t)h(x)$ has the uniqueness property.

8. Suppose that $f(t,x)$ is defined for all (t,x). Prove that f can be written in the form $f(t,x) = g(t)h(x)$ if and only if $f(a,b)f(c,d) = f(a,d)f(c,b)$ for all numbers a, b, c, d.

3. Linear Equations

The solutions of a linear first-order equation

$$\frac{dx}{dt} = a(t)x + b(t) \tag{3.1}$$

can be expressed by means of a formula involving integral signs. This formula, due to Leibniz, can be derived in a variety of ways. The derivation given here is not the shortest (compare it with the derivations given in the Exercises), but it has the advantage that the basic idea can be applied also to linear equations of higher order, and to linear systems. We suppose throughout that $a(t)$ and $b(t)$ are defined and continuous on an interval I. Consequently, the functions $f(t,x) = a(t)x + b(t)$ and $\partial f(t,x)/\partial x = a(t)$ are continuous and the equation has the uniqueness property.

Recall that (3.1) is said to be *homogeneous* if $b(t)$ is identically zero, and *non-homogeneous* otherwise. The homogeneous equation

$$\frac{dx}{dt} = a(t)x \tag{3.2}$$

is solvable by separation of variables. The solutions are given by

$$x = ce^{\int a(t)\, dt} \tag{3.3}$$

and are therefore defined on I.

In discussing the non-homogeneous equation, we consider simultaneously the homogeneous equation obtained by dropping the b term.

The homogeneous equation is in this connection called the *reduced* equation; the non-homogeneous equation itself is referred to as the *complete* equation. Let u be any solution of the reduced equation other than the zero solution. We attempt to find a function v so that $v(t)u(t)$ is a solution of the complete equation. Assuming such a v exists, substitution of vu for x in (3.1) yields

$$(v(t)u(t))' = a(t)v(t)u(t) + b(t),$$

or

$$v'(t)u(t) + u'(t)v(t) = a(t)v(t)u(t) + b(t).$$

Because $u'(t) = a(t)u(t)$, the last equation becomes

$$v'(t)u(t) + a(t)u(t)v(t) = a(t)u(t)v(t) + b(t),$$

or

$$v'(t)u(t) = b(t).$$

Because $u(t)$ does not vanish,

$$v'(t) = \frac{b(t)}{u(t)}$$

and therefore

$$v(t) = \int \frac{b(t)}{u(t)}\, dt + c.$$

We thus obtain the formula

$$x = u(t)\left(\int \frac{b(t)}{u(t)}\, dt + c\right). \tag{3.4}$$

It is a simple matter to check that this formula does indeed give us a solution of (3.1) for every c:

$$\frac{dx}{dt} = u'(t)\left(\int \frac{b(t)}{u(t)}\, dt + c\right) + u(t)\frac{b(t)}{u(t)}$$

$$= a(t)u(t)\left(\int \frac{b(t)}{u(t)} + c\right) + b(t)$$

$$= a(t)x + b(t).$$

We have thus succeeded in obtaining infinitely many solutions of (3.1). Indeed, we have found *all* solutions. To check this, let us denote the indefinite integral $\int \dfrac{b(t)}{u(t)}\, dt$ by $Q(t)$, so that the formula reads $x = u(t)(Q(t) + c)$. Setting $t = t_0$, $x = x_0$, we have $c = \dfrac{x_0}{u(t_0)} - Q(t_0)$, and therefore we can fit any initial condition. Note that all solutions are defined on I.

To complete our derivation of the promised formula, we replace the function $u(t)$ in (3.4) by $e^{\int a(t)\,dt}$. We state the result as a theorem:

Theorem 3.1 (Leibniz's Formula). *The solutions of (3.1) are given by*

$$x = e^{\int a(t)\,dt}\left(\int e^{-\int a(t)\,dt} b(t)\,dt + c\right), \tag{3.5}$$

where t is in I and c is arbitrary.

The indefinite integral $\int a(t)\,dt$, which occurs twice in (3.5), may be chosen at pleasure, but it must be the same in both places.

The reader will easily verify that the solution of (3.1) which satisfies $u(t_0) = x_0$ is given by

$$u(t) = e^{\int_{t_0}^{t} a(s)\,ds}\left[\int_{t_0}^{t} e^{-\int_{t_0}^{s} a(\tau)\,d\tau}\, b(s)\,ds + x_0\right]. \tag{3.6}$$

EXAMPLE. $\qquad\qquad\dfrac{dx}{dt} = 2x + e^t.$

Here $a(t) \equiv 2$ and $b(t) = e^t$. Substitute in (3.5):

$$\begin{aligned}
x &= e^{\int 2\,dt}\left(\int e^{-\int 2\,dt} e^t\,dt + c\right) \\
  &= e^{2t}\left(\int e^{-2t} e^t\,dt + c\right) \\
  &= e^{2t}\left(\int e^{-t}\,dt + c\right) \\
  &= ce^{2t} - e^t. \qquad \square
\end{aligned}$$

The method by which we derived (3.4) is known as *variation of constants*. In effect, a solution of the complete equation is obtained by allowing the constant c in the formula $x = cu(t)$ for the solutions of the reduced equation to *vary*, i.e., to be a function v of t.

Homogeneous linear equations have a property which is obvious from formula (3.3): If u is a solution, then for any constant c, cu is also a solution, and, unless $u = 0$, every solution is obtained in this way. In the language of linear algebra, the solutions from a *one-dimensional vector space*. The following theorem shows that this property is not shared by other equations.

Theorem 3.2. *Suppose that for every t_0 in an interval I and every x_0, the equation $dx/dt = f(t,x)$ has a solution u satisfying $u(t_0) = x_0$. If every multiple of a solution is itself a solution, then there exists a function a defined on I such that $f(t,x) = a(t)x$.*

Proof. Let t_0 be in I, let c be a real number, and let u be a solution satisfying $u(t_0) = 1$. By hypothesis, $cu(t)$ is also a solution, so $cu'(t) =$

$f(t,cu(t))$, or $cf(t,u(t)) = f(t,cu(t))$. Set $t = t_0$: then $cf(t_0,1) = f(t_0,c)$, or, writing x instead of c and t instead of t_0, we have $f(t,x) = f(t,1)x$. Thus, $a(t) = f(t,1)$ is the desired function. $\square$

Periodic Functions.   In the theory and applications of differential equations, continuous periodic functions play an extremely important role. Let us recall that a function f is *periodic* if it is defined for all t and there is a positive number p such that $f(t + p) = f(t)$ for all t. Such a number p is called a *period* of f. Every constant function is, of course, periodic, because every positive number is a period of it. If f is continuous and periodic but not constant, it has a least period, which is sometimes called *the* period of f. For example, the least periods of $\sin t$ and $\sin^2 t$ are 2π and π, respectively. Note that if p is a period of f and n is an integer, then also $f(t + np) = f(t)$ for all t. Conversely, every period of f is an integral multiple of the least period.

A function continuous on a closed bounded interval assumes a largest and a smallest value in the interval. Let M and m be the largest and smallest values of $f(t)$ for t in the interval $[0,p]$, where p is a period of f. Because f assumes all of its values in this interval, it follows that M and m are the largest and smallest values of $f(t)$ for all t. Consequently, a continuous periodic function is bounded.

The following theorems deal with integrals of periodic functions, and with the behavior of the solutions of a linear homogeneous equation $dx/dt = a(t)x$, where $a(t)$ is periodic.

Theorem 3.3.   *Let f be continuous and of period p, and let $F(t) = \int_0^t f(s)\, ds$. A necessary and sufficient condition that F has period p is that*

$$\int_0^p f(s)\, ds = 0.$$

Proof.   Define the function G by $G(t) = F(t + p)$, so that F has period p if and only if $G = F$. Because $G(t) = \int_0^{t+p} f(s)\, ds$, the fundamental theorem of calculus and the periodicity of f yield $G'(t) = f(t)$. But $F'(t) = f(t)$ also, so that $G(t) - F(t) = $ constant. Setting $t = 0$ yields $G(t) - F(t) = G(0)$. Hence, $G = F$ if and only if $G(0) = 0$. But $G(0) = \int_0^p f(s)\, ds$, which proves the theorem. $\square$

Theorem 3.4.   *Let f be continuous and of period p, and let $f_0(t) = f(t) - \dfrac{1}{p}\int_0^p f(s)\, ds$. Then f_0 is continuous and of period p, and*

$$\int_0^p f_0(s)\, ds = 0.$$

The reader is asked to supply the simple proof (Exercise 6). The number $\dfrac{1}{p}\displaystyle\int_0^p f(t)\,dt$ is called the *mean value* of f, and will be denoted by $\bar f$. Theorem 3.4 thus asserts that every continuous periodic function has the representation $f(t) = \bar f + f_0(t)$, where $\bar f_0 = 0$. An example is $\cos^2 t = \tfrac{1}{2} + \tfrac{1}{2}\cos 2t$.

Theorem 3.5.　　　*Let f be continuous and of period p, and let $F(t) = \displaystyle\int_0^t f(s)\,ds$.*

    (a)　*If $\bar f = 0$, $F(t)$ is periodic.*
    (b)　*If $\bar f > 0$, $F(t) \to \infty$ as $t \to \infty$.*
    (c)　*If $\bar f < 0$, $F(t) \to -\infty$ as $t \to \infty$.*

Proof.　Part (a) follows directly from Theorem 3.3. To prove the rest, we write $f(t) = \bar f + f_0(t)$, so that $F(t) = \displaystyle\int_0^t (\bar f + f_0(s))\,ds = \bar f t + G(t)$, where $G(t) = \displaystyle\int_0^t f_0(s)\,ds$ is periodic by part (a). Because G is bounded, F becomes positively or negatively large for large positive t according as $\bar f$ is positive or negative.　□

Theorem 3.6.　　　*Let $a(t)$ be continuous and of period p, and let u be a non-zero solution of $dx/dt = a(t)x$.*

    (a)　*If $\bar a = 0$, u is periodic.*
    (b)　*If $\bar a < 0$, $u(t) \to 0$ as $t \to \infty$.*
    (c)　*If $\bar a > 0$, $|u(t)| \to \infty$ as $t \to \infty$.*

Proof.　$u(t) = x_0 e^{\int_0^t a(s)\,ds}$, where $x_0 = u(0)$. If $\bar a = 0$, the integral is periodic by the preceding theorem. Because the exponential of a periodic function is periodic, the exponential of the integral is periodic.

If $\bar a < 0$, the integral tends to $-\infty$ as $t \to \infty$, and its exponential therefore tends to 0.

If $\bar a > 0$, the integral tends to ∞, and therefore its exponential does also. Hence, $u(t)$ tends to ∞ or $-\infty$ according as $x_0 > 0$ or $x_0 < 0$.

    □

The case of a non-homogeneous equation $dx/dt = a(t)x + b(t)$ with a and b periodic is discussed in Exercises 8 and 9.

EXERCISES

1.　Derive formula (3.5) by representing a in the form $a(t) = -\phi'(t)/\phi(t)$. Let u be a solution, and note that u satisfies $\phi u' +$

$\phi'u = \phi b$. Integrate, solve for u, and substitute a suitable function for ϕ.

2. Derive formula (3.5) by assuming a solution u and multiplying the equation $u' - au = b$ by $e^{-\int a}$. This has the same effect as the procedure in Exercise 1, the equation being now integrable.

3. Solve the following equations:

$$\text{(a)} \quad \frac{dx}{dt} = x + \cos t \qquad\qquad \text{(b)} \quad \frac{dx}{dt} = \frac{1}{t}x + t^2, \quad t > 0$$

$$\text{(c)} \quad \frac{dx}{dt} = -\frac{1}{t}x + \log t, \quad t > 0$$

$$\text{(d)} \quad \frac{dx}{dt} = \frac{1}{t}x + \frac{1}{t}, \quad t > 0$$

$$\text{(e)} \quad \frac{dx}{dt} = \frac{1}{1+t}x + t, \quad t > -1$$

$$\text{(f)} \quad \frac{dx}{dt} = -2x + e^{-t} \qquad\qquad \text{(g)} \quad \frac{dx}{dt} = -2x + e^{-2t}$$

4. Prove:

(a) The difference of two solutions of the complete equation is a solution of the reduced equation.

(b) The sum of a solution of the complete equation and a solution of the reduced equation is a solution of the complete equation.

(c) If u is a non-zero solution of the reduced equation and v is a solution of the complete equation, then all the solutions of the complete equation are given by $cu(t) + v(t)$, where c is arbitrary.

5. Prove: If u is a solution of $dx/dt = a(t)x + b(t)$ and v is a solution of $dx/dt = a(t)x + c(t)$, then $u + v$ is a solution of $dx/dt = a(t)x + b(t) + c(t)$.

6. Prove Theorem 3.4.

7. Find the periodic solutions, if any, of the following equations.

$$\text{(a)} \quad \frac{dx}{dt} = (\cos^2 t)x \qquad\qquad \text{(b)} \quad \frac{dx}{dt} = (\cos t)x$$

$$\text{(c)} \quad \frac{dx}{dt} = x + \sin t \qquad\qquad \text{(d)} \quad \frac{dx}{dt} = (\cos t)x + e^{\sin t}$$

$$\text{(e)} \quad \frac{dx}{dt} = e^{\sin t} x$$

8. Consider the equation $dx/dt = a(t)x + b(t)$, where a and b are continuous and of period p. Let u be a solution. (Note that all solutions are defined for all t.) Prove:

(a) The function v defined by $v(t) = u(t + p)$ is also a solution. Interpret this geometrically.

(b) u has period p if and only if $u(p) = u(0)$. [To show that $u(p) = u(0)$ implies $u(t + p) = u(t)$ for all t, note that $v(0) = u(p)$ and use uniqueness.]

9. Use Exercise 8 and formula (3.6) with $t_0 = 0$ to prove:

(a) If $\bar{a} \neq 0$, the equation has a unique solution of period p.

(b) If $\bar{a} = 0$ and $\int_0^p e^{-\int_0^s a(\tau)\,d\tau} b(s)\,ds = 0$, all solutions have period p.

(c) If $\bar{a} = 0$ and $\int_0^p e^{-\int_0^s a(\tau)\,d\tau} b(s)\,ds \neq 0$, there is no solution of period p.

10. Consider the equation $dx/dt = a(t)x + b(t)$, a and b continuous for all t, and a periodic. Let u and v be any two solutions. Prove: If $\bar{a} < 0$, then $\lim_{t \to \infty} [u(t) - v(t)] = 0$. [Hint: Exercise 4(a) and Theorem 3.6.]

11. Let f be continuous and of period p. Prove: For any a,
$$\int_a^{a+p} f(s)\,ds = \int_0^p f(s)\,ds.$$

12. Let f be continuous and of period p. Prove:

$$\lim_{T \to \infty} \frac{1}{T} \int_0^T f(s)\,ds = \frac{1}{p} \int_0^p f(s)\,ds.$$

13. Let f be differentiable and of period p. Prove that f' has period p.

14. (a) Find all solutions of $dx/dt = (\sin t)x$ and sketch the solution curves.

(b) Denote the maximum and minimum values of a solution by u_M and u_m, respectively. Compute $u_M - u_m$ in terms of $u(0)$.

15. Identify the periodic solutions of the following nonlinear equations by their values at $t = 0$.

(a) $\dfrac{dx}{dt} = (\sin t)x^2$ (b) $\dfrac{dx}{dt} = \sin t \cos^2 x$

(c) $\dfrac{dx}{dt} = (\cos t)e^x$

4. Equations with Parameters

The presence of parameters in a differential equation introduces no new computational problems; the parameters simply reappear in the formulas for the solutions. For example, the solutions of $dx/dt = ax$

are given by $x = ce^{at}$. Sometimes, however, the solutions are given by one formula for one range of values of the parameters, and by another formula for a different range. For example, the solutions of

$$\frac{dx}{dt} = ax + 1 \tag{4.1}$$

are given by

$$x = -\frac{1}{a} + ce^{at} \tag{4.2}$$

if $a \neq 0$, and by

$$x = t + c \tag{4.3}$$

if $a = 0$. Similarly, the solutions of

$$\frac{dx}{dt} = e^{ax} \tag{4.4}$$

are given by

$$x = -\frac{1}{a} \log (c - at) \tag{4.5}$$

if $a \neq 0$, and by $x = t + c$ if $a = 0$. In both cases, the solutions undergo a marked change at $a = 0$. The change in the solutions of (4.4) is particularly noteworthy; if $a \neq 0$, no solution is defined for all t, whereas, if $a = 0$, every solution is defined for all t.

The purpose of this section is to point out that such apparently abrupt transitions in reality represent continuous changes in the solutions.

We consider a first-order differential equation

$$\frac{dx}{dt} = f(t,x,a) \tag{4.6}$$

containing a parameter a, where f and $\partial f/\partial x$ are continuous for all (t,x,a). The fundamental existence and uniqueness theorem guarantees that, for each value of a and each initial condition, the differential equation has a unique solution. Geometrically, the equation represents a continuous direction field for each value of a, the field changing continuously as a changes. On geometric grounds, we expect the solution curves to change continuously with a also. To formulate this idea precisely, let us choose a point (t_0,x_0) and consider the solution of (4.6) that passes through this point. This solution is a function of a as well as of t; we denote it by $u(t,a)$. Thus, $u'(t,a) = f(t,u(t,a),a)$, where the prime, as always, denotes the derivative of u with respect to t. (It is actually a partial derivative now.) The crucial property of the function u is that it satisfies the same initial condition for all a: $u(t_0,a) = x_0$. Our geometric expectation may now be formulated by saying that the function u thus defined is a continuous function of a. A little more care is required, because the t-interval on which u is defined may depend on a. Let us denote it by $I(a)$. Then we can say precisely:

Given a number a_0 and a number t in $I(a_0)$, there exists an $\epsilon > 0$ such that if $|a - a_0| < \epsilon$, then t is in $I(a)$. Furthermore, $\lim_{a \to a_0} u(t,a) = u(t,a_0)$.

Our concern here is not to prove this statement but to illustrate it. Let us begin with Eq. (4.1). Setting $x = x_0$ and $t = t_0$ in (4.2), we find that $c = e^{-at_0}(x_0 + 1/a)$. Thus the solution of (4.1) satisfying the initial condition (t_0,x_0) is given by

$$u(t,a) = -\frac{1}{a} + \left(x_0 + \frac{1}{a}\right) e^{a(t-t_0)} \tag{4.7}$$

if $a \neq 0$, while, obviously,

$$u(t,0) = t + x_0 - t_0. \tag{4.8}$$

The continuity of u is obvious at every value of a except $a = 0$. To show that $\lim_{a \to 0} u(t,a) = u(t,0)$, we write (4.7) in the form

$$u(t,a) = \frac{(ax_0 + 1)e^{a(t-t_0)} - 1}{a}.$$

Because the numerator and denominator of this fraction both tend to zero as a tends to 0, it is natural to have recourse to L'Hospital's rule. The derivative of the numerator with respect to a is $x_0 e^{a(t-t_0)} + (ax_0 + 1)(t - t_0)e^{a(t-t_0)}$, while the derivative of the denominator is 1. Now we set $a = 0$ and conclude that $\lim_{a \to 0} u(t,a) = x_0 + t - t_0 = u(t,0)$.

The continuity of the solutions of (4.1) as functions of the parameter a is thus established. Let us take another look at the formula $x = -1/a + ce^{at}$. Clearly x tends to no limit at all as $a \to 0$. The reason is not far to seek. By holding c fixed as we let a tend to zero, we are in fact varying the initial condition, and varying it in such a way that the absolute value of the solution at $t = 0$ tends to infinity.

Turning now to Eq. (4.4), we find that

$$u(t,a) = -(1/a) \log (a(t_0 - t) + e^{-ax_0})$$

if $a \neq 0$, while $u(t,0) = t - t_0 + x_0$. The interval of definition $I(a)$ is $t < t_0 + (1/a)e^{-ax_0}$ if $a > 0$, $t > t_0 + (1/a)e^{-ax_0}$ if $a < 0$, and

R if $a = 0$. Its endpoint for $a \neq 0$ is a continuous function of a, and as $a \to 0$, $I(a) \to R$. The continuity of $u(t,a)$ being obvious for $a \neq 0$, it only remains to show that $\lim_{a \to 0} u(t,a) = u(t,0)$. This follows as before by L'Hospital's rule.

These examples might suggest that $I(a)$ always varies continuously with a. Unfortunately this is not the case. An example to the contrary is given in Exercise 4.

We have denoted the solution whose value at t_0 is x_0 by $u(t,a)$, thereby suppressing its dependence on t_0 and x_0. Our reason for doing so was simply that for our purposes, we could regard t_0 and x_0 as fixed. But t_0 and x_0 are, after all, arbitrary, which is only another way of saying that they are variables, and a better notation is $u(t,t_0,x_0,a)$. The same notation is also useful for the solutions of a differential equation without parameters, $u(t,t_0,x_0)$ denoting the value at t of the solution whose value at t_0 is x_0. For example, if the equation is $dx/dt = x$, then $u(t,t_0,x_0) = x_0 e^{t-t_0}$. In general, if the equation $dx/dt = f(t,x)$ has the uniqueness property, there is a unique function u (of three variables) satisfying $u'(t,t_0,x_0) = f(t,u(t,t_0,x_0))$ and $u(t_0,t_0,x_0) = x_0$, the prime denoting the derivative of u with respect to t. We may call this function the *general* solution of $dx/dt = f(t,x)$, and any function (of t alone) obtained from it by replacing t_0 and x_0 by numbers, a *particular* solution. Similarly, if the equation $dx/dt = f(t,x,a)$ has the uniqueness property for every a (in some range), there is a unique function u (of four variables) satisfying $u'(t,t_0,x_0,a) = f(t,u(t,t_0,x_0,a),a)$ and $u(t_0,t_0,x_0,a) = x_0$. Expressing the solutions of a differential equation in terms of initial conditions (t_0,x_0), rather than in terms of other "arbitrary constants," is frequently advantageous.

This brief discussion was confined to equations with one parameter. However, everything that we have said is equally true for equations with any number of parameters.

EXERCISES

1. Prove that the solutions of the following equations are continuous functions of the parameters.

(a) $\dfrac{dx}{dt} = ax^2$ (b) $\dfrac{dx}{dt} = x + e^{at}$

(c) $\dfrac{dx}{dt} = ax + t$ (d) $\dfrac{dx}{dt} = ax + e^{bt}$

2.   Let $u(t,a)$ be the solution of $dx/dt = \cos^2 ax$ satisfying the initial condition $u(0,a) = 0$.

   (a)   Compute u and prove that it is continuous in a at $a = 0$.

   (b)   Compute $\lim_{t\to\infty} u(t,a)$ for $a \neq 0$.

   (c)   Sketch the curve $x = u(t,a)$ for $a = \dfrac{\pi}{10}, \dfrac{\pi}{100}, 0$.

3.   Let $f(t,a) = \sin at/a$ if $a \neq 0$, and let $f(t,0) = t$. (Note that f is continuous.) Let $u(t,a)$ be the solution of $dx/dt = f(t,a)$ satisfying the initial condition $u(0,a) = 0$. Compute u and prove that it is continuous in a at $a = 0$.

4.   Let $u(t,a)$ be the solution of $dx/dt = 2(t + 1)(x + a^2 - 1)^2$ satisfying the initial condition $u(0,a) = 0$. Show that $u(t,a)$ is defined for all t if $-1 \leq a \leq 1$ but $a \neq 0$, while $u(t,0)$ is defined only for $t > -1$. What can you say about $\lim_{a\to 0} u(t,a)$ if

(a) $t > -1$ 　　　　 (b) $t = -1$ 　　　　 (c) $t < -1$

5.   Physical Examples; Simple Linear Systems

We now make a thorough study of two physical systems described by linear differential equations. The first example illustrates, among other things, how differential equations are used to determine physical constants which cannot be measured directly. The second example leads to a simple system of two first-order linear differential equations and to some of the questions connected with such systems, giving a preview of the subject matter of later chapters.

1.   *Mixing.*   We return to the mixing problem already described in Chapter 1. A vessel contains a volume v of salt solution. A solution of concentration c runs into the vessel at the rate r. The solution in the vessel is kept perfectly mixed and flows out of the vessel at the rate r.

Let m be the mass of salt in the vessel. The rate of change of m with respect to time is equal to the rate at which salt enters the vessel, which is rc, minus the rate at which salt leaves the vessel, which is $r\dfrac{m}{v}$. Thus

$$\frac{dm}{dt} = rc - \frac{r}{v}m. \tag{5.1}$$

Denote the concentration in the vessel by x, so that $x = m/v$. Then $\dfrac{dx}{dt} = \dfrac{1}{v}\dfrac{dm}{dt}$ and therefore

$$\frac{dx}{dt} = \frac{rc}{v} - \frac{r}{v}\,x. \tag{5.2}$$

The solutions are easily found to be

$$x = c + (x_0 - c)e^{-(r/v)t}, \tag{5.3}$$

where $x = x_0$ when $t = 0$. [Since every solution of (5.2) is defined at $t = 0$, it is convenient to set $t_0 = 0$. The choice of time origin is, of course, arbitrary.] We note in particular the constant solution $x \equiv c$ produced by the initial value $x_0 = c$.

Because r and v are positive, $e^{-(r/v)t} \to 0$ as $t \to \infty$, and therefore every solution approaches the constant solution as $t \to \infty$. Note that every solution is monotonic, and that x approaches c from above or from below according as $x_0 > c$ or $x_0 < c$. The geometry of the solution curves reflects obvious physical features of the model.

If r, v and c are known, Eq. (5.3) can be used to predict the value of x on the basis of a single measurement. (Measurements of x can be made on the outflow.) Suppose that the volume v is unknown and impossible to measure directly. Then Eq. (5.3) enables us to calculate it by means of an additional measurement of x provided that $x_0 \neq c$. If $x = x_1$ when $t = t_1$, so that $x_1 = c + (x_0 - c)e^{-(r/v)t_1}$, we find that

$$v = \frac{rt_1}{\log\left(\dfrac{x_0 - c}{x_1 - c}\right)}. \tag{5.4}$$

If v is known, then this formula can be used to calculate r.

This technique finds application in medicine. The rate of flow of blood through the heart (r) can be measured in this way, a dye injected into the heart taking the place of the salt in our example. In this application, $c = 0$; the positive x_0 is achieved by the dye injection. The same technique is used to measure lung volume, with inhaled pure oxygen constituting the inflow (at $c = 0$), and the residue of nitrogen from previously inhaled air furnishing the positive x_0.

Because the differential equation (5.2) depends only on the ratio of r to v, one of these two quantities can be thus determined if the other is known. Although, in principle, two measurements of x suffice to determine the ratio r/v, the presence of experimental error suggests the desirability of averaging a number of measurements. Let us suppose that $c = 0$ and $x_0 > 0$. Then $\log x = \log x_0 - (r/v)t$, so that if x is plotted against t on semi-logarithmic graph paper,* the graph ought to be a line of slope $-r/v$. Therefore, the slope of a line fitted to the experimental data furnishes a measure of this ratio. Another averaging technique is suggested in Exercise 2.

2. *Radioactivity.* A radioactive isotope decays at a rate which is proportional to the mass of the isotope present. Thus, if x denotes the mass of the isotope,

$$\frac{dx}{dt} = -\lambda x, \tag{5.5}$$

where λ is a positive constant called the *decay constant* of the isotope. Obviously

$$x = x_0 e^{-\lambda t}. \tag{5.6}$$

The time in which the mass is halved is called the *half-life* of the isotope. Denoting the half-life by T and setting $x = \frac{1}{2}x_0$, and $t = T$ in (5.6), we find $T = \log 2/\lambda = 0.69/\lambda$.

Many radioactive isotopes are themselves the decay products of other radioactive isotopes. If an isotope A is transformed as a result of radioactive decay into an isotope B, we say that B is the *daughter* of A and that A is the *parent* of B; we symbolize this relationship by $A \to B$. A chain of transformations $A \to B \to C \to \cdots$ terminating in a stable (non-radioactive) isotope is called a *radioactive series*. Clearly Eq. (5.5) is valid only if either the isotope has no parent, or if the parent substance is not present. We now investigate the behavior of an isotope B whose changes in mass are the result both of its own decay and the decay of its parent, A. We suppose that A has no parent. Denote the decay constant of A by λ and that of B by μ. If x is the mass of A, then $dx/dt = -\lambda x$. Denote the mass of B by y. The rate of change of y is equal to the rate at which B gains mass by the decay of A, which is λx,† minus the rate at which B loses mass by its own decay, which is μy. We thus obtain the linear homogeneous system with constant coefficients

$$\frac{dx}{dt} = -\lambda x, \qquad \frac{dy}{dt} = \lambda x - \mu y. \tag{5.7}$$

This system is of a particularly simple type. We can solve the first equation, which involves only x, and substitute the solution into the second equation. The second equation then becomes a linear equation

---

* In effect, $\log x$ is plotted against t.

† We ignore the fact that only part of the mass lost by A appears in the mass of B, i.e., we ignore the mass of the radiation. If it is taken into account, the term λx becomes $\lambda \rho x$, where ρ is the ratio of the atomic weight of B to the atomic weight of A.

in y, and can also be solved. The solutions of the first equation are given by $x = c_1 e^{-\lambda t}$, where c_1 is arbitrary. (Of course, only solutions corresponding to $c_1 \geq 0$ have physical significance.) We substitute into the second equation and obtain

$$\frac{dy}{dt} = -\mu y + c_1 \lambda e^{-\lambda t}. \tag{5.8}$$

The solutions are given by

$$y = e^{-\mu t}[-c_1 \lambda \int e^{(\mu - \lambda) t} \, dt + c_2], \tag{5.9}$$

where c_2 is arbitrary. The integral in this formula is equal to $\dfrac{1}{\mu - \lambda} e^{(\mu - \lambda) t}$ or to t according as $\mu \neq \lambda$ or $\mu = \lambda$, so that these two cases must be discussed separately. We shall take the case $\mu \neq \lambda$, leaving the case of equality as an exercise for the reader. After some simplification, (5.9) becomes

$$y = c_1 \frac{\lambda}{\mu - \lambda} e^{-\lambda t} + c_2 e^{-\mu t}. \tag{5.10}$$

The solutions of the system (5.7) are uniquely determined by initial conditions $x = x_0$, $y = y_0$, $t = t_0$; because all the solutions are defined at $t = 0$, we may take $t_0 = 0$. Setting $t = 0$, $x = x_0$, $y = y_0$ in the formulas for x and y, we find $x_0 = c_1$, $y_0 = c_1 \dfrac{\lambda}{\mu - \lambda} + c_2$. Thus $c_1 = x_0$ and $c_2 = y_0 - \dfrac{\lambda}{\mu - \lambda} x_0$. In terms of initial conditions at $t = 0$, the solutions of (5.7) are therefore given by

$$x = x_0 e^{-\lambda t}$$
$$y = x_0 \frac{\lambda}{\mu - \lambda} e^{-\lambda t} + \left(y_0 - \frac{\lambda}{\mu - \lambda} x_0 \right) e^{-\mu t}. \tag{5.11}$$

Because μ and λ are both assumed different from zero, and indeed positive, there is just one constant solution, namely $x \equiv 0$, $y \equiv 0$, and every solution tends to this constant solution as $t \to \infty$. Throughout the remainder of the discussion, we suppose $x_0 \geq 0$, $y_0 \geq 0$, as required by the physical context.

Let us find the conditions under which y has a relative maximum. If $x_0 = 0$, then $y = y_0 e^{-\mu t}$; unless also $y_0 = 0$, y is a strictly decreasing function with neither relative maxima nor relative minima. Henceforth, suppose $x_0 > 0$. We show first that y cannot have any critical points other than relative maxima, by showing that if $\dfrac{dy}{dt} = 0$, then $\dfrac{d^2 y}{dt^2} < 0$.

Suppose that $dy/dt = 0$ for $t = t_1$, and let $x = x_1$ and $y = y_1$, when $t = t_1$. Differentiating the second of the Eqs. (5.7), we have that

$$\frac{d^2y}{dt^2} = \lambda\frac{dx}{dt} - \mu\frac{dy}{dt} = \lambda(-\lambda x) = -\lambda^2 x_1 \text{ when } t = t_1. \text{ Because } x_1 > 0,$$

$\dfrac{d^2y}{dt^2} < 0$ at t_1, which proves the assertion. The search for relative

maxima is thereby reduced to a search for solutions of $dy/dt = 0$. Computing dy/dt from (5.11) and setting the derivative equal to zero yields the equation

$$\mu\,\frac{\lambda x_0 - (\mu - \lambda)y_0}{\lambda^2 x_0}\,e^{(\lambda - \mu)t} = 1. \tag{5.12}$$

This equation has a solution t if and only if the coefficient of $e^{(\lambda-\mu)t}$ is positive, that is, if and only if

$$\lambda x_0 - (\mu - \lambda)y_0 > 0. \tag{5.13}$$

If $u - \lambda < 0$, this condition is satisfied by all solutions (5.11). We shall set this case aside, and henceforth suppose that $\mu > \lambda$. We may then write (5.13) in the form

$$\frac{y_0}{x_0} < \frac{\lambda}{\mu - \lambda}. \tag{5.14}$$

If this inequality is satisfied, y has a relative maximum, which moreover is unique. Because y has no relative minimum, this relative maximum is an absolute maximum.

If $\dfrac{y_0}{x_0} \geqq \dfrac{\lambda}{\mu - \lambda}$, y is a strictly decreasing function. If equality holds, i.e., if

$$\frac{y_0}{x_0} = \frac{\lambda}{\mu - \lambda}, \tag{5.15}$$

then (5.11) yields $y = \dfrac{\lambda x_0}{\mu - \lambda}\,e^{-\lambda t}$, and therefore $\dfrac{y}{x} = \dfrac{\lambda}{\mu - \lambda}$. Hence, the ratio y/x is constant for all time.

The only way to get a grip on these and subsequent results and gather them together into a coherent whole is to interpret (5.11) as parametric equations of curves in the xy-plane. These curves are shown in Fig. 9. The reader should follow the discussion by referring to this diagram. If (5.15) is satisfied, the curve is simply the portion of the line $y = \dfrac{\lambda}{\mu - \lambda}\,x$ lying inside the first quadrant $x > 0$, $y > 0$. If $x_0 = y_0 = 0$, the curve degenerates to the point $x = 0$, $y = 0$, i.e., the origin. If $x_0 = 0$ and $y_0 > 0$, the curve is the positive y-axis.

Let us show that every curve which satisfies (5.14) enters the first quadrant (the only region of physical interest) by crossing the positive x-axis—in other words, that all y-solutions which have a maximum must vanish for some t. We find from (5.11) that $y = 0$ if and only if

$$\frac{x_0 \lambda}{\mu - \lambda} + \left(y_0 - \frac{\lambda}{\mu - \lambda} x_0 \right) e^{(\lambda - \mu)t} = 0.$$ Since $\mu - \lambda > 0$, a solution t exists if and only if $y_0 - \dfrac{\lambda}{\mu - \lambda} x_0 < 0$, which is precisely (5.14).

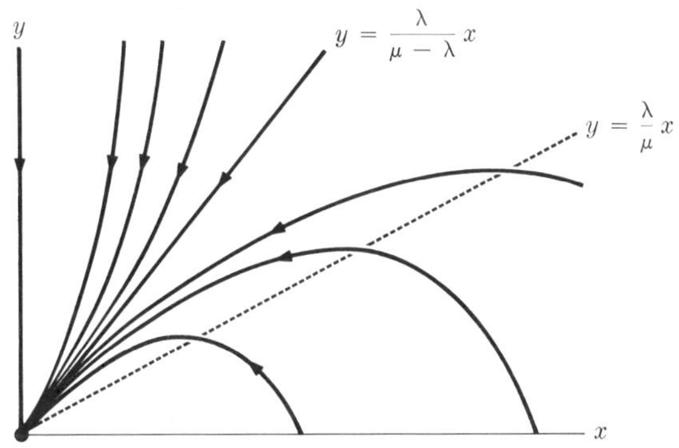

FIG. 9. Radioactive decay, $\mu > \lambda > 0$.

Returning to (5.7), we see that $dy/dt = 0$ if and only if $\lambda x = \mu y$. Hence, the point at which y reaches its maximum lies on the line $y = (\lambda/\mu)x$. This is the dashed line in Fig. 9.

We already know that all of the curves (5.11) approach the origin as $t \to \infty$. To complete the analysis summarized by Fig. 9, we show that $\lim\limits_{t \to \infty} v'(t)/u'(t) = \lambda/(\mu - \lambda)$ for every solution $x = u(t)$, $y = v(t)$ for which $x_0 > 0$. Indeed, we find from (5.11) that if $x_0 > 0$,

$$\frac{v'(t)}{u'(t)} = \frac{\lambda}{\mu - \lambda} + \mu \left(\frac{y_0}{\lambda x_0} - \frac{1}{\mu - \lambda} \right) e^{(\lambda - \mu)t}.$$

Since $\lambda - \mu < 0$, the result is obvious. We conclude that all the curves for which $x_0 > 0$ approach the origin tangentially to the line $y = \dfrac{\lambda}{\mu - \lambda} x$. The exceptional curves are given by $x_0 = y_0 = 0$, i.e., the origin itself, and $x_0 = 0$, $y_0 > 0$, the positive y-axis.

Figure 9 is called the *phase diagram*, or *phase portrait*, of the system (5.7). The reader will find it instructive to extend the diagram to the other three quadrants by lifting the restriction $x \geq 0$, $y \geq 0$. If we interpret the system (5.7) as prescribing the x and y components of a velocity-vector field in the xy-plane, then the phase diagram becomes simply a picture of the flow-lines, with the origin a stagnation point. A systematic introduction to this type of geometric analysis is given in Chapter 6.

EXERCISES

1.   We derive a result which will be needed in some of the following exercises.

(a)   Show that $u(t) = te^{at}$, $a < 0$, attains its maximum at some $t_0 > 0$.

(b)   Let $\lambda > \mu > 0$. Use (a) to show that there is a number k such that $te^{-\lambda t} \leq ke^{-\mu t}$ for all t.

(c)   Conclude that if $\lambda > 0$, $\lim_{t \to \infty} te^{-\lambda t} = 0$. [This also follows from L'Hospital's rule. However, the inequality derived in (b) tells us more, viz., that $te^{-\lambda t}$ tends to zero as quickly as an exponential function.]

Mixing

2.   Consider Eq. (5.2) with $c = 0$, and let u be a solution, $u(0) = x_0 > 0$. Show that $\int_0^\infty u(t)\, dt = \dfrac{v}{r} x_0$. [Hence, the ratio v/r can be computed from the area under the experimentally obtained curve $x = u(t)$.]

3.   Suppose that the concentration of the inflow fluctuates periodically, say $c = c_0 + \sin t$, $c_0 \geq 1$. For simplicity, let $r/v = 1$. Then (5.2) becomes

$$\frac{dx}{dt} = c_0 + \sin t - x.$$

(a)   Compute the solution $x = u(t)$ satisfying $u(0) = x_0$.

(b)   Find the value of x_0 for which u is periodic.

(c)   Show that every solution approaches this periodic solution as $t \to \infty$.

(d)   Let x_0 be chosen as in (b), so that the concentration in the vessel fluctuates periodically. Do the concentrations in the vessel and in the inflow reach their peak values simultaneously? Are these peak values the same? Discuss.

4.   Let the inflow and outflow rates in the mixing problem be r and R, respectively. To simplify the problem, suppose also that $c = 0$.

(a)   Show that the mass m and the volume v satisfy the system

$$\frac{dm}{dt} = -\frac{R}{v}\,m, \qquad \frac{dv}{dt} = r - R.$$

(b)   Assuming that $r > R$, find the solution $m = m(t)$, $v = v(t)$ satisfying the initial condition $m(0) = m_0$, $v(0) = v_0$, where $v_0 > 0$. What is the interval on which the solution is defined?

(c)   Find the solution satisfying the same initial condition on the assumption that $r < R$, and specify its interval of definition.

(d)   To exhibit the dependence of these solutions on the parameters r and R, let us denote them more explicitly by $m = m(t,r,R)$, $v = v(t,r,R)$. Prove that $\lim_{R \to r} m(t,r,R) = m(t,r,r)$.

5.   We are given two vessels of volume v_1 and v_2 filled with salt solutions (Fig. 10). A solution of concentration c runs into the first

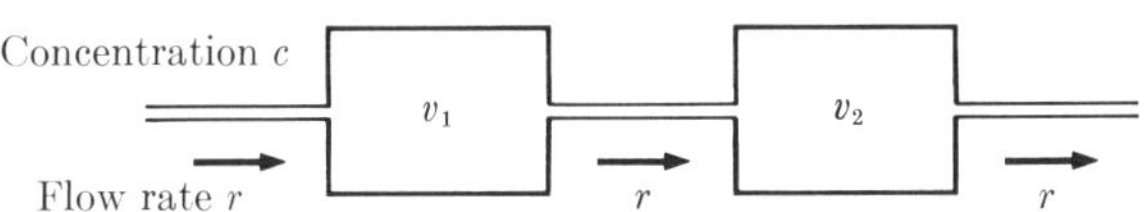

FIG. 10.   See Exercise 5.

vessel at the rate r. The solution runs out of the first vessel and into and out of the second at the same rate. The concentration is kept uniform throughout each vessel. Let m, M denote the mass of salt in the two vessels, and $x = m/v_1$, $y = M/v_2$ the concentrations.

(a)   Show that

$$\frac{dm}{dt} = rc - \frac{r}{v_1}\,m, \qquad \frac{dM}{dt} = \frac{r}{v_1}\,m - \frac{r}{v_2}\,M$$

and hence that

$$\frac{dx}{dt} = \frac{rc}{v_1} - \frac{r}{v_1}\,x, \qquad \frac{dy}{dt} = \frac{r}{v_2}\,x - \frac{r}{v_2}\,y.$$

(b)   Henceforth, suppose that $v_1 = v_2 = r = 1$. The concentrations are thus governed by

$$\frac{dx}{dt} = c - x, \qquad \frac{dy}{dt} = x - y.$$

Compute the solution $x = u(t)$, $y = v(t)$ satisfying $u(0) = x_0$, $v(0) = y_0$, where x_0 and y_0 are non-negative.

(c)   Show that $\lim\limits_{t \to \infty} u(t) = \lim\limits_{t \to \infty} v(t) = c$.

(d)   Let $c = 0$. Show that if $x_0 > y_0 > 0$, v has a relative maximum at some $t > 0$, and compute this value of t.

Radioactivity

6.   Set μ equal to λ in Eq. (5.7), obtaining

$$\frac{dx}{dt} = -\lambda x, \qquad \frac{dy}{dt} = \lambda x - \lambda y.$$

(a)   Compute the solution $x = u(t)$, $y = v(t)$ satisfying $u(0) = x_0$, $v(0) = y_0$. Show that u and v are the limits of the solutions (5.11) as $\mu \to \lambda$.

(b)   Show that if $x_0 > 0$, v has a maximum. Show that when v attains its maximum, $u(t) = v(t)$.

(c)   Show that $\lim\limits_{t \to \infty} u(t) = \lim\limits_{t \to \infty} v(t) = 0$, and that

$$\lim_{t \to \infty} \frac{u'(t)}{v'(t)} = 0.$$

(d)   Interpret (b) and (c) by sketching the curves $x = u(t)$, $y = v(t)$ in the first quadrant of the xy-plane, noting in particular that all approach the origin tangentially to the y-axis.

(e)   Describe how your sketch could be obtained from Fig. 9 by letting μ approach λ in that figure.

7.   To complete the analysis of (5.7), we consider the case $\mu < \lambda$. As before, write $x = u(t)$, $y = v(t)$. It has already been shown [see Eq. (5.13)] that v has a maximum if $x_0 > 0$. The conclusion that the maximum is attained when $y = (\lambda/\mu)\, x$ remains valid when $\mu < \lambda$, as does the conclusion that $u(t)$ and $v(t)$ tend to zero as $t \to \infty$. Show that $\lim\limits_{t \to \infty} u'(t)/v'(t) = 0$, so that the phase portrait in the first quadrant looks exactly like that in the case $\mu = \lambda$.

8.   It is obvious from Fig. 9 that each curve $x = u(t)$, $y = v(t)$ for which $x_0 > 0$ is of the form $y = f(x)$. The function f can be obtained in two ways: by eliminating t from the solutions, i.e., solving the equations $x = u(t)$, $y = v(t)$ for y as a function of x, or by eliminating dt from the differential equation (5.7) itself.

(a)   Eliminate t from (5.11), assuming $x_0 > 0$, and thus show that the curves in Fig. 9 are given by

$$y = \frac{\lambda}{\mu - \lambda} x + \left(y_0 - \frac{\lambda}{\mu - \lambda} x_0 \right) \left(\frac{x}{x_0} \right)^{\frac{\mu}{\lambda}}.$$

(b) Restricting (5.7) to positive values of x, we can divide:

$$\frac{dy/dt}{dx/dt} = \frac{\lambda x - \mu y}{-\lambda x} = \frac{\mu}{\lambda x} y - 1.$$

Because $\dfrac{dy/dt}{dx/dt} = \dfrac{dy}{dx}$, we have

$$\frac{dy}{dx} = \frac{\mu}{\lambda x} y - 1, \qquad x > 0,$$

which is a linear differential equation. Solve the equation on the assumption that $\mu \neq \lambda$. This will again yield the formula obtained in (a).

6. Autonomous Equations; Stability of Equilibrium

One of the principal objects of the theory of differential equations is to obtain information about the solutions of a differential equation when, as is usually the case, it is impossible to represent the solutions by formulas. In the simple case of a first-order autonomous equation

$$\frac{dx}{dt} = f(x) \tag{6.1}$$

it is quite easy to obtain a comprehensive picture of the solutions. Most of the important features of the solutions can indeed be read off more easily from the differential equation itself than from formulas.

We assume that f' is continuous for all x, so that (6.1) has a unique solution u for every initial condition $u(t_0) = x_0$. Throughout this discussion, the term "solution" means a maximal solution of (6.1).

The direction field of (6.1) is independent of t. Hence, if a solution curve is moved parallel to the t-axis, another solution curve is obtained. This primary characteristic of autonomous equations furnishes our first theorem.

Theorem 6.1. *Let u be a solution of 6.1, and let c be a number. Define the function v by $v(t) = u(t + c)$. Then v is also a solution.*

Proof. $v'(t) = u'(t + c) = f(u(t + c)) = f(v(t)).$ $\square$

Theorem 6.2. *Let u be a solution of 6.1. If for some t_0 we have $u'(t_0) = 0$, then u is a constant solution.*

Proof. If $u'(t_0) = 0$, then $f(u(t_0)) = 0$. Define v by $v(t) = u(t_0)$ for all t. Then $v'(t) = 0$ and $f(v(t)) = 0$ for all t, so that v is a solution.

Because $v(t_0) = u(t_0)$, uniqueness implies $v = u$. Hence $u(t) = u(t_0)$ for all t. $\square$

It follows that no solution of (6.1) can have a relative maximum or minimum, unless it is, in fact, a constant solution. We thus have the following characterization of the solutions of (6.1).

Corollary 6.3.   *A solution of* (6.1) *is either constant, strictly increasing, or strictly decreasing.*

We note in particular that (6.1) cannot have any oscillatory solutions. In physical terms, the solutions represent either a growth or a decay process, or—in the case of a constant solution—an equilibrium. The constant solutions are in every way the most important, as we shall see.

The reader is reminded of Theorem 6.3, Chapter 1, according to which the interval of definition of a solution can have an endpoint only if the solution tends to $\pm\infty$ as the endpoint is approached. It follows that if u is a solution defined at t_0 and bounded for all $t \geq t_0$ $(t \leq t_0)$ in its interval of definition, then u is defined for all $t \geq t_0$ $(t \leq t_0)$. We now show that if a solution remains bounded, it must actually be asymptotic to a constant solution.

Theorem 6.4.   *Let u be a solution such that $|u(t)| \leq K$ for $t \geq t_0$. Then $\lim\limits_{t\to\infty} u(t)$ exists. If $\lim\limits_{t\to\infty} u(t) = a$, then $f(a) = 0$.*

Proof.   If u is itself a constant solution, there is nothing to prove. We know that the only other possibilities are that u is an increasing or a decreasing function. A fundamental principle of analysis asserts that a monotonic and bounded function tends to a limit, so that in either case $\lim\limits_{t\to\infty} u(t)$ exists. We denote it by a. It remains to show that $f(a) = 0$.

Since f is continuous,
$$\lim_{t\to\infty} f(u(t)) = f(a),$$
and since
$$u'(t) = f(u(t)),$$
it follows that
$$\lim_{t\to\infty} u'(t) = f(a).$$

Suppose $f(a) > 0$. Then for t sufficiently large, say $t \geq T$,
$$u'(t) > \frac{f(a)}{2}.$$

Hence, integrating from T to t, where $t > T$, we obtain

$$u(t) - u(T) > \frac{f(a)}{2}(t - T),$$

or

$$u(t) > u(T) + \frac{f(a)}{2}(t - T), \qquad t > T.$$

The right side of this last inequality tends to ∞ as $t \to \infty$, and therefore the left side does also. This contradiction shows that $f(a)$ cannot be positive. The assumption that $f(a) < 0$ leads in the same way to the contradiction that $u(t) \to -\infty$ as $t \to \infty$. We conclude that $f(a) = 0$. $\quad\square$

We leave it to the reader to show by a similar argument that if $|u(t)| \leq K$ for $t \leq t_0$, then

$$\lim_{t \to -\infty} u(t) = b$$

exists, and $f(b) = 0$. Briefly, a solution bounded to the left approaches a constant solution as $t \to -\infty$.

Theorem 6.5. *If* (6.1) *has no constant solution, then every solution assumes all real values.*

Proof. No solution can be bounded either to the right or to the left because if it were, it would be asymptotic to a constant solution. Because the solutions are monotonic, they have neither an upper nor a lower bound. $\quad\square$

It may be worthwhile to emphasize a point which so far has not been discussed. By virtue of the uniqueness theorem, the mere existence of constant solutions gives information about the intervals of definition of the other solutions. It is clear, for example, that as long as there is at least one constant solution, all of the solutions have unbounded intervals of definition. It is equally clear that if there are two constant solutions, the solutions which lie between them are defined for all t.

The results which we have obtained enable us to get a very good idea of the general behavior of the solutions of an autonomous equation. Let us take as an example the equation

$$\frac{dx}{dt} = x - x^3.$$

A sketch of the solution curves is given in Fig. 11. The constant solutions are $x = -1$, $x = 0$, and $x = 1$. Because $f(x) = x - x^3 < 0$ if $-1 < x < 0$, the solutions which lie between $x = -1$ and $x = 0$ are decreasing; they tend to -1 as $t \to \infty$ and to 0 as $t \to -\infty$. If $0 < x < 1$, $f(x) > 0$; hence the solutions in this strip are increasing, approaching 1 as $t \to \infty$ and 0 as $t \to -\infty$. Note that all solutions u such that $-1 \leqq u(t) \leqq 1$ for some t are defined for all t. Because $f(x) < 0$ if $x > 1$, the solutions with initial values > 1 are decreasing and tend to 1 as $t \to \infty$. Finally, $f(x) > 0$ if $x < -1$ and therefore the solutions which lie below the line $x = -1$ are increasing and approach -1 as $t \to \infty$. If we are given a solution curve in each of the four strips $x < -1$, $-1 < x < 0$, $0 < x < 1$, and $x > 1$, all the other curves in that strip are simply translates of the given curve. Note that the curves in the strips $-1 < x < 0$ and $0 < x < 1$ are represented in Fig. 11 as having just one inflection point each. The justification of this feature of the sketch is left as an exercise.

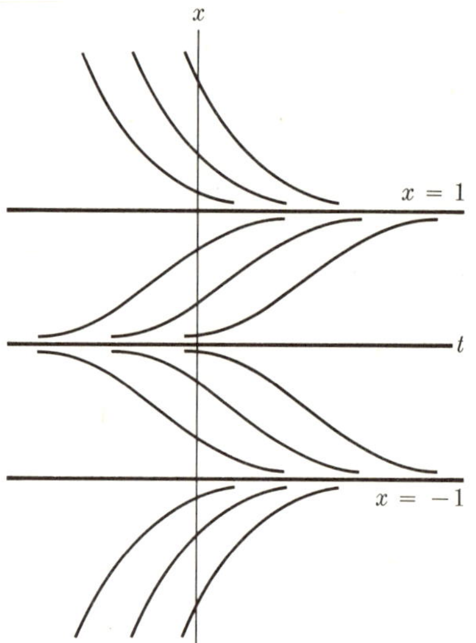

Fig. 11.   Solution curves of $dx/dt = x - x^3$

Stability. We have already mentioned that a constant solution has the physical meaning of an equilibrium, literally "equal balance." The term comes from mechanics, where it carries the connotation of the state of rest of a body acted upon by mutually cancelling forces. We are not yet in a position to give a mathematical discussion of mechanical systems, because such systems are governed by second-order differential equations. However, a glance at the familiar example of a pendulum will help to motivate the mathematical definitions to follow.

A pendulum has two positions of equilibrium. In the first, the bob is directly below the point of support; in the second, it is directly above it. If the pendulum is displaced from the first equilibrium, it will swing back and forth about the equilibrium position. The amplitude of its swing can be made as small as desired by making the displacement small. Let us suppose for the moment that there is neither air resistance nor internal friction to retard the motion of the pendulum. The pendulum then persists in its small oscillations—its motion is periodic and of small amplitude. The situation is quite different with respect to disturbances of the second equilibrium. If the pendulum is displaced, however slightly, from this equilibrium, it will execute large periodic swings about the other equilibrium position; it will depart widely from the original equilibrium, no matter how small the initial displacement. We say that the second equilibrium is *unstable*, and that the first is *stable*. The distinction arises when we consider not merely the equilibrium itself, but the motions which originate *near* the equilibrium. If, now, we consider the effect of friction on the motion of the pendulum, we see that any oscillation about the stable equilibrium will gradually die out; in each swing some energy is lost, and the amplitude of each swing is somewhat smaller than that of the preceding one. Thus, a small displacement of the pendulum from the position of stable equilibrium produces a motion which not only remains near the equilibrium, but actually approaches it. We say that the equilibrium is *asymptotically stable*. This type of equilibrium is possible only in the presence of frictional forces.

If the reader was inclined to dismiss the unstable equilibrium of the pendulum as a theoretical fiction without physical significance, his instinct set him on the right track. It is precisely because a state of unstable equilibrium cannot persist in nature, where the minute disturbances which suffice to demolish it are always present, that the distinction between stability and instability is of such very great importance.

For the remainder of this section, we shall use the terms "equilibrium solution" and "constant solution" interchangeably.

Definition 6.6.   Let $x = c$ be an equilibrium solution of (6.1). We say that c is *stable* if for every $\epsilon > 0$, there is a $\delta > 0$ such that if u is a solution satisfying $|u(t_0) - c| < \delta$ for some t_0, then $|u(t) - c| < \epsilon$ for all $t \geqq t_0$. If δ can be chosen so that $|u(t_0) - c| < \delta$ implies $\lim_{t \to \infty} u(t) = c$, we say that c is *asymptotically stable*. If c is not stable, we say that it is *unstable*.

Note that all solutions coming sufficiently close to a stable solution for some value of t are defined for all greater values of t.

Referring to Fig. 11, we see that $x = -1$ and $x = 1$ are asymptotically stable, whereas $x = 0$ is unstable. A suitable δ for either of the former is obviously $\delta = 1$; it is, in fact, the largest such.

A necessary and sufficient condition that $x = c$ is asymptotically stable is clearly that $f(x) > 0$ for all x near c and smaller than c, and $f(x) < 0$ for all x near c and larger than c—in other words, that $f(x)$ changes sign from positive to negative at c. The following simple theorem is stated here not for its own sake, but in anticipation of its important generalization to systems of differential equations.

Theorem 6.7.   *Let $f(c) = 0$. If $f'(c) < 0$, c is asymptotically stable. If $f'(c) > 0$, c is unstable.*

Proof.   If $f'(c) < 0$, then f is decreasing at c, i.e., it changes sign from positive to negative. Hence, c is asymptotically stable. If $f'(c) > 0$, f changes sign from negative to positive at c. Let $r > 0$ be a number such that $c - r < x < c$ implies $f(x) < 0$ and $c < x < c + r$ implies $f(x) > 0$. Every solution u satisfying $c - r < u(t_0) < c$ for some t_0 is, therefore, a decreasing function, and is either asymptotic to a constant solution $x = a \leqq c - r$, or tends to $-\infty$. Likewise, every solution u satisfying $c < u(t_0) < c + r$ for some t_0 is an increasing function and is, therefore, either asymptotic to a constant solution $x = b \geqq c + r$, or tends to ∞. Hence, if $\epsilon < r$, there is no $\delta > 0$ satisfying the definition of stability.   □

If $f'(c) = 0$, the solution $x = c$ may be stable, unstable, or even asymptotically stable. For example, every solution of $dx/dt = 0$ is a stable equilibrium, the zero solution of $dx/dt = x^2$ is unstable, and the zero solution of $dx/dt = -x^3$ is asymptotically stable. When we want to assert that a solution is stable but not asymptotically stable, we say that it is *weakly stable*. A solution $x = c$ of (6.1) is weakly stable if and only if every interval about c contains infinitely many solutions of $f(x) = 0$. The proof of this fact is left as an exercise.

A Population Model.   Consider an animal species living in an ideal environment with an unlimited food supply and unlimited living space. The simplest hypothesis concerning the growth rate of such an animal population is that the number of births and the number of deaths per unit time are each proportional to x, the size of the population. The rate of change of the population is thus given by $dx/dt = ax$, where a (= birth rate − death rate) must be positive if the species is viable. Hence, $x = x_0 e^{at}$ and the population increases exponentially. In some circumstances, this is a good approximation to the truth, but it is obviously inadequate for large values of x or t. We can improve our model by postulating a maximum population $M > 0$ which the environment can support, and by assuming that the growth of the population is proportional not only to x, but also to $M - x$. We then obtain a differential equation of the form $dx/dt = b(M - x)x$, where $b > 0$ and $M > 0$. A last refinement takes into account that a very scarce species is likely to die out, so that we postulate a minimum population m, where $0 < m < M$, needed for survival. We assume that the death rate exceeds the birth rate if $x < m$ and obtain an equation of the form $dx/dt = c(M - x)(x - m)x$, $c > 0$. It has three equilibrium solutions, $x = M$, m, 0; the last corresponds to extinction. The reader should verify that $x = M$ and $x = 0$ are asymptotically stable, while $x = m$ is unstable.

This crude attempt to account for population changes is highly unsatisfactory in many respects, but it is about the best that can be done in terms of a single first-order equation. A more adequate model must take into account the interaction between a species and its environment. We shall discuss the simplest such model in Chapter 6.

EXERCISES

1. (a) Show that a non-constant solution u of (6.1) has an inflection point at t_0 if and only if f has a strict relative maximum or minimum at $u(t_0)$.

(b) Let u be a bounded solution of (6.1). Show that $u''(t) = 0$ for at least one t.

2. Show that the inflection points of the solution curves of $dx/dt = x - x^3$ (Fig. 11) lie on the lines $x = 1/\sqrt{3}$ and $x = -1/\sqrt{3}$.

3. Describe and sketch the solution curves of the following equations.

(a) $\dfrac{dx}{dt} = -xe^{-x^2}$ (b) $\dfrac{dx}{dt} = x^2 - 1$

(c) $\dfrac{dx}{dt} = -x^3 + x^2 + x - 1$ (d) $\dfrac{dx}{dt} = \arctan(x^2 + x)$

4. Sketch the solution curve of the given equation passing through the given point (t_0, x_0).

(a) $\dfrac{dx}{dt} = x^2 - x$; $(0, \tfrac{1}{2})$ (b) $\dfrac{dx}{dt} = 5x - x^5$; $(0, 1)$

5. Find the constant solutions of each of the following differential equations and identify them as weakly stable, asymptotically stable, or unstable.

(a) $\quad \dfrac{dx}{dt} = \sin x$

(b) $\quad \dfrac{dx}{dt} = \sin^2 x$

(c) $\quad \dfrac{dx}{dt} = x^2 e^{-x}$

(d) $\quad \dfrac{dx}{dt} = f(x)$, where $f(x) = x^3 \sin \dfrac{1}{x}$ for $x \neq 0$, $f(0) = 0$

(e) $\quad \dfrac{dx}{dt} = f(x)$, where $f(x) = \begin{cases} 0 \text{ for } x \leqq 0 \\ x^2 - x^3 \text{ for } x \geqq 0 \end{cases}$

6.   Let u be the solution of $dx/dt = x(x-2)(x-3)$ which satisfies $u(0) = 1$. Find $\lim\limits_{t \to \infty} u(t)$ and $\lim\limits_{t \to -\infty} u(t)$.

7.   Find the constant solutions of $dx/dt = (x-1)(x-a)$ and discuss how their stability or instability depends on the parameter a.

7. Euler Polygons

In the preceding section, we gave an elementary introduction to the qualitative or geometric theory of differential equations. The purpose of the qualitative theory is to characterize the solutions of a differential equation "in the large," that is, to describe the shape of individual solution curves, and even more importantly, to describe the geometric relationship among a whole family of solution curves. Questions of boundedness and stability are typical qualitative problems.

An entirely different kind of problem is to compute the numerical values of a solution. If we are given a differential equation and want to know the value $u(t_1)$ of a solution u satisfying a given initial condition $u(t_0) = x_0$, the only means at our disposal so far is to find a formula for $u(t)$—that is, to express $u(t)$ in terms of functions whose values are easily computed or are available in tables. If this is impossible, we must make our own table for u. Methods for constructing such tables are called "numerical methods." The study of numerical methods properly belongs to numerical analysis rather than the theory of differential equations and we shall not pursue it in any detail. The following discussion is limited to a particularly simple numerical method, which also happens to be the oldest historically. It is due to Euler, and goes by the name of "Euler polygons."

We consider the equation $dx/dt = f(t,x)$ and suppose for simplicity that f and $\partial f / \partial x$ are continuous for all (t,x). We are given an initial condition $u(t_0) = x_0$ and we want to compute the values $u(t)$ to within a prescribed degree of accuracy. Euler's method is nothing more than a systematic exploitation of the notion of the direction field. We draw a line segment from (t_0,x_0) with slope $f(t_0,x_0)$. Denoting the right-hand end of the segment by (t_1,x_1), we continue the segment by means of another segment of slope $f(t_1,x_1)$, and so on. We thus obtain a polygon starting at (t_0,x_0), and it is reasonable to hope that if its segments are sufficiently short, the polygon will closely approximate the curve $x = u(t)$. If we are interested in values of $u(t)$ for $t < t_0$, we simply continue the polygon to the left of (t_0,x_0). We now describe this procedure somewhat more formally.

Let it be required to approximate $u(t)$ for t in a prescribed interval $t_0 \leq t \leq b$. (We make the tacit assumption that u is defined at b.) We subdivide the interval $[t_0,b]$ into n equal subintervals of length $h = (b - t_0)/n$ by means of the points $t_k = t_0 + kh$, $k = 0, 1, \ldots, n$. Let $x_1 = x_0 + f(t_0,x_0)h$, $x_2 = x_1 + f(t_1,x_1)h$, and, in general, $x_{k+1} = x_k + f(t_k,x_k)h$, $k = 0, 1, \ldots, n - 1$. Then the points (t_k,x_k) are the vertices of the Euler polygon corresponding to our partition of $[t_0,b]$. The number h is called the *step size* of the polygon. The polygonal function itself is given by v_h, where $v_h(t) = x_k + f(t_k,x_k)(t - t_k)$ for $t_k \leq t \leq t_{k+1}$, $k = 0, 1, \ldots, n - 1$. It can be shown that for any $\epsilon > 0$, there is an h such that $|u(t) - v_h(t)| < \epsilon$ for all $t_0 \leq t \leq b$.

From the numerical point of view, Euler's method yields a table of values (t_k,x_k), where x_k is an approximation to $u(t_k)$. The accuracy of x_k depends in general on k, on the step size h, on the number of decimal places carried in the calculations, on the accuracy of the values $f(t_k,x_k)$ used, and on the solution being approximated. We naturally expect that the accuracy of x_k decreases with increasing k. However, if the solutions near u tend toward u with increasing t, the reverse may be the case, because the differential equation will tend to correct the deviations of the polygon from the true solution curve. Qualitative considerations are therefore important even in connection with purely numerical questions.

The method of Euler polygons can also be applied to systems of first-order equations. We leave this simple generalization to the reader (Exercise 3). Since any differential equation or system can be written as a system of first-order equations, Euler's method is generally applicable.

The numerical methods in actual use on computers are essentially refinements of Euler's method. While in a sense Euler's method does everything that can be asked of a numerical method, more complex methods are frequently more efficient, i.e., they require less computer time to achieve a given degree of accuracy. This greater efficiency, or speed, is important not only because computers are expensive to use, but because in many cases the results of the computations are useless unless they are obtained very quickly. It should be pointed out that the numerical solution of even a very complicated system of differential equations is among the easiest things for a computer to do, and usually requires only a few minutes of computer time. For an extensive and rigorous discussion of numerical methods the reader is referred to: Henrici, *Discrete Variable Methods in Ordinary Differential Equations*, Wiley, New York, 1962.

EXERCISES

1. (a) Use Euler's method to approximate the solution u of $dx/dt = x$ satisfying $u(0) = 1$, on the interval [0,1]. Use the step size $h = 0.1$ and make a table of the values t_k, x_k, $0 \leq k \leq 10$. Carry four decimal places in the calculations. Use a table of e^t to compute the errors $e^{0.1k} - x_k$ to the nearest one-hundredth.

(b) Show that $x_k = (1 + 0.1)^k$, so that $x_{10} = (1 + \frac{1}{10})^{10}$.

(c) Show that if $h = 1/n$, the approximation x_n to e is $\left(1 + \dfrac{1}{n}\right)^n$. Hence, $\lim\limits_{n \to \infty} x_n = e$.

2. Approximate the solution u of $dx/dt = t - 2x$ satisfying $u(0) = 0.5$ on the interval [0,1]. Use the step size $h = 0.1$. What is the error in the value of x_{10}?

3. Apply the method of Euler polygons to a system of first-order equations $\dfrac{dx_i}{dt} = f_i(t,x_1, \ldots ,x_n), \quad i = 1, \ldots , n.$

4. Let u be the solution of $\dfrac{d^2x}{dt^2} = -x$ satisfying the initial condition $u(0) = 0$, $u'(0) = 1$. Show how one can construct a table of values for u using Euler polygons. (Hint: Write the equation as a system of first-order equations.)

8. General Comments on the Theory of Differential Equations

By now the reader has a fairly good idea what the subject of differential equations is all about. The great French mathematician Henri Poincaré, the founder of the modern theory of differential equations, divided the subject into a *qualitative* and a *quantitative* part. This distinction is of inestimable value if we wish to keep our goals steadily before our eyes, and avoid confusing means and ends. We already touched upon this distinction on page 19, where we noted that it is—at least implicitly—familiar to the reader from his study of the elementary functions of calculus.

The solutions of an nth-order system of differential equations are represented by curves filling out a space of $n + 1$ dimensions. To describe the shape of these curves is the qualitative problem posed by the system. The physical or mathematical origin of the system almost invariably directs the inquiry to certain special questions concerning the shape of these curves. Frequently we are interested only in those curves which pass through a prescribed region of the space, i.e., which satisfy certain restrictions on the initial conditions, and we may wish to know if these curves are bounded, if they all approach a particular curve as $t \to \infty$, if among them there is a curve which repeats itself periodically, and so on. We may also need to know certain numbers associated with these curves. For example, if we have established the existence of a periodic solution we may also want to have an accurate estimate of its period, and if a solution tends to a limit as $t \to \infty$, we may wish to know the value of this limit. The determination of such numbers is still a part of the qualitative problem posed by the differential equation, because the significance of these numbers lies in the information they supply about the curve as a whole.

It is easy to see that the quantitative problem of making a table of values, which we discussed in the preceding section, amounts to locating a finite number of points on one or several solution curves. The growth of numerical analysis and computer technology has largely rendered the construction of numerical tables a matter of clerical routine to be entrusted to a computer programmer. The reader who wishes to gain familiarity with the uses and limitations of computers, and has access to one, should make a practice of supplementing the exercises given in this book by selecting simple differential equations or systems for careful qualitative analysis followed by construction of tables or curves for a number of solutions by means of the computer.

The difference between qualitative and quantitative problems and methods is forcefully demonstrated by the most famous problem in the history of differential equations, that of predicting the motion of the planets on the basis of Newtonian mechanics. The positions of the planets are routinely predicted with great accuracy for many years ahead by means of numerical methods and computers. However, we are unable to prove that the earth will never escape from the solar system, or that it will never collide with the sun. Obviously we cannot hope to settle such qualitative questions by means of tables. Indeed, most advances toward proving the "stability of the solar system" have involved ideas and techniques which are essentially geometric. Of course, our nearly total ignorance about the long-term behavior of the solar system is not a matter of practical concern, and numerical methods give us all the information we need. It must be said, however, that nothing has contributed more to the development of the theory of differential equations than the persistent effort of mathematicians over several centuries to solve this difficult astronomical problem.

The distinction between qualitative and quantitative problems does not exclude the possibility that quantitative information may be helpful in the qualitative analysis of a differential equation, or that qualitative information may be needed before reliance can be placed on numerical results. Here we must mention that the engineer and the applied mathematician have still another string to their bow: their scientific or technical knowledge of the physical system described by the differential equation. All major advances toward the solution of mathematical problems posed by science and technology were made by men who were intimately acquainted with the empirical background of the problems. Sometimes, understanding a mathematical problem in physical terms is the first step toward its solution. Without question, many subtle mathematical concepts are most easily understood in terms of a physical example. The purpose of the numerous physical examples included in this book is not to teach the reader physics or to train him to apply mathematics to physical problems. On the contrary, their purpose is to teach him to use his physical intuition in thinking about differential equations, and to introduce him to new mathematical concepts in a simple physical setting.

MISCELLANEOUS EXERCISES FOR CHAPTER 2

1.   Show that if u is a solution of $dx/dt = f(x)$ and $v(t) = u(at)$, then v is a solution of $dx/dt = af(x)$.

2.   Let f' be continuous for all x and suppose that every solution of $dx/dt = f(x)$ is defined for all t. Denote the solution whose value at $t = 0$ is η by $u(t,\eta)$, so $u(0,\eta) = \eta$. Prove that $u(t_1, u(t_0,\eta)) = u(t_0 + t_1, \eta)$ for all t_0, t_1, η.

3.   Let u be a solution of $dx/dt = f(t,x)$, $u(t_0) = x_0$. Show that if $|f(t,x)| \leq M$ for all (t,x), then $|u(t) - x_0| \leq M|t - t_0|$. Conclude that u is defined for all t.

4. Consider the equation $dx/dt = f(x)$, where f' is continuous for all x, f has period p, and $f(x) > 0$ for all x. By Theorem 6.5, page 59, the range of every solution is R, the set of all reals, and because a continuous periodic function is bounded, it follows from the preceding exercise that the domain is also R. Let u be the solution satisfying $u(0) = 0$ and define the number ω by $u(\omega) = p$. Prove:

(a) $\quad \omega = \displaystyle\int_0^p \frac{dx}{f(x)}$

(b) $\quad u(t + \omega) = u(t) + p \qquad$ for all t

The following exercises deal with qualitative properties of solutions of non-autonomous equations.

5. Let $f(t,x)$ be continuous for all (t,x), and suppose there are numbers a, b with $a < b$ such that $f(t,a) > 0$, $f(t,b) < 0$ for all $t \geq 0$. Let u be a solution of $dx/dt = f(t,x)$ such that $a \leq u(0) \leq b$. Show that $a \leq u(t) \leq b$ for $t \geq 0$. It follows that u is defined for all $t \geq 0$.

6. Let $g_1(t)$, $g_2(t)$ be continuous and satisfy $a < g_1(t) < b < g_2(t)$ for some a, b and all $t \geq 0$. Prove that if u is a solution of $dx/dt = (x - g_1(t))(x - g_2(t))$ and $a \leq u(0) \leq b$, then $a \leq u(t) \leq b$ for $t \geq 0$. Use Exercise 5.

7. Let $f(t,x)$ have period p in t, that is, $f(t + p, x) = f(t,x)$ for all (t,x), and let f and $\partial f/\partial x$ be continuous. Let u be a solution of $dx/dt = f(t,x)$. Prove:

(a) If $v(t) = u(t + p)$, then v is also a solution

(b) If u is defined on $[0,p]$ and $u(0) = u(p)$, then u is defined for all t and has period p.

8. Let ϕ be a continuous function which satisfies $a \leq \phi(x) \leq b$ for $a \leq x \leq b$.

(a) Let $g(x) = x - \phi(x)$ for $a \leq x \leq b$. Use the intermediate value theorem to show that there is at least one number x in $[a,b]$ for which $g(x) = 0$.

(b) Conclude that $\phi(x) = x$ for some x in $[a,b]$.

9. Let $f(t,x)$ satisfy the hypotheses of Exercises 5 and 7. We show that $dx/dt = f(t,x)$ has a solution of period p.

(a) Let $u(t,\eta)$ be the solution whose value at $t = 0$ is η, that is $u(0,\eta) = \eta$. Show that if $a \leq \eta \leq b$, then $a \leq u(t,\eta) \leq b$ for all $t \geq 0$.

(b) Let $\phi(\eta) = u(p,\eta)$ for $a \leq \eta \leq b$. It is easy to prove that ϕ is continuous. Accepting the continuity of ϕ, use Exercise 8 to show that there is an η_0 such that $\phi(\eta_0) = \eta_0$.

(c) Prove that $u(t + p, \eta_0) = u(t,\eta_0)$ for all t.

10.   Show by means of the preceding exercise that each of the following equations has a solution of period 2π.

(a) $\dfrac{dx}{dt} = \sin t - x^3$

(b) $\dfrac{dx}{dt} = (x + \sin^2 t)(x - e^{\sin t})$

11.   Let $f(t,x)$ have period p in t, that is, $f(t + p, x) = f(t,x)$. The question arises whether a periodic solution of $dx/dt = f(t,x)$ must necessarily have period p. The answer is yes, provided that the differential equation has the uniqueness property.

(a)   Let u be a continuous function of period ω, and let p be a positive number. Let $w(t) = u(t + p) - u(t)$, so that w is also continuous and has period ω. Show that the mean value of w is zero.

(b)   Let u be a solution of period ω of $dx/dt = f(t,x)$, where f has period p in t. Let $v(t) = u(t + p)$, so that v is also a solution and has period ω. Show that $u(t_0) = v(t_0)$ for some t_0. Conclude that if the differential equation has the uniqueness property then u has period p.

12.   In Exercise 10 we showed that

$$\frac{dx}{dt} = \sin t - x^3 \tag{1}$$

has at least one solution of period 2π. We now show that it has exactly one periodic solution.

(a)   Let u and v be distinct solutions of (1) defined on $[0,2\pi]$. Show that $\displaystyle\int_0^{2\pi} u^3(t) \neq \int_0^{2\pi} v^3(t)\, dt$.

(b)   Let u be a periodic solution of (1), so that by Exercise 11 u has period 2π. Show that $\displaystyle\int_0^{2\pi} u^3(t)\, dt = 0$. Conclude that (1) has exactly one periodic solution. (It can be shown by a more sophisticated argument that all solutions of (1) tend toward the periodic solution as $t \to \infty$. The reader will find it instructive to plot, or construct tables for, a few solutions of this equation by means of a computer.)

13.   (a)   Let u be the periodic solution of (1), and let $v(t) = -u(t + \pi)$. Show that v is a solution of (1).

(b)   Conclude that $u = v$, so that $u(t) = -u(t + \pi)$. (Hence the graph of u has the same type of symmetry as the graphs of the sine and cosine functions: it is carried into itself by a reflection in the t-axis followed by a translation through π units parallel to the t-axis.)

(c)   Show that the mean value of u is zero.

3

LINEAR EQUATIONS

The only prerequisite for this chapter is Section 3 of Chapter 2. In turn, the only portion of the present chapter which will be needed in later chapters is Section 1, which is an introduction to complex numbers and complex-valued functions.

1. Preliminaries

Complex Numbers. A *complex number* is a number of the form $a = \alpha + i\beta$ where α and β are real numbers and $i^2 = -1$. If $\beta = 0$, a is real; hence every real number is a complex number. If $\alpha = 0$, we say that a is *imaginary*. If $\alpha = 0$ and $\beta \neq 0$, we say that a is *purely imaginary*. We say that α is the *real part*, and β the *imaginary part*, of a, and write $\alpha = \operatorname{Re} a$, $\beta = \operatorname{Im} a$. Equality of complex numbers means equality of their real and imaginary parts, i.e., $a = b$ if and only if $\operatorname{Re} a = \operatorname{Re} b$ and $\operatorname{Im} a = \operatorname{Im} b$. Complex numbers obey the same arithmetic rules as do real numbers. Thus the sum and product of two complex numbers are given by $(\alpha + i\beta) + (\gamma + i\delta) = \alpha + \gamma + i\beta + i\delta = \alpha + \gamma + i(\beta + \delta)$ and $(\alpha + i\beta)(\gamma + i\delta) = \alpha\gamma + i\alpha\delta + i\beta\gamma - \beta\delta = \alpha\gamma - \beta\delta + i(\alpha\delta + \beta\gamma)$. The reciprocal of $\alpha + i\beta$ is found by writing

$$\frac{1}{\alpha + i\beta} = \frac{1}{\alpha + i\beta} \cdot \frac{\alpha - i\beta}{\alpha - i\beta} = \frac{\alpha - i\beta}{\alpha^2 + \beta^2} = \frac{\alpha}{\alpha^2 + \beta^2} - i\frac{\beta}{\alpha^2 + \beta^2}.$$

If $a = \alpha + i\beta$, we write $\bar{a} = \alpha - i\beta$; $\bar{a}$ is called the *complex conjugate* of a. Note that $\bar{a} = a$ if and only if a is real. It is readily verified that $\overline{a + b} = \bar{a} + \bar{b}$ and that $\overline{ab} = \bar{a}\,\bar{b}$.

The relations of greater and less are not defined for strictly complex numbers. Whenever we write $a < b$, $a \leq b$, etc., it is understood that a and b are real. Similarly the terms "positive" and "negative" apply only to real numbers. The *absolute value* of $\alpha + i\beta$ is defined by $|\alpha + i\beta| = \sqrt{\alpha^2 + \beta^2}$ and is therefore a non-negative real number. If $\beta = 0$, this definition reduces to the familiar definition of the absolute value of a real number. Note that $|a| = 0$ if and only if $a = 0$.

Complex numbers have a simple geometric interpretation as points in the plane. With the usual x and y coordinate axes we assign the number $x + iy$ to the point (x,y). Every real number is thus assigned to its usual place on the x-axis, while an imaginary number iy is assigned to the point $(0,y)$ on the y-axis. In this context, the x-axis is called the *real axis*, the y-axis the *imaginary axis*, and the plane the *complex plane*. If we define addition of points by $(x,y) + (x',y') = (x + x', y + y')$, which is simply addition of their position vectors, then the point corresponding to the sum of two complex numbers is the sum of the corresponding points. If we go a step further and define the product of two points by $(x,y)(x',y') = (xx' - yy', xy' + yx')$, then products of points and products of numbers correspond also. The distinction between complex numbers and pairs of real numbers, i.e., points of the plane, is thus essentially a matter of notation. To make the correspondence complete, note that we can multiply a point (x,y) by a real number c according to the usual rule $c(x,y) = (cx,cy)$, so that $(x,0) = x(1,0)$, $(0,y) = y(0,1)$ and thus $(x,y) = (x,0) + (0,y) = x(1,0) + y(0,1)$. Since $(1,0)$ corresponds to 1 and $(0,1)$ to i, it only remains to write $1 = (1,0)$ and $i = (0,1)$, and we have $(x,y) = x + iy$.

If $a = \alpha + i\beta \neq 0$, then

$$a = \sqrt{\alpha^2 + \beta^2}\left(\frac{\alpha}{\sqrt{\alpha^2 + \beta^2}} + i\frac{\beta}{\sqrt{\alpha^2 + \beta^2}}\right) = |a|\left(\frac{\alpha}{|a|} + i\frac{\beta}{|a|}\right).$$

Since the absolute value of the number in parentheses is 1, we have represented a as the product of a positive number and a number of absolute value 1. This is called the *polar form* of a, for the following reason. If (r,θ) are polar coordinates of $a = (\alpha,\beta)$, then $r = |a|$, $\cos\theta = \alpha/|a|$, and $\sin\theta = \beta/|a|$, and therefore the polar form reduces to $a = r(\cos\theta + i\sin\theta)$. Note that $r = |a|$ is a uniquely determined positive number, while θ is determined only to within integer multiples of 2π. θ is called an *argument* of a, and we write $\theta = \arg a$. If $0 \leq \theta < 2\pi$, we say that θ is the *principal* argument of a and write $\theta = \text{Arg } a$.

Given two complex numbers $a = r(\cos\theta + i\sin\theta)$ and $b = \rho(\cos\varphi + i\sin\varphi)$, we have for their product

$$ab = r\rho[\cos\theta\cos\varphi - \sin\theta\sin\varphi + i(\cos\theta\sin\varphi + \sin\theta\cos\varphi)]$$
$$= r\rho[\cos(\theta + \varphi) + i\sin(\theta + \varphi)].$$

Thus $|ab| = |a|\,|b|$ and $\arg(ab) = \arg a + \arg b$, the latter equation having the meaning that the sum of an argument of a and an argument of b is an argument of ab. We now have a geometric interpretation of complex multiplication: the product of a and b is obtained by multiplying the lengths of their position vectors and adding their arguments. Note in particular that the effect of multiplying a number by i is to rotate it through an angle of $90°$ about the origin.

If $a = b$, the formula for the product reduces to $a^2 = r^2(\cos 2\theta + i\sin 2\theta)$. By induction it follows that for any positive integer n,

$$a^n = r^n(\cos n\theta + i\sin n\theta). \tag{1.1}$$

We leave it to the reader to show that this formula is in fact valid for all integers n. If $r = 1$, we have DeMoivre's formula:

$$(\cos\theta + i\sin\theta)^n = \cos n\theta + i\sin n\theta. \tag{1.2}$$

Polynomials. A *polynomial of degree n* is a function of the form $f(x) = a_0x^n + a_1x^{n-1} + \cdots + a_n$, where n is a non-negative integer, the coefficients a_k are complex numbers, and $a_0 \neq 0$. We allow the variable x to assume complex values. If $f(c) = 0$, we say that c is a *root* of $f(x)$. The following theorem is the sole reason that complex numbers occur in this book.

Theorem 1.1 (Fundamental Theorem of Algebra). *Every polynomial of positive degree has a root.*

A proof will be found in any text on the theory of functions of a complex variable. Note that the theorem would be false if only real numbers were admitted.

If $f(c) = 0$, then $x - c$ is a factor of $f(x)$; i.e., there is a polynomial $g(x)$ such that $f(x) = (x - c)g(x)$. This follows from the identity $f(x) - f(c) = a_0(x^n - c^n) + a_1(x^{n-1} - c^{n-1}) + \cdots + a_{n-1}(x - c)$. Since every term on the right has the factor $x - c$, we have $f(x) - f(c) = (x - c)g(x)$, where $g(x)$ is a polynomial whose coefficients depend on c. Since $f(c) = 0$, we have the desired factorization. If $g(x)$ has positive degree, i.e., if $n > 1$, then $g(x)$ has a root c_2. Hence $g(x) = (x - c_2)h(x)$ for some polynomial $h(x)$, and therefore, writing

c_1 for c, $f(x) = (x - c_1)(x - c_2)h(x)$. Continuing this process shows that we can factor $f(x)$ in the form $f(x) = a_0(x - c_1)(x - c_2) \cdots (x - c_n)$. Since $f(c_k) = 0$ for $k = 1, \ldots, n$, we may say that every polynomial of degree n has n roots. However, it need not be the case that all the c_k are different. In fact, they may all be equal, as is shown by the example $f(x) = x^n$.

Let $f(x)$ have s distinct roots, and label them $r_1, r_2, \ldots, r_s$. If the factor $x - r_k$ occurs exactly m_k times in the factorization of $f(x)$, we say that r_k is a *root of multiplicity* m_k. We can then write the factorization in the form $f(x) = a_0(x - r_1)^{m_1}(x - r_2)^{m_2} \cdots (x - r_s)^{m_s}$, where $r_i \neq r_j$ if $i \neq j$ and $\sum_{k=1}^{s} m_k = n$. If $m_k = 1$, we say that r_k is a *simple* root, and if $m_k > 1$, that r_k is a *repeated* root. Clearly $f(x)$ has n distinct roots if and only if all the roots are simple. With the understanding that each (distinct) root is to be counted a number of times equal to its multiplicity, it is true that every polynomial of degree $n \geq 1$ has n roots.

The reader is no doubt aware that the problem of computing the roots of a polynomial is difficult at best if n is large, and that in most cases one must have recourse to the methods of numerical analysis. However, the special problem of solving $x^n - a = 0$, i.e., of finding the nth roots of a, can be reduced to trigonometry. Let $a = r(\cos \theta + i \sin \theta)$; then also $a = r[\cos (\theta + 2\pi k) + i \sin (\theta + 2\pi k)]$ for any integer k. We know that if b is an nth root of a, then $|b|^n = |a|$ and $\arg a = \arg (b^n) = n \arg b$. Thus $|b| = \sqrt[n]{|a|} = \sqrt[n]{r}$ and $\arg b = (1/n) \arg a$. Hence

$$b = \sqrt[n]{r}\left(\cos \frac{\theta + 2\pi k}{n} + i \sin \frac{\theta + 2\pi k}{n}\right),$$

where as always $\sqrt[n]{r}$ denotes the positive nth root of r. This formula actually yields n distinct numbers corresponding to the n values $k = 0, 1, \ldots, n - 1$, since no two of the angles θ/n, $\theta/n + 2\pi/n$, $\ldots, \theta/n + 2\pi(n - 1)/n$ differ by a multiple of 2π. Of course, $k = n$ yields the same value for b as $k = 0$; $k = n + 1$ yields the same value as $k = 1$; and so on. We summarize the results in a theorem.

Theorem 1.2.   *Every non-zero complex number* $a = r(\cos \theta + i \sin \theta)$ *has n distinct nth roots, which are given by*

$$\sqrt[n]{r}\left(\cos \frac{\theta + 2\pi k}{n} + i \sin \frac{\theta + 2\pi k}{n}\right), \qquad k = 0, 1, \ldots, n - 1.$$

Complex Functions.   By a *complex function* we shall mean a function whose domain is an interval of real numbers and whose range consists of complex numbers. If f is a complex function with domain I, and t is in I, then $f(t)$ is a complex number. Denoting the real and imaginary parts of $f(t)$ by $g(t)$ and $h(t)$, we have $f(t) = g(t) + ih(t)$. Conversely, if g and h are any real functions defined on an interval I, the function f defined by $f(t) = g(t) + ih(t)$ is a complex function. Hence a complex function is completely determined by two real functions, its real and imaginary parts. We say that f is continuous, differentiable, integrable, etc., according as its real and imaginary parts are. The derivative of f is given by $f'(t) = g'(t) + ih'(t)$. Similarly $\int f(t)\,dt = \int g(t)\,dt + i\int h(t)\,dt$. We say that $\lim_{t\to\infty} f(t) = a = \alpha + i\beta$ if $\lim_{t\to\infty} g(t) = \alpha$ and $\lim_{t\to\infty} h(t) = \beta$. The conjugate of f is defined by $\bar{f}(t) = g(t) - ih(t)$.

Let $a = \alpha + i\beta$ be a complex number. We define the exponential function e^{at} by

$$e^{at} = e^{\alpha t}\cos \beta t + ie^{\alpha t}\sin \beta t$$
$$= e^{\alpha t}(\cos \beta t + i\sin \beta t).$$

To motivate this definition we would have to anticipate some later developments. Let us instead simply verify that with this definition all the familiar properties of exponentials are preserved.

1. *For any complex numbers a and b, $e^a e^b = e^{a+b}$.*
    Let $a = \alpha + i\beta$, $b = \gamma + i\delta$. Then

$$\begin{aligned}
e^a e^b &= e^{\alpha+i\beta}e^{\gamma+i\delta} = e^{\alpha}(\cos \beta + i\sin \beta)e^{\gamma}(\cos \delta + i\sin \delta)\\
&= e^{\alpha+\gamma}[\cos \beta \cos \delta - \sin \beta \sin \delta + i(\cos \beta \sin \delta + \sin \beta \cos \delta)]\\
&= e^{\alpha+\gamma}[\cos (\beta + \delta) + i\sin (\beta + \delta)]\\
&= e^{\alpha+\gamma+i(\beta+\delta)}\\
&= e^{a+b}.
\end{aligned}$$

2. *For any complex number a and any integer n, $(e^a)^n = e^{na}$.*
    The proof is left as an exercise.

3. $\dfrac{d}{dt}\, e^{at} = ae^{at}.$

    Let $g(t) = e^{\alpha t}\cos \beta t$ and $h(t) = e^{\alpha t}\sin \beta t$. Then $g'(t) = \alpha e^{\alpha t}\cos \beta t - \beta e^{\alpha t}\sin \beta t$ and $h'(t) = \alpha e^{\alpha t}\sin \beta t + \beta e^{\alpha t}\cos \beta t$. Hence $g'(t) + ih'(t) = e^{\alpha t}[\alpha \cos \beta t - \beta \sin \beta t + i(\alpha \sin \beta t + \beta \cos \beta t)]$. On the other hand,

$$\begin{aligned}
ae^{at} &= (\alpha + i\beta)e^{\alpha t}(\cos \beta t + i\sin \beta t)\\
&= e^{\alpha t}[\alpha \cos \beta t - \beta \sin \beta t + i(\alpha \sin \beta t + \beta \cos \beta t)].
\end{aligned}$$

Thus $g'(t) + ih'(t) = ae^{at}$.

It follows that $\int e^{at}\,dt = (1/a)e^{at}$ if $a \neq 0$. If $a = 0$, $e^{at} = 1$ so that the integral is equal to t.

Note that the conjugate of e^{at} is $e^{\bar{a}t}$:

$$\overline{e^{at}} = e^{\alpha t} \cos \beta t - i e^{\alpha t} \sin \beta t = e^{\alpha t} \cos (-\beta)t + i e^{\alpha t} \sin (-\beta)t$$
$$= e^{(\alpha - i\beta)t} = e^{\bar{a}t},$$

where we used the identities $\cos (-\theta) = \cos \theta$ and $\sin (-\theta) = -\sin \theta$. From the identities

$$e^{it} = \cos t + i \sin t \qquad \text{and} \qquad e^{-it} = \cos t - i \sin t,$$

we obtain by elimination the equivalent identities

$$\cos t = \frac{1}{2} (e^{it} + e^{-it}) \qquad \text{and} \qquad \sin t = \frac{1}{2i} (e^{it} - e^{-it}).$$

Simultaneous Linear Equations and Determinants. We conclude these preliminaries with a few words about simultaneous linear algebraic equations and their connection with determinants, a subject to which we shall return again later. Here we confine ourselves to two equations in two unknowns:

$$a_{11}x_1 + a_{12}x_2 = b_1$$
$$a_{21}x_1 + a_{22}x_2 = b_2. \tag{1.3}$$

The numbers a_{ij} and b_i are given complex numbers. A solution of (1.3) is by definition a pair of complex numbers x_1, x_2 which satisfies both equations. If $b_1 = b_2 = 0$, we say that (1.3) is *homogeneous*, and otherwise that it is *non-homogeneous*. Note that a homogeneous system always has the solution $x_1 = x_2 = 0$, which we refer to as the *zero solution*. The number $a_{11}a_{22} - a_{12}a_{21}$ is called the *determinant* of the coefficients a_{ij} and will be denoted by Δ.

Theorem 1.3.   (a) *If $\Delta \neq 0$, the system (1.3) has a unique solution.*

(b) *If $\Delta = 0$, then there exist numbers b_1 and b_2 such that (1.3) has no solution. If b_1 and b_2 are such that (1.3) does have a solution, then it has infinitely many solutions.*

Proof.   (a)   Let $x_1 = \dfrac{1}{\Delta} (b_1 a_{22} - b_2 a_{12})$ and $x_2 = \dfrac{1}{\Delta} (b_2 a_{11} - b_1 a_{21})$. Substitution into (1.3) shows that x_1, x_2 is a solution. Conversely, if x_1, x_2 is a solution of (1.3), multiply the first equation by a_{22} and the second by a_{12}. Subtracting the second from the first yields $\Delta x_1 = b_1 a_{22} - b_2 a_{12}$, so $x_1 = \dfrac{1}{\Delta} (b_1 a_{22} - b_2 a_{12})$. In the same way, we find that $x_2 = \dfrac{1}{\Delta} (b_2 a_{11} - b_1 a_{21})$. Thus (a) is proved.

(b)   Now let $\Delta = 0$. If all the a_{ij} are zero, the system obviously has no solution if some b_i is non-zero, while any numbers x_1, x_2 form a solution if $b_1 = b_2 = 0$. The case in which not all the a_{ij} vanish is adequately treated by supposing $a_{11} \neq 0$. Then $x_1 = -\dfrac{a_{12}}{a_{11}} x_2 + \dfrac{b_1}{a_{11}}$, and substituting into the second equation, we get $\Delta x_2 = b_2 a_{11} - b_1 a_{21}$. Since $\Delta = 0$, there is no solution if $b_2 a_{11} - b_1 a_{21} \neq 0$. If $b_2 a_{11} - b_1 a_{21} = 0$, then for any c, $x_1 = -\dfrac{a_{12}}{a_{11}} c + \dfrac{b_1}{a_{11}}$, $x_2 = c$ is a solution. Hence (1.3) has infinitely many solutions or no solution according as $b_2 a_{11} - b_1 a_{21}$ is or is not equal to zero. $\square$

We shall sometimes use the standard notation

$$\Delta = \begin{vmatrix} a_{11} & a_{12} \\ a_{21} & a_{22} \end{vmatrix}.$$

EXERCISES

1.   Prove that formula (1.1) is valid for all integers n.

2.   (a)   Prove that $\overline{a + b} = \bar{a} + \bar{b}$ and $\overline{ab} = \bar{a}\,\bar{b}$.

   (b)   Prove that $\overline{a_1 + a_2 + \cdots + a_n} = \overline{a_1} + \overline{a_2} + \cdots + \overline{a_n}$ and $\overline{a_1 a_2 \cdots a_n} = \overline{a_1}\,\overline{a_2} \cdots \overline{a_n}$.

   (c)   Prove that $\overline{(a^{-1})} = (\bar{a})^{-1}$ and hence that $\overline{(a^n)} = (\bar{a})^n$ for all integers n.

3.   Prove:

   (a) $a\bar{a} = |a|^2$ and hence $|a| = \sqrt{a\bar{a}}$

   (b) $|ab| = |a||b|$ by using the identity $|a|^2 = a\bar{a}$

   (c) $a + \bar{a} = 2\,\mathrm{Re}\,a$

   (d) $|a| = |\bar{a}|$

   (e) $|\mathrm{Re}\,a| \leq |a|$ and $|\mathrm{Im}\,a| \leq |a|$

4.   Prove:

   (a) $a\bar{b} + \bar{a}b \leq 2|a||b|$

   (b) $|a + b| \leq |a| + |b|$ [Hint: Expand the right side of the identity $|a + b|^2 = (a + b)\overline{(a + b)}$ and use part (a).]

5.   Prove that $(e^a)^n = e^{na}$ for all integers n.

6.   Find all solutions of $x^n = 1$ for (a) $n = 3$; (b) $n = 4$; (c) $n = 6$.

7.   Find the cube roots of -1.

8.   Find the square roots of i.

9.   Prove that the roots of a polynomial with *real* coefficients occur in conjugate pairs. More precisely: If $f(x) = a_0x^n + a_1x^{n-1} + \cdots + a_n$ where a_k is real, $\ k = 0, 1, \ldots, n$, and if $f(c) = 0$, then $f(\bar{c}) = 0$. (Hint: Show that $\overline{f(x)} = f(\bar{x})$.)

10.   Solve the first-order linear differential equations

$$\text{(a)} \quad \frac{dx}{dt} = ix \qquad \text{(b)} \quad \frac{dx}{dt} = ix + e^{it}$$

$$\text{(c)} \quad \frac{dx}{dt} = x + ie^t$$

11.   Let the roots of $x^2 + a_1x + a_2$ be r_1 and r_2. Prove that $a_1 = -(r_1 + r_2)$ and $a_2 = r_1r_2$.

2. Second-Order Equations

The general linear differential equation of order two has the form

$$\frac{d^2x}{dt^2} + a_1(t)\frac{dx}{dt} + a_2(t)x = b(t).$$

We shall write $x' = \dfrac{dx}{dt}$, $x'' = \dfrac{d^2x}{dt^2}$, so that the equation becomes

$$x'' + a_1(t)x' + a_2(t)x = b(t). \tag{2.1}$$

For some purposes it is desirable that a_1, a_2 and b, as well as the solutions, be allowed to assume complex values. Henceforth a_1, a_2 and b are complex functions defined and continuous for all (real) t. A solution of (2.1) is a complex function u defined on an interval I which satisfies

$$u''(t) + a_1(t)u'(t) + a_2(t)u(t) = b(t)$$

for all t in I.

We now state the fundamental existence and uniqueness theorem for the solutions of (2.1).

Theorem 2.1.   *Given a real number t_0 and complex numbers x_0, y_0, there exists a unique solution u of (2.1) defined for all t such that $u(t_0) = x_0$ and $u'(t_0) = y_0$.*

The proof is given in Chapter 7.

Since (2.1) is a second-order equation, the initial conditions specify not only the value of u at t_0 but also the value of u'. Note that every solution is defined for all t. (If we had assumed that a_1, a_2 and b are continuous on some interval I, we could conclude that the solutions are also defined on I.) This is a consequence of the linearity of (2.1).

If $b(t) \equiv 0$, we say that (2.1) is *homogeneous*, and otherwise that it is *non-homogeneous*. The theory of a non-homogeneous equation (2.1) involves the associated homogeneous equation

$$x'' + a_1(t)x' + a_2(t)x = 0. \tag{2.2}$$

We shall refer to (2.2) as the *reduced* equation corresponding to the *complete* equation (2.1).

It will be convenient to use the abbreviation $L(x) = x'' + a_1(t)x' + a_2(t)x$, so that (2.1) becomes $L(x) = b(t)$. In this notation u is a solution of (2.1) if $L(u(t)) = b(t)$ and of (2.2) if $L(u(t)) = 0$.

Theorem 2.2. *For any twice-differentiable functions u_k and constants c_k, $k = 1, \ldots, m$,*

$$L(c_1 u_1(t) + c_2 u_2(t) + \cdots + c_m u_m(t))$$
$$= c_1 L(u_1(t)) + c_2 L(u_2(t)) + \cdots + c_m L(u_m(t)).$$

The proof is left as an exercise.

Theorem 2.3. *The difference of two solutions of the complete equation is a solution of the reduced equation.*

Proof. Let $L(u_k(t)) = b(t)$, $k = 1, 2$. Then, using the preceding theorem, $L(u_1(t) - u_2(t)) = L(u_1(t)) - L(u_2(t)) = b(t) - b(t) = 0$. $\square$

Theorem 2.4. *The sum of a solution of the complete equation and a solution of the reduced equation is a solution of the complete equation.*

Proof. Let $L(u(t)) = b(t)$ and $L(v(t)) = 0$. Then $L(u(t) + v(t)) = L(u(t)) + L(v(t)) = b(t) + 0 = b(t)$. $\square$

Theorem 2.5. *Let w be a solution of the complete equation. Then every solution of the complete equation is of the form $w + u$, where u is a solution of the reduced equation.*

Proof. If v is a solution of the complete equation, then, by Theorem 2.3, $v - w$ is a solution u of the reduced equation. Hence $v = w + u$. $\square$

The totality of solutions of the complete equation thus coincides with the totality of functions of the form $w + u$, where w is some one solution of the complete equation, and u ranges over the solutions of the reduced equation. The problem of finding all solutions of the complete equation may therefore be split into two parts (i) finding a single solution of the complete equation, and (ii) finding all solutions of the reduced equation. The principal burden of the theory lies therefore on the homogeneous equation (2.2), to which we now turn.

Theorem 2.6. *If u is a solution of (2.2) such that for some t_0, $u(t_0) = 0$ and $u'(t_0) = 0$, then $u(t) \equiv 0$.*

Proof. The function $v(t) \equiv 0$ is a solution of (2.2). Since $v(t_0) = u(t_0)$ and $v'(t_0) = u'(t_0)$, Theorem 2.1 yields that $v = u$. $\square$

Definition 2.7. Let the functions u_1, u_2, . . . , u_m be defined for all t. We say that they are *linearly dependent* if there exist complex numbers c_1, c_2, . . . , c_m, not all zero, such that $c_1 u_1(t) + c_2 u_2(t) + \cdots + c_m u_m(t) = 0$ for all t. If, on the other hand, this equation is satisfied only if all the c_k vanish, we say that the u_k are *linearly independent*. By a *linear combination* of u_1, u_2, . . . , u_m we mean a function u of the form

$$u(t) = \sum_{k=1}^{m} c_k u_k(t).$$

Theorem 2.8. *Any linear combination of solutions of (2.2) is a solution.*

Proof. This is an immediate consequence of Theorem 2.2. $\square$

Given two solutions u_1 and u_2 of (2.2), we form the determinant function

$$\varphi(t) = \begin{vmatrix} u_1(t) & u_2(t) \\ u_1'(t) & u_2'(t) \end{vmatrix} = u_1(t)u_2'(t) - u_2(t)u_1'(t).$$

φ is called the *Wronskian* of u_1 and u_2.

Theorem 2.9. *Two solutions of (2.2) are linearly dependent if and only if their Wronskian vanishes for some value of t.*

Proof. Let u_1, u_2 be linearly dependent solutions of (2.2), and let $c_1 u_1(t) + c_2 u_2(t) = 0$ for all t, where c_1 and c_2 are not both zero. Differentiation yields that $c_1 u_1'(t) + c_2 u_2'(t) = 0$ for all t. Fix t, say $t = t_0$, and consider the system of simultaneous linear equations

$$\begin{aligned} u_1(t_0)x_1 + u_2(t_0)x_2 &= 0 \\ u_1'(t_0)x_1 + u_2'(t_0)x_2 &= 0. \end{aligned} \tag{2.3}$$

Since $x_1 = c_1$, $x_2 = c_2$ is a non-zero solution, the determinant of the coefficient vanishes. But this determinant is the Wronskian of u_1 and u_2 evaluated at t_0. Note that we have actually proved that the Wron-

skian of *any* two linearly dependent differentiable functions vanishes
for *all* t. We did not use the hypothesis that u_1, u_2 are solutions of
(2.2), and t_0 was arbitrary.

To prove the converse, let u_1 and u_2 be solutions of (2.2), and suppose
that their Wronskian vanishes at $t = t_0$. Then the system (2.3) has
a non-zero solution $x_1 = c_1$, $x_2 = c_2$. Let $v(t) = c_1 u_1(t) + c_2 u_2(t)$. By
Theorem 2.8, v is a solution of (2.2). By the choice of c_1 and c_2, $v(t_0) = 0$
and $v'(t_0) = 0$. It follows from Theorem 2.6 that $v(t) = 0$. Hence u_1
and u_2 are linearly dependent.   □

Theorem 2.10.   *The Wronskian of two solutions of (2.2) is either
zero for all t or for no t. In the first case the solutions are linearly dependent, in the second case they are independent.*

Proof.   If the Wronskian vanishes for some t_0, then the solutions are
linearly dependent, and it was shown in the proof of Theorem 2.9
that the Wronskian of linearly dependent functions vanishes identically. If the Wronskian vanishes for no t, then by Theorem 2.9 the
solutions are not dependent, i.e., they are independent.   □

Theorem 2.11.   *Equation (2.2) has two linearly independent
solutions.*

Proof.   Let u_1, u_2 be the solutions of (2.2) satisfying the initial
conditions

$$u_1(0) = 1, \qquad u_1'(0) = 0$$
$$u_2(0) = 0, \qquad u_2'(0) = 1.$$

Then the Wronskian of u_1, u_2 has the value 1 at $t = 0$.   □

Theorem 2.12.   *Let u_1, u_2 be linearly independent solutions of (2.2).
Then every solution of (2.2) is a linear combination of u_1 and u_2.*

Proof.   Let u be a solution. The system

$$u_1(0)x_1 + u_2(0)x_2 = u(0)$$
$$u_1'(0)x_1 + u_2'(0)x_2 = u'(0)$$

has a non-vanishing determinant, and therefore has a unique solu-
tion $x_1 = c_1$, $x_2 = c_2$. Let $v(t) = c_1 u_1(t) + c_2 u_2(t)$. Then v is a solution of (2.2). Since $v(0) = u(0)$ and $v'(0) = u'(0)$, $v = u$. Thus $u(t) =
c_1 u_1(t) + c_2 u_2(t)$.   □

This is the principal theorem about the homogeneous equation. It
tells us that the solutions are given by $c_1 u_1(t) + c_2 u_2(t)$, where c_1 and
c_2 are arbitrary constants and u_1, u_2 are linearly independent solutions. Thus to find all the solutions, it suffices to find two linearly

independent solutions. The next theorem follows directly from Theorems 2.5 and 2.12.

Theorem 2.13.   *The solutions of the complete equation* (2.1) *are given by* $w + c_1u_1 + c_2u_2$, *where* w *is a solution of the complete equation,* u_1, u_2 *are linearly independent solutions of the reduced equation, and* c_1, c_2 *are arbitrary constants.*

A pair of linearly independent solutions u_1, u_2 of (2.2) is called a *basis* for the solutions of (2.2). If u_1, u_2 are real, we call it a *real* basis. In general, the only real solution of (2.2) is $u \equiv 0$. However, if (2.2) is real, i.e., if a_1 and a_2 are real functions, the basis constructed in the proof of Theorem 2.11 is real because the initial values are real. We first prove that if (2.2) is real, the real and imaginary parts of a solution are themselves solutions.

Theorem 2.14.   *If* (2.2) *is real and* $u = v + iw$ *is a solution* (v, w *real*), *then* v *and* w *are also solutions.*

Proof. $L(v + iw) = L(v) + iL(w)$ by Theorem 2.2. Since $L(v)$ and $L(w)$ are real functions, and since a complex number is zero only if its real and imaginary parts are zero, $L(v) + iL(w) = 0$ implies $L(v) = 0$ and $L(w) = 0$. $\square$

Theorem 2.15.   *If* (2.2) *is real and* u *is a solution such that* $u(0)$ *and* $u'(0)$ *are real, then* u *is real for all* t.

Proof.   Let $u = v + iw$. By the preceding theorem, w is a solution. By hypothesis, $w(0) = 0$ and $w'(0) = 0$. Hence by Theorem 2.6, $w \equiv 0$. $\square$

If u_1, u_2 is a real basis, then obviously the real solutions of (2.2) are given by $c_1u_1 + c_2u_2$ where c_1, c_2 are real.

We direct the reader's attention to Exercises 6 to 8 below, which describe a method of extracting a linearly independent set of real functions from a linearly independent set of complex functions. Exercise 9 is also important and will be used later on.

EXERCISES

1.   Prove Theorem 2.2.

2.   Prove that m functions $u_1, \ldots, u_m$ are linearly dependent if and only if it is possible to express one of them, say u_j, as a linear

combination of the others: $u_j(t) = \displaystyle\sum_{k=1,k\neq j}^{m} c_k u_k(t)$. (This may be taken as an alternative definition of linear dependence. In most cases the text definition is easier to apply.)

3.   Test for linear dependence:

(a) $u_1(t) = t$, $u_2(t) = t^2 + 1$, $u_3(t) = t^2 - t + 1$
(b) $u_1(t) = t^2 + 1$, $u_2(t) = t^2 - 1$, $u_3(t) = t^2 + t + 1$
(c) $u_1(t) = \sin^2 t$, $u_2(t) = \cos^2 t$
(d) $u_1(t) = t^2$, $u_2(t) = t^2 + t$, $u_3(t) = t + 1$, $u_4(t) = t^2 - 1$
(e) $u_1(t) = e^{it}$, $u_2(t) = e^{-it}$, $u_3(t) = \cos t$

4.   (a)   Prove that if one of the functions $u_1, \ldots, u_m$ is identically zero, then the functions are linearly dependent.

(b)   Prove that if two of the functions $u_1, \ldots, u_m$ are equal, then the functions are linearly dependent.

5.   Let $u_1, \ldots, u_m$ be real functions. Prove that they are linearly dependent if and only if there are *real* constants $c_1, \ldots, c_m$ not all zero such that $\displaystyle\sum_{k=1}^{m} c_k u_k = 0$.

6.   Let $u = v + iw$.

(a)   Show that $v = \frac{1}{2}(u + \bar{u})$, $w = -\dfrac{i}{2}(u - \bar{u})$.

(b)   Prove that if u and $\bar{u}$ are linearly independent, then v and w are linearly independent.

7.   Let $u_k = v_k + iw_k$, $k = 1, \ldots, m$. Prove: If the $2m$ functions u_k, $\bar{u}_k$ are linearly independent, so are the $2m$ real functions v_k, w_k. (Use the method of Exercise 6.)

8.   Let $u_k = v_k + iw_k$, $k = 1, \ldots, m$, and let $u_k = v_k$ be real for $k = m + 1, \ldots, m + r$. Prove: If the $2m + r$ functions u_k, $\bar{u}_k$, u_j ($k = 1, \ldots, m$, $j = m + 1, \ldots, m + r$) are linearly independent, so are the $2m + r$ real functions v_k, w_k, u_j.

9.   Suppose that $b(t)$ in Eq. (2.1) is a sum of m functions, $b(t) = \displaystyle\sum_{k=1}^{m} b_k(t)$. Prove that if u_k is a solution of $L(x) = b_k(t)$, then $v = \displaystyle\sum_{k=1}^{m} u_k$ is a solution of $L(x) = b(t)$. Thus the terms may be dealt with separately. This is known as the *superposition principle*.

10.   It is remarkable that the Wronskian of two solutions of (2.2) can be computed even if neither solution can. From the formula for the Wronskian, the fact that it is either zero for all t or for no t is obvious. This formula is due to Abel.

Let the solutions be u_1 and u_2, and denote their Wronskian by φ, $\varphi = u_1 u_2' - u_1' u_2$.

    (a)   Compute φ'.

    (b)   Eliminate the second-derivative terms from the formula for φ' by using (2.2).

    (c)   Conclude that $\varphi(t) = \varphi(0)e^{-\int_0^t a_1(s)\,ds}$.

3.   Second-Order Equations with Constant Coefficients

We now consider the equation

$$x'' + a_1 x' + a_2 x = b(t), \tag{3.1}$$

where a_1 and a_2 are constants. This equation can be solved by the successive solution of two first-order equations, a process which leads to a formula involving two integrations. The reduction of (3.1) to an equivalent system of first-order equations is effected by the factorization of a differential operator, a concept which we now define.*

Just as a function f of the type studied in calculus assigns to each number t of a given domain another number $f(t)$, an operator Q assigns to each function f of some set of functions another function $Q(f)$. An operator Q is thus defined by prescribing $Q(f)$ for each f in the domain of Q. Two operators P and Q are equal if they have the same domain, and $P(f) = Q(f)$ for all f in the domain. We shall consider only functions f defined for all t, and operators Q such that $Q(f)$ is also defined for all t.

Given two operators Q_1 and Q_2, we define their sum $Q_1 + Q_2$ by $(Q_1 + Q_2)(f) = Q_1(f) + Q_2(f)$, which is just the way that the sum of ordinary functions is defined. It follows from the definition that $Q_1 + Q_2 = Q_2 + Q_1$ for any operators Q_1, Q_2. The product of Q_1 and Q_2 is defined by composition, that is, $Q_1 \cdot Q_2$ is the operator which assigns to a function f the function $Q_1(Q_2(f))$: $(Q_1 \cdot Q_2)(f) = Q_1(Q_2(f))$. Consequently it is generally not true that $Q_1 \cdot Q_2 = Q_2 \cdot Q_1$, just as in composing two functions f and g, it is generally not true that $f(g(t)) = g(f(t))$. If $Q_1 \cdot Q_2 = Q_2 \cdot Q_1$, we say that Q_1 and Q_2 *commute*. All of the operators which we shall consider will commute with each other. We usually drop the dot and write $Q_1 \cdot Q_2 = Q_1 Q_2$.

The operators with which we deal are combinations of the differentiation operator D, which assigns to each differentiable function f its

---

*The equivalent system of first-order equations as defined in Chapter 1, Section 2, is of no use to us here, since we cannot solve it starting with either of the equations.

derivative, $D(f) = f'$, and operators which multiply f by a constant c. We denote the latter simply by c, so that $c(f) = cf$. The product DD, which according to the general rule of operator multiplication is defined by $(DD)(f) = D(D(f)) = D(f') = f''$, and thus assigns to each twice-differentiable function its second derivative, will be denoted by D^2. In general, if Q is an operator and n a positive integer, Q^n means $QQ \cdots Q$, with n factors. Thus $D^n(f) = f^{(n)}$. Note that $Q^n Q^m = Q^{n+m}$ for any Q. The rule $(cf)' = cf'$ implies $Dc = cD$, i.e., differentiation and multiplication by a constant commute. Combining the operator D and the constant operators by means of operator addition and multiplication leads to operators of the form

$$L = c_0 D^n + c_1 D^{n-1} + \cdots + c_{n-1} D + c_n.$$

Clearly $L(f) = c_0 f^{(n)} + c_1 f^{(n-1)} + \cdots + c_{n-1} f' + c_n f$. Since powers of D commute with each other and with constants, all operators of the form L also commute with each other. Indeed, the algebra of these operators is the same as the algebra of integers or of polynomials, in that all of the commutative, associative, and distributive laws are satisfied. (The cancellation law, $L_1 L_2 = 0$ implies $L_1 = 0$ or $L_2 = 0$, is also satisfied, but we do not need this fact. See Exercise 1.) Hereafter we shall write Lf instead of $L(f)$.

Using operators, we can write (3.1) in the form,

$$(D^2 + a_1 D + a_2)x = b(t),$$

a solution being a function u such that $(D^2 + a_1 D + a_2)u = b$. We now proceed to factor the operator $D^2 + a_1 D + a_2$. The polynomial $p(\lambda) = \lambda^2 + a_1\lambda + a_2$ is called the *characteristic polynomial* of (3.1). Let its roots be λ_1 and λ_2, so that $p(\lambda) = (\lambda - \lambda_1)(\lambda - \lambda_2)$. The case $\lambda_1 = \lambda_2$ is not excluded. λ_1 and λ_2 are called the *characteristic roots* of (3.1). It follows from the algebra of operators that

$$D^2 + a_1 D + a_2 = (D - \lambda_1)(D - \lambda_2).$$

However, let us verify this directly. For any twice-differentiable function u,

$$\begin{aligned}
(D - \lambda_1)(D - \lambda_2)u &= (D - \lambda_1)(u' - \lambda_2 u) \\
&= u'' - \lambda_2 u' - \lambda_1 u' + \lambda_1\lambda_2 u \\
&= u'' - (\lambda_1 + \lambda_2)u' + \lambda_1\lambda_2 u \\
&= (D^2 - (\lambda_1 + \lambda_2)D + \lambda_1\lambda_2)u \\
&= (D^2 + a_1 D + a_2)u,
\end{aligned}$$

the last step following from the relations $\lambda_1 + \lambda_2 = -a_1$, $\lambda_1\lambda_2 = a_2$ (see Exercise 11, Section 1).

We can therefore write (3.1) in the form

$$(D - \lambda_1)(D - \lambda_2)x = b(t). \tag{3.2}$$

Let

$$y = (D - \lambda_2)x. \tag{3.3}$$

Then (3.2) becomes

$$(D - \lambda_1)y = b(t). \tag{3.4}$$

Now notice that (3.3) and (3.4) constitute a system of two first-order linear differential equations,

$$\frac{dx}{dt} = \lambda_2 x + y$$
$$\frac{dy}{dt} = \lambda_1 y + b(t). \tag{3.5}$$

The last equation contains only y and can be solved by Leibniz's formula. The solutions are

$$y = e^{\lambda_1 t}[\int e^{-\lambda_1 t}b(t)\,dt + c_1].$$

Substituting into the first equation of the system (3.5) yields

$$\frac{dx}{dt} = \lambda_2 x + e^{\lambda_1 t}[\int e^{-\lambda_1 t}b(t)\,dt + c_1].$$

This is a first-order linear differential equation for x and is also solved by Leibniz's formula. The solutions are

$$x = e^{\lambda_2 t}[\int e^{(\lambda_1 - \lambda_2)\,t}[\int e^{-\lambda_1 t}b(t)\,dt + c_1]\,dt + c_2]. \tag{3.6}$$

This is the desired formula, giving the solutions of any second-order linear differential equation with constant coefficients.

We evaluate (3.6) for the case $b(t) = 0$, i.e., we solve the homogeneous equation

$$x'' + a_1 x' + a_2 x = 0. \tag{3.7}$$

Setting $b(t) = 0$ in (3.6) yields

$$x = c_1 e^{\lambda_2 t}\int e^{(\lambda_1 - \lambda_2)t}\,dt + c_2 e^{\lambda_2 t}. \tag{3.8}$$

Since

$$\int e^{(\lambda_1 - \lambda_2)t}\,dt = \frac{1}{\lambda_1 - \lambda_2} e^{(\lambda_1 - \lambda_2)t} \qquad \text{if } \lambda_1 \neq \lambda_2$$
$$= t \qquad\qquad\quad \text{if } \lambda_1 = \lambda_2$$

there are two cases.

If $\lambda_1 \neq \lambda_2$,

$$x = \frac{c_1}{\lambda_1 - \lambda_2} e^{\lambda_1 t} + c_2 e^{\lambda_2 t}.$$

Since $\dfrac{c_1}{\lambda_1 - \lambda_2}$ is an arbitrary constant, we simply rename it c_1, so that

$$x = c_1 e^{\lambda_1 t} + c_2 e^{\lambda_2 t}.$$

If $\lambda_1 = \lambda_2$ is a repeated root, we denote it by λ. Then

$$x = c_1 t e^{\lambda t} + c_2 e^{\lambda t}.$$

We state these results as a theorem.

Theorem 3.1. *If the characteristic polynomial has distinct roots* λ_1, λ_2, *the solutions of* (3.7) *are*

$$x = c_1 e^{\lambda_1 t} + c_2 e^{\lambda_2 t}. \tag{3.9}$$

If the characteristic polynomial has a repeated root λ, *the solutions are*

$$x = (c_1 + c_2 t) e^{\lambda t} \tag{3.10}$$

where c_1, c_2 *are arbitrary constants.*

If a_1 and a_2 are *real*, we can distinguish three cases, according as $p(\lambda)$ has

(1) two distinct real roots,
(2) a repeated root, which is necessarily real, or
(3) strictly complex roots, which are necessarily complex conjugates λ, $\bar{\lambda}$.

In cases (1) and (2) the real solutions correspond to real constants c_1, c_2 in formulas (3.9) and (3.10). In case (3), however, the basis $e^{\lambda t}$, $e^{\bar{\lambda} t}$ is complex, and the real solutions correspond to values of c_1, c_2 in (3.9) which are complex conjugates of each other. It is desirable to rewrite (3.9) in this case, so that real solutions again correspond to real constants.

Theorem 3.2. *If* a_1 *and* a_2 *are real and the characteristic polynomial has complex roots* $\lambda = \alpha + i\beta$ *and* $\bar{\lambda} = \alpha - i\beta$, *then the solutions of* (3.7) *are*

$$x = (c_1 \cos \beta t + c_2 \sin \beta t) e^{\alpha t}. \tag{3.11}$$

Proof. The solutions are given by (3.9), where now $\lambda_1 = \alpha + i\beta$, $\lambda_2 = \alpha - i\beta$. Thus

$$\begin{aligned}
x &= c_1 e^{(\alpha + i\beta) t} + c_2 e^{(\alpha - i\beta) t} \\
&= [(c_1 + c_2) \cos \beta t + i(c_1 - c_2) \sin \beta t] e^{\alpha t} \\
&= (k_1 \cos \beta t + k_2 \sin \beta t) e^{\alpha t},
\end{aligned}$$

where $k_1 = c_1 + c_2$, $k_2 = i(c_1 - c_2)$. Since k_1, k_2 may be chosen arbitrarily, determining in turn the values of c_1, c_2, it only remains to change notation, and we have (3.11). $\square$

Note that Theorem 3.2 also follows from Theorem 2.14 and Exercise 6, Section 2.

To sum up, if a_1 and a_2 are real, the solutions of (3.7) are given in terms of real bases by $c_1 e^{\lambda_1 t} + c_2 e^{\lambda_2 t}$ if λ_1, λ_2 are real distinct roots, by $(c_1 + c_2 t)e^{\lambda t}$ if λ is a repeated root, and by $(c_1 \cos \beta t + c_2 \sin \beta t)e^{\alpha t}$ if $\alpha \pm i\beta$ are complex conjugate roots.

We now return to the fundamental formula (3.6). If the formula is multiplied out, it is seen to have the form

$$x = w(t) + c_1 u_1(t) + c_2 u_2(t),$$

where

$$
\begin{aligned}
w(t) &= e^{\lambda_2 t} \textstyle\int e^{(\lambda_1 - \lambda_2)t}[\int e^{-\lambda_1 t} b(t)\, dt]\, dt && (3.12)\\
u_1(t) &= e^{\lambda_2 t} \textstyle\int e^{(\lambda_1 - \lambda_2)t}\, dt && (3.13)\\
u_2(t) &= e^{\lambda_2 t}, && (3.14)
\end{aligned}
$$

thus illustrating Theorem 2.13. Let us prove that u_1, u_2 are linearly independent. If $c_1 u_1(t) + c_2 u_2(t) = 0$ for all t, i.e., if

$$c_1 e^{\lambda_2 t} \textstyle\int e^{(\lambda_1 - \lambda_2)t}\, dt + c_2 e^{\lambda_2 t} = 0,$$

or

$$e^{\lambda_2 t}(c_1 \textstyle\int e^{(\lambda_1 - \lambda_2)t}\, dt + c_2) = 0,$$

then, since $e^{\lambda_2 t}$ does not vanish,

$$c_1 \textstyle\int e^{(\lambda_1 - \lambda_2)t}\, dt + c_2 = 0.$$

Differentiation yields

$$c_1 e^{(\lambda_1 - \lambda_2)t} = 0$$

whence $c_1 = 0$, since $e^{(\lambda_1 - \lambda_2)t}$ does not vanish. The preceding equation now yields $c_2 = 0$.

The easiest way to solve the non-homogeneous equation is first to obtain the general solution of the reduced equation by using formula (3.9), (3.10), or (3.11), and then to compute a particular solution of the complete equation by means of (3.12). If a_1, a_2 and $b(t)$ are real and λ_1, λ_2 are complex conjugates, then (3.12) produces a real function $w(t)$ provided the integration constants are chosen properly. We omit the proof because this fact is not important. If $w(t)$ turns out to be complex, its imaginary part is obviously a solution of the reduced equation, and may be discarded.

The question arises if the operator method can be used to solve the equation

$$x'' + a_1(t)x' + a_2(t)x = b(t) \qquad (3.15)$$

if a_1 and a_2 are not constant. The answer is: in general no, for the simple reason that differentiation and multiplication by a non-constant function do not commute. Let us try it. Let the equation $\lambda^2 + a_1(t)\lambda + a_2(t) = 0$ have the solutions $\lambda = \lambda_1(t)$ and $\lambda = \lambda_2(t)$, so that

$$\lambda^2 + a_1(t)\lambda + a_2(t) = (\lambda - \lambda_1(t))(\lambda - \lambda_2(t)).$$

The method would work if it were true that

$$(D - \lambda_1(t))(D - \lambda_2(t)) = D^2 - (\lambda_1(t) + \lambda_2(t))D + \lambda_1(t)\lambda_2(t), \qquad (3.16)$$

but, as the reader will verify, this is true only if $\lambda_2'(t) \equiv 0$. Hence the most general case to which the method applies is the case in which

$$\lambda^2 + a_1(t)\lambda + a_2(t) = (\lambda - \lambda_1(t))(\lambda - \lambda_2).$$

We leave it to the reader to solve Eq. (3.15) in this case. A formula for the solutions of (3.15) does not exist for arbitrary functions a_1, a_2, since it is known for example that

$$x'' + tx = 0$$

cannot be solved by means of a formula. Further results will be found in the Miscellaneous Exercises at the end of this chapter.

EXERCISES

1. (a) Let $L = a_0D^n + a_1D^{n-1} + \cdots + a_{n-1}D + a_n$. Prove that if $Lf = 0$ for all f, then $a_k = 0$, $k = 0, 1, \ldots, n$. (The domain of L may be taken to be the set of functions f which have derivatives of all orders.)

(b) It follows from part (a) that 0, i.e., multiplication by 0, is the only operator which sends every function into zero. Prove that if $L_1L_2 = 0$, then either $L_1 = 0$ or $L_2 = 0$.

2. Prove that (3.1) and (3.5) are equivalent, i.e., if $x = u(t)$, $y = v(t)$ is a solution of (3.5), then $u(t)$ is a solution of (3.1), and if $u(t)$ is a solution of (3.1), then $x = u(t)$, $y = u'(t) - \lambda_2 u(t) = v(t)$ is a solution of (3.5).

3. Show by substituting into (3.7) that $e^{\lambda t}$ is a solution if and only if λ is a characteristic root, and that $te^{\lambda t}$ is a solution if and only if λ is a repeated characteristic root.

4. Find all solutions of

(a) $x'' + x' - 2x = 0$ (b) $x'' + 2x' + x = 0$
(c) $x'' - 2ix' - x = 0$ (d) $x'' + x' = 0$
(e) $x'' = 0$ (f) $x'' - x = 0$
(g) $x'' - 2ix = 0$ (h) $x'' + 3x' + 2x = 0$
(i) $x'' - 2x' + x = 0$

5. Find the solutions of the following equations in real form:

(a) $x'' + x = 0$ (b) $x'' + 2x' + 2x = 0$
(c) $x'' + x' + x = 0$ (d) $x'' + 2x' + 3x = 0$

6. Show that

(a) $e^{\alpha t} \cos \beta t$ and $e^{\alpha t} \sin \beta t$ are linearly independent if and only if $\beta \neq 0$

(b) $e^{\lambda_1 t}$ and $e^{\lambda_2 t}$ are linearly independent if and only if $\lambda_1 \neq \lambda_2$

(c) $e^{\lambda t}$ and $te^{\lambda t}$ are linearly independent for all λ

7. Find all solutions of

(a) $x'' - x = e^{-t}$ (b) $x'' + 4x = e^{it}$
(c) $x'' - 3x' + 2x = \cos t$ (d) $x'' - 2x' + x = t$

8. Let $u(t,x_0,y_0,\lambda_1,\lambda_2)$ be the solution of $(D - \lambda_1)(D - \lambda_2)x = 0$ satisfying $u(0,x_0,y_0,\lambda_1,\lambda_2) = x_0$, $u'(0,x_0,y_0,\lambda_1,\lambda_2) = y_0$, where λ_1 and λ_2 are real. Prove that $\lim\limits_{\lambda_2 \to \lambda_1} u(t,x_0,y_0,\lambda_1,\lambda_2) = u(t,x_0,y_0,\lambda_1,\lambda_1)$.

9. Prove that (3.16) holds only if $\lambda_2(t)$ is constant.

10. Show that the solutions of

$$(D - \lambda_1(t))(D - \lambda_2)x = b(t)$$

are obtained by replacing the exponents $\lambda_1 t$ in formula (3.6) by $\int \lambda_1(t)\, dt$. Use the method by which (3.6) is derived in the text.

11. Solve the following equations. Non-elementary integrals must naturally be left unevaluated.

(a) $x'' + (\cos^2 t)x' - (\sin^2 t)x = 0$
(b) $x'' + (1 - t)x' - tx = 0$
(c) $x'' + \left(\dfrac{t - 1}{t}\right)x' - \dfrac{1}{t}x = 0, \qquad t > 0$

12. In solving Eq. (3.1), we did not appeal to the theorems of Section 2. In particular, no use was made of Theorem 2.1, which is so far unproved. Hence, by using our formula, we are in a position to give an independent proof of the existence and uniqueness theorem for (3.1). Prove that, given t_0, x_0 and y_0 there exists exactly one solution u of (3.1) which satisfies $u(t_0) = x_0$, $u'(t_0) = y_0$. (Hint: t_0, x_0, y_0 determine c_1, c_2 uniquely.)

Note that this type of proof is possible only in the presence of formulas known to include all solutions, a circumstance in which the theorem loses most of its interest.

4. The Method of Undetermined Coefficients; The Variation-of-Constants Formula

Integration by parts shows that

$$\int t^k e^{at} dt = \frac{1}{a} t^k e^{at} - \frac{k}{a} \int t^{k-1} e^{at}\, dt \qquad \text{if } a \neq 0$$

$$= \frac{1}{k+1} t^{k+1} \qquad \text{if } a = 0.$$

Hence, if $f(t)$ is a polynomial of degree d,

$$\int f(t)e^{at}\, dt = g(t)e^{at},$$

where $g(t)$ is a polynomial of degree d if $a \neq 0$, and of degree $d + 1$ if $a = 0$. We apply this fact to the equation

$$x'' + a_1 x' + a_2 x = f(t)e^{\mu t}, \tag{4.1}$$

where $f(t)$ is a polynomial of degree d. Letting $b(t) = f(t)e^{\mu t}$ in formula (3.12) yields the particular solution

$$w(t) = e^{\lambda_2 t}\int e^{(\lambda_1 - \lambda_2) t}[\int f(t)e^{(\mu - \lambda_1) t}\, dt]\, dt. \tag{4.2}$$

Now

$$\int f(t)e^{(\mu - \lambda_1) t}\, dt = g(t)e^{(\mu - \lambda_1) t},$$

where d_1, the degree of $g(t)$, is equal to d if $\mu \neq \lambda_1$ and to $d + 1$ if $\mu = \lambda_1$. Consequently

$$\begin{aligned}
w(t) &= e^{\lambda_2 t}\int g(t)e^{(\mu - \lambda_2) t}\, dt \\
&= e^{\lambda_2 t} h(t)e^{(\mu - \lambda_2) t} \\
&= h(t)e^{\mu t},
\end{aligned}$$

where d_2, the degree of $h(t)$, is equal to d_1 if $u \neq \lambda_2$ and to $d_1 + 1$ if $\mu = \lambda_2$. Thus d_2 is equal to d, $d + 1$, or $d + 2$ according as μ is equal to neither of λ_1, λ_2, exactly one of them, or both of them.

We may therefore compute $w(t)$ by setting it equal to the product of $e^{\mu t}$ and a polynomial $h(t)$ of suitable degree and unknown coefficients, substituting into (4.1), and determining the coefficients by equating coefficients of like powers of t in (4.1).

The integration constants in (4.2) should be set equal to zero, since they only give rise to solutions of the reduced equations. If this is done,

$h(t)$ has no constant term in case μ equals either λ_1 or λ_2, and neither a constant nor a first-degree term in case $\mu = \lambda_1 = \lambda_2$. To summarize these results, let us agree to say that μ is a characteristic root of multiplicity 0 if μ is not a characteristic root. We then have the following theorem.

Theorem 4.1.   *Let $f(t)$ be a polynomial of degree d. The equation*

$$x'' + a_1 x' + a_2 x = f(t)e^{\mu t}$$

has a solution of the form

$$w(t) = h(t)e^{\mu t},$$

where $h(t) = t^m r(t)$; m is the multiplicity of μ as a characteristic root, and $r(t)$ is a polynomial of degree d.

Theorem 4.1 may also be proved by means of the identity

$$L(h(t)e^{\mu t}) = [p(\mu)h(t) + p'(\mu)h'(t) + h''(t)]e^{\mu t}, \tag{4.3}$$

where $L(x) = x'' + a_1 x' + a_2 x$, $h(t)$ is an arbitrary function, and $p(\lambda)$ is the characteristic polynomial. Note that μ is a simple or repeated root according as $p(\mu) = 0$ and $p'(\mu) \neq 0$, or $p(\mu) = p'(\mu) = 0$. It is easy to see that if $h(t)$ is a suitably chosen polynomial, the term in brackets is equal to $f(t)$. The derivation of (4.3) is left as an exercise.

EXAMPLE. $\qquad\qquad\qquad x'' - x = t^2 e^t. \tag{4.4}$

The solutions of the reduced equation are $c_1 e^t + c_2 e^{-t}$. Since $d = 2$ and $m = 1$, we take $h(t)$ to be a polynomial of degree 3 without constant term:

$$h(t) = at^3 + bt^2 + ct.$$

The coefficients a, b, c are computed by setting $x = h(t)e^t$ in (4.4) and equating coefficients of like powers of t. The task of substituting is simplified by using (4.3), which yields

$$\begin{aligned}
L(h(t)e^t) &= [p'(1)h'(t) + h''(t)]e^t \\
&= [2h'(t) + h''(t)]e^t \\
&= [6at^2 + (4b + 6a)t + 2c + 2b]e^t.
\end{aligned}$$

a, b, c must satisfy

$$6at^2 + (4b + 6a)t + 2c + 2b = t^2,$$

whence

$$\begin{aligned}
6a &= 1 \\
4b + 6a &= 0 \\
2c + 2b &= 0.
\end{aligned}$$

We find $a = \frac{1}{6}$, $b = -\frac{1}{4}$, $c = \frac{1}{4}$. Thus

$$w(t) = (\tfrac{1}{6}t^3 - \tfrac{1}{4}t^2 + \tfrac{1}{4}t)e^t.$$

The general solution of (4.4) is therefore

$$c_1e^t + c_2e^{-t} + (\tfrac{1}{6}t^3 - \tfrac{1}{4}t^2 + \tfrac{1}{4}t)e^t.$$

We remind the reader of the superposition principle (Exercise 9, Section 2): *If $u_k(t)$ is a solution of $Lx = b_k(t)$, then $\sum_{k=1}^{N} u_k(t)$ is a solution of $Lx = \sum_{k=1}^{N} b_k(t)$. This principle gives us the following generalization of Theorem 4.1.*

Theorem 4.2. *If $f_k(t)$ is a polynomial of degree d_k and μ_k is a characteristic root of multiplicity m_k, then the equation*

$$x'' + a_1x' + a_2x = \sum_{k=1}^{N} f_k(t)e^{\mu_k t}$$

has a solution of the form

$$w(t) = \sum_{k=1}^{N} t^{m_k}r_k(t)e^{\mu_k t},$$

where $r_k(t)$ is a polynomial of degree d_k.

By virtue of the identities

$$\cos \omega t = \tfrac{1}{2}(e^{i\omega t} + e^{-i\omega t})$$
$$\sin \omega t = \frac{1}{2i}(e^{i\omega t} - e^{-i\omega t}),$$

Theorem 4.2 includes the case in which the right side of the equation contains terms of the form $f(t)e^{\gamma t}\cos \omega t$ or $f(t)e^{\gamma t}\sin \omega t$. However, if a_1 and a_2 are real, there is an easier way to deal with such terms. Consider the equation

$$\begin{aligned} Lx &= f(t)e^{(\gamma + i\omega)t} \\ &= f(t)e^{\gamma t}\cos \omega t + if(t)e^{\gamma t}\sin \omega t. \end{aligned} \tag{4.5}$$

If $u = v + iw$ is a solution, then, assuming $f(t)$ is a real polynomial, v is a solution of

$$Lx = f(t)e^{\gamma t}\cos \omega t \tag{4.6}$$

and w is a solution of

$$Lx = f(t)e^{\gamma t}\sin \omega t. \tag{4.7}$$

Thus, to solve either of the equations (4.6) or (4.7), we solve instead Eq. (4.5). The real part of the solution will be a solution of (4.6), and the imaginary part will be a solution of (4.7).

EXAMPLE. $$x'' + 4x = \sin 2t.$$

The solutions of the reduced equation are $c_1 \cos 2t + c_2 \sin 2t$. To find a solution of the complete equation we solve

$$x'' + 4x = e^{2it}$$

instead, and then select the imaginary part. Since $2i$ is a simple characteristic root, we look for a solution $w(t) = cte^{2it}$. Since $Lw(t) = 4ice^{2it}$, we must have $4ic = 1$, whence $c = -i/4$. Thus

$$w(t) = -\frac{i}{4} te^{2it}$$

$$= \tfrac{1}{4}t \sin 2t - \frac{i}{4} t \cos 2t.$$

It follows that $-\tfrac{1}{4}t \cos 2t$ is a solution of $x'' + 4x = \sin 2t$, and the general solution is

$$c_1 \cos 2t + c_2 \sin 2t - \tfrac{1}{4}t \cos 2t.$$

The Variation-of-Constants Formula.   We return once more to formulas (3.12) to (3.14):

$$w(t) = e^{\lambda_2 t}\int e^{(\lambda_1-\lambda_2)t}[\int e^{-\lambda_1 t}b(t)\,dt]\,dt$$
$$u_1(t) = e^{\lambda_2 t}\int e^{(\lambda_1-\lambda_2)t}\,dt$$
$$u_2(t) = e^{\lambda_2 t}.$$

Our object is to rewrite the formula for $w(t)$ so as to express $w(t)$ in terms of $u_1(t)$ and $u_2(t)$. Let

$$v(t) = \frac{u_1(t)}{u_2(t)} = \int e^{(\lambda_1-\lambda_2)t}\,dt.$$

Then

$$w(t) = u_2(t)\int v'(t)[\int e^{-\lambda_1 t}b(t)\,dt]\,dt.$$

Integration by parts yields

$$\int v'(t)[\int e^{-\lambda_1 t}b(t)\,dt]\,dt = v(t)\int e^{-\lambda_1 t}b(t)\,dt - \int v(t)e^{-\lambda_1 t}b(t)\,dt.$$

Hence

$$w(t) = u_2(t)v(t)\int e^{-\lambda_1 t}b(t)\,dt - u_2(t)\int v(t)e^{-\lambda_1 t}b(t)\,dt$$

$$= u_1(t)\int \frac{u_2(t)b(t)}{e^{(\lambda_1+\lambda_2)t}}\,dt - u_2(t)\int \frac{u_1(t)b(t)}{e^{(\lambda_1+\lambda_2)t}}. \qquad (4.8)$$

It remains to express the denominators $e^{(\lambda_1+\lambda_2)t}$ in terms of $u_1(t)$ and $u_2(t)$. Let φ be the Wronskian of u_1, u_2. Then

$$\varphi(t) = \begin{vmatrix} e^{\lambda_2 t}\int e^{(\lambda_1-\lambda_2)t}\,dt & e^{\lambda_2 t} \\ \lambda_2 e^{\lambda_2 t}\int e^{(\lambda_1-\lambda_2)t}\,dt + e^{\lambda_1 t} & \lambda_2 e^{\lambda_2 t} \end{vmatrix}$$
$$= -e^{(\lambda_1+\lambda_2)t}.$$

Thus $e^{(\lambda_1+\lambda_2)t} = -\varphi(t)$, and (4.8) can be written

$$w(t) = u_2(t) \int \frac{u_1(t)b(t)}{\varphi(t)}\,dt - u_1(t) \int \frac{u_2(t)b(t)}{\varphi(t)}\,dt. \qquad (4.9)$$

We shall now show that this formula furnishes a solution $w(t)$ of the complete equation if u_1, u_2 are any linearly independent solutions of the reduced equation, even if the coefficients a_1, a_2 in the differential equation are not constant.

Theorem 4.3.   *Let u_1, u_2 be linearly independent solutions of*

$$x'' + a_1(t)x' + a_2(t)x = 0 \qquad (4.10)$$

and let $\varphi(t)$ be their Wronskian. Then (4.9) is a solution of

$$x'' + a_1(t)x' + a_2(t)x = b(t). \qquad (4.11)$$

Proof.   Differentiating (4.9) yields

$$w'(t) = u_2'(t) \int \frac{u_1(t)b(t)}{\varphi(t)}\,dt + u_2(t)\frac{u_1(t)b(t)}{\varphi(t)}$$
$$- u_1'(t) \int \frac{u_2(t)b(t)}{\varphi(t)}\,dt - u_1(t)\frac{u_2(t)b(t)}{\varphi(t)}$$
$$= u_2'(t) \int \frac{u_1(t)b(t)}{\varphi(t)}\,dt - u_1'(t) \int \frac{u_2(t)b(t)}{\varphi(t)}\,dt.$$

Differentiating once more, and using the relation $\varphi = u_1 u_2' - u_1' u_2$, we find

$$w''(t) = u_2''(t) \int \frac{u_1(t)b(t)}{\varphi(t)}\,dt - u_1''(t) \int \frac{u_2(t)b(t)}{\varphi(t)}\,dt + b(t).$$

A simple computation now shows that

$$w''(t) + a_1(t)w'(t) + a_2(t)w(t) = b(t). \qquad \square$$

Formula (4.9) is called the *variation-of-constants* formula, because it results if we seek a solution of (4.11) in the form

$$w(t) = c_1(t)u_1(t) + c_2(t)u_2(t),$$

that is, if we attempt to convert the general solution of the reduced equation into a solution of the complete equation by replacing the constants c_1, c_2 by suitable functions $c_1(t)$, $c_2(t)$. We shall derive generalized variation-of-constants formula in just this way in Chapter 5.

For theoretical purposes there is an advantage in using definite integrals in (4.9). The lower limits of integration are of course arbitrary, and we take both to be zero. We then obtain

$$w(t) = u_2(t) \int_0^t \frac{u_1(s)b(s)}{\varphi(s)}\, ds - u_1(t) \int_0^t \frac{u_2(s)b(s)}{\varphi(s)}\, ds \qquad (4.12)$$

$$= \int_0^t \frac{u_1(s)u_2(t) - u_1(t)u_2(s)}{\varphi(s)}\, b(s)\, ds. \qquad (4.13)$$

The choice of zero as the lower limit of integration has the consequence that w satisfies the initial conditions $w(0) = w'(0) = 0$.

EXERCISES

1. Derive formula (4.3).
2. Solve:

    (a) $x'' + x' - 2x = (t^2 + 1)e^{-t}$
    (b) $x'' + 2x' + x = (t + 1)e^{-t}$
    (c) $x'' + 3x' + 2x = t^2 e^{-2t}$
    (d) $x'' + x' = t$
    (e) $x'' + x = \sin 2t$
    (f) $x'' - x' + x = t \cos t$
    (g) $x'' - x = e^t \sin t$

3. Show that the function $w(t)$ defined by (4.12) satisfies $w(0) = w'(0) = 0$.

4. Let $u(t,t_0)$ be the solution of $x'' + a_1(t)x' + a_2(t)x = 0$ which satisfies the initial conditions $u(t_0,t_0) = 0$, $u'(t_0,t_0) = 1$.
    (a)   Compute $u(t,t_0)$ in terms of a basis u_1, u_2.
    (b)   Show that in terms of this function, formula (4.13) becomes

$$w(t) = \int_0^t b(s)u(t,s)\, ds.$$

[$u(t,s)$ is called the *Green's function* associated with (4.11) and the initial conditions $w(0) = w'(0) = 0$.]

5.   Oscillations

Up to now our discussion of linear equations has been purely formal. We now investigate the qualitative behavior of the real solutions of $x'' + a_1x' + a_2x = b(t)$, where $a_1 \geq 0$, $a_2 > 0$, and $b(t)$ is real and periodic. As we shall see, this equation governs a spring-mass system subject to a periodic force, as well as a linear electric circuit subject to a periodic electromotive force.

We begin by describing the qualitative properties of the solutions of

$$x'' + a_1x' + a_2x = 0, \tag{5.1}$$

assuming only that a_1, a_2 are real. We denote the characteristic roots by λ_1, λ_2.

If λ_1, λ_2 are real and distinct, the solutions are $c_1e^{\lambda_1 t} + c_2e^{\lambda_2 t}$. If both λ_1 and λ_2 are negative, all the solutions tend to zero as $t \to \infty$. If one root is zero and the other negative, then all the solutions are bounded for $t \geq 0$ but do not all tend to zero. If at least one root is positive, then there is a family of solutions which are unbounded for $t \geq 0$.

If λ is a repeated root, the solutions are $c_1e^{\lambda t} + c_2te^{\lambda t}$. If λ is negative, all the solutions tend to zero as $t \to \infty$. (See Chapter 2, Section 5, Exercise 1.) If $\lambda \geq 0$, there is a family of solutions which are unbounded for $t \geq 0$.

If $\alpha \pm i\beta$ are complex conjugate roots, the solutions are

$$(c_1 \cos \beta t + c_2 \sin \beta t)e^{\alpha t}.$$

If α is negative, $e^{\alpha t} \to 0$ as $t \to \infty$. Since $\cos \beta t$ and $\sin \beta t$ are bounded, it follows that all solutions tend to zero as $t \to \infty$. If $\alpha > 0$, all nonzero solutions are unbounded for $t \geq 0$. If $\alpha = 0$, the solutions are periodic with period $2\pi/\beta$, and are therefore bounded for all t.

We summarize these facts as follows.

Theorem 5.1.   *A necessary and sufficient condition that all solutions of (5.1) tend to zero as $t \to \infty$ is that the characteristic roots have negative real part.*

If some root has positive real part, or if zero is a repeated root, there is a family of solutions which are unbounded for $t \geq 0$.

In the remaining cases (one root zero and one negative, or both purely imaginary) the solutions are bounded for $t \geq 0$ but do not all tend to zero as $t \to \infty$.

These same facts may be stated in terms of the stability of the zero solution:

The zero solution (5.1) is asymptotically stable if and only if the characteristic roots have negative real part. If some root has positive real part, or if zero is a repeated root, the zero solution is unstable. In the remaining cases the zero solution is weakly stable.

We consider in some detail the case of purely imaginary roots $a_1 = 0$, $a_2 > 0$, which leads to periodic solutions. Setting $\omega = \sqrt{a_2}$, the equation becomes

$$x'' + \omega^2 x = 0, \qquad \omega > 0 \tag{5.2}$$

and the solutions are $c_1 \cos \omega t + c_2 \sin \omega t$. We shall consider only real solutions, i.e., real constants c_1, c_2. If c_1 and c_2 are not both zero, we have

$$c_1 \cos \omega t + c_2 \sin \omega t = \sqrt{c_1^2 + c_2^2}\left(\frac{c_1}{\sqrt{c_1^2 + c_2^2}} \cos \omega t + \frac{c_2}{\sqrt{c_1^2 + c_2^2}} \sin \omega t\right)$$

$$= A \sin (\omega t + \phi), \tag{5.3}$$

where $A = \sqrt{c_1^2 + c_2^2}$ and ϕ satisfies $\sin \phi = \dfrac{c_1}{\sqrt{c_1^2 + c_2^2}}$, $\cos \phi = \dfrac{c_2}{\sqrt{c_1^2 + c_2^2}}$. Thus the solutions of (5.2) may be expressed in the form (5.3), where A and ϕ are arbitrary constants which may always be chosen so that $A \geqq 0$ and $0 \leqq \phi < 2\pi$.

A function of the form (5.3) is called a *sinusoidal* function. It is periodic with least period $p = 2\pi/\omega$. The coefficient A is called its *amplitude*, and ϕ is called its *phase*. The reciprocal of the period, $\omega/2\pi$, is called its *frequency*, and ω itself is called its *angular frequency*. Note that a large ω means a high frequency and therefore a small period. Note also that the mean value of $A \sin(\omega t + \phi)$ is zero. Its maximum is A and its minimum is $-A$. Because $\sin (\omega t + \phi) = \sin \omega(t + \phi/\omega)$, the curve $x = A \sin (\omega t + \phi)$ is obtained from the curve $x = A \sin \omega t$ by translating the latter ϕ/ω units to the left.

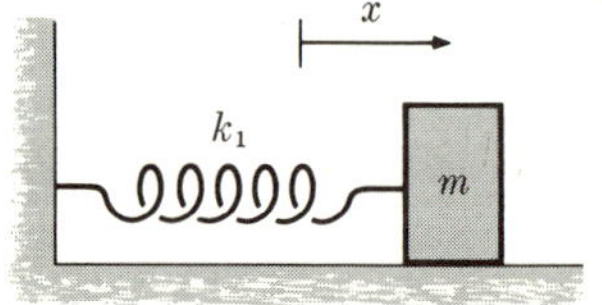

FIG. 12. Spring-mass system.

Consider a mass m attached to the end of a spring which exerts on it a force proportional to the extension of the spring (Fig. 12). Let x be the displacement of the mass and $k_1 > 0$ the spring constant, so that the force on the mass is $-k_1 x$. If the system is conservative, i.e., if there is no loss of mechanical energy through friction or internal heating of the spring, Newton's second law yields the equation of motion

$$m\frac{d^2x}{dt^2} = -k_1 x.$$

If the system is not conservative, we account for the energy loss by postulating a force which is a function of the velocity of the mass. We assume that this force, like the spring force, is linear, and is thus equal to $-k_2\, dx/dt$, where $k_2 > 0$. The equation of motion under friction or damping is then

$$m\frac{d^2x}{dt^2} = -k_1 x - k_2 \frac{dx}{dt}.$$

We are particularly interested in the effect of a force which is wholly external to and independent of the state of the system, i.e., of x and dx/dt, being arbitrarily imposed on it from without. Such, for example, is the force exerted by sound waves on a microphone membrane, of which a spring-mass system is a crude model. For the time being, we make no assumption about the nature of this force, except that it is a continuous periodic function $f(t)$ of the time. The equation of motion of the mass under the action of such an external force thus becomes

$$m\frac{d^2x}{dt^2} = -k_1 x - k_2 \frac{dx}{dt} + f(t)$$

or

$$mx'' + k_2 x' + k_1 x = f(t).$$

We divide by m and let $k_2/m = 2k$, $k_1/m = \omega_0^2$, $(1/m)f(t) = b(t)$, so that

$$x'' + 2kx' + \omega_0^2 x = b(t). \tag{5.4}$$

We shall always suppose $\omega_0 > 0$, $k \geq 0$. The cases $k = 0$ (no damping) and $k > 0$ (positive damping) will need to be discussed separately.

Equation (5.4) is a second-order linear differential equation with constant coefficients; as we shall see, the restrictions on the coefficients are precisely those which guarantee that the solutions of the reduced equation are either asymptotic to zero ($k > 0$) or periodic ($k = 0$). While a spring-mass system has (for most of us) a superior intuitive value as a physical model of the differential equation, the physical system which most closely obeys such an equation is a linear electric circuit. If the inductance of the circuit is L, its resistance R, and its capacity C, then the charge x on the condenser satisfies the equation

$$Lx'' + Rx' + \frac{1}{C}x = f(t),$$

where $f(t)$ is an electromotive force due, for example, to a generator or to radio waves.

Either of these physical models enables us to consider the reduced equation as a physical system, and the complete equation as this system acted upon by an external force $f(t)$.

The Damped Oscillator. We consider first the homogeneous equation

$$x'' + 2kx' + \omega_0^2 x = 0$$

with $k > 0$. The characteristic roots are $-k \pm \sqrt{k^2 - \omega_0^2}$. If $k > \omega_0$, they are real and distinct, and since $k > \sqrt{k^2 - \omega_0^2}$, they are both negative. If $k = \omega_0$, $-k$ is a repeated root. Finally, if $k < \omega_0$, the roots are complex with real part $-k$. Thus in all cases the roots have negative real part, and all solutions tend to zero as $t \to \infty$. If $k \geqq \omega_0$ so that the damping is large, the solutions are monotonic for large t: the mass "creeps" toward the equilibrium $x = 0$. If $k < \omega_0$, the motion may be described as oscillatory with steadily diminishing amplitude. Note that the mass passes through the equilibrium position at time intervals of length $\pi/\sqrt{\omega_0^2 - k^2}$.

We now turn to the non-homogeneous equation

$$x'' + 2kx' + \omega_0^2 x = b(t), \tag{5.5}$$

where $k > 0$ and $b(t)$ is a continuous function of period $p > 0$. We ask whether this equation has any solutions of period p.

If u is a solution of (5.5) and $v(t) = u(t + p)$, then v is also a solution, since $Lv(t) = Lu(t + p) = b(t + p) = b(t)$. Suppose $u(0) = u(p)$ and $u'(0) = u'(p)$. Then $v(0) = u(0)$ and $v'(0) = u'(0)$, so that by Theorem 2.1, $v = u$. But this means $u(t + p) = u(t)$. Hence u has period p if and only if $u(0) = u(p)$ and $u'(0) = u'(p)$. [Note that this is also true if $b(t) \equiv 0$.] We must therefore look for a solution with this property. It turns out that there is exactly one. We could show this by using the formulas for the solutions of the reduced equation, considering separately each of the cases $k > \omega_0$, $k = \omega_0$, $k < \omega_0$. A shorter and more instructive argument is based on the following lemma.

Lemma 5.2. *Let u_1, u_2 be a basis for $x'' + a_1x' + a_2x = 0$, and let $p > 0$. The equation has a solution of period p (other than $x \equiv 0$) if and only if*

$$\begin{vmatrix} u_1(0) - u_1(p) & u_2(0) - u_2(p) \\ u_1'(0) - u_1'(p) & u_2'(0) - u_2'(p) \end{vmatrix} = 0.$$

Proof. Let u be a solution, so $u(t) = c_1u_1(t) + c_2u_2(t)$. u has period p if and only if $u(0) = u(p)$ and $u'(0) = u'(p)$, or

$$c_1u_1(0) + c_2u_2(0) = c_1u_1(p) + c_2u_2(p)$$
$$c_1u_1'(0) + c_2u_2'(0) = c_1u_1'(p) + c_2u_2'(p).$$

Collecting terms in c_1 and c_2 we obtain

$$[u_1(0) - u_1(p)]c_1 + [u_2(0) - u_2(p)]c_2 = 0$$
$$[u_1'(0) - u_1'(p)]c_1 + [u_2'(0) - u_2'(p)]c_2 = 0,$$

which is a homogeneous linear system for c_1 and c_2. The system has a non-zero solution c_1, c_2 if and only if the determinant of the coefficients vanishes. The lemma is therefore proved. $\qquad \square$

Theorem 5.3. *Equation (5.5), with $k > 0$, has a unique solution u of period p, and all other solutions tend toward u as $t \to \infty$.*

Proof. Let u_1, u_2 be a basis for the reduced equation and w a solution of the complete equation. A solution $u = c_1u_1 + c_2u_2 + w$ of the complete equation has period p if and only if $u(0) = u(p)$ and $u'(0) = u'(p)$, that is, if and only if c_1 and c_2 satisfy

$$c_1u_1(0) + c_2u_2(0) + w(0) = c_1u_1(p) + c_2u_2(p) + w(p)$$
$$c_1u_1'(0) + c_2u_2'(0) + w'(0) = c_1u_1'(p) + c_2u_2'(p) + w'(p).$$

Collecting terms in c_1 and c_2 we have

$$[u_1(0) - u_1(p)]c_1 + [u_2(0) - u_2(p)]c_2 = w(p) - w(0)$$
$$[u_1'(0) - u_1'(p)]c_1 + [u_2'(0) - u_2'(p)]c_2 = w'(p) - w'(0).$$

Since the reduced equation has no solution of period p other than $x \equiv 0$, we conclude from the lemma that the determinant of the coefficients of this system does not vanish. Hence the system has a unique solution c_1, c_2. It follows that (5.5) has a unique solution of period p. If v is any other solution, then $u - v$ is a solution of the reduced equation. It follows that $\lim_{t \to \infty} [u(t) - v(t)] = 0.$ $\qquad \square$

The engineer refers to the periodic solution as the "steady state" solution, and to all other solutions (or the difference between them and the periodic solution) as "transients." This terminology reflects

the fact that the asymptotic stability of the equilibrium solution of
the reduced equation is transferred to the periodic solution of the
complete equation, as well as the fact that periodic states play a role
very similar to equilibrium states both in physical reality and in
mathematical theory.

In many engineering systems the approach to the periodic solution
is so rapid that for all practical purposes no other solution need be
taken into account, i.e., the initial conditions become irrelevant. The
nature of the periodic solution depends in a rather complicated way
on the external force. Before we turn to this problem, we point out
that to the designer of machinery it is important to minimize vibrations altogether, while to the designer of sound equipment it is important that the oscillations resemble the external force as closely as
possible.

Consider the case $b(t) = A \sin \omega t$, i.e., the equation

$$x'' + 2kx' + \omega_0^2 x = A \sin \omega t.$$

To find a solution we replace the right side by $A e^{i\omega t}$, set $w(t) = c e^{i\omega t}$,
and compute c by substituting into the equation. The imaginary part
of $w(t)$ then yields the solution

$$u(t) = \frac{A}{\sqrt{4k^2\omega^2 + (\omega_0^2 - \omega^2)^2}} \sin (\omega t + \phi),$$

where $\pi < \phi < 2\pi$; $\phi = 3\pi/2$ if $\omega = \omega_0$ and $\tan \phi = 2k\omega/(\omega^2 - \omega_0^2)$ if
$\omega \neq \omega_0$. The coefficient

$$\frac{1}{\sqrt{4k^2\omega^2 + (\omega_0^2 - \omega^2)^2}}$$

is the ratio of the amplitude of the forced oscillation to the amplitude
of the external force, and is called the *magnification factor* M of the
oscillation. It depends not only on k and ω_0, the parameters of the
system, but also on the frequency of the external force. For fixed
values of k and ω_0 it defines a function

$$F(\omega) = \frac{1}{\sqrt{4k^2\omega^2 + (\omega_0^2 - \omega^2)^2}}$$

which measures the response of the system to sinusoidal forces of
different frequencies. Engineers are interested in the shape of the
curves $M = F(\omega)$ for different values of k, i.e., for various degrees of
damping. These curves, called *response curves*, are shown in Fig. 13.
We leave it to the reader to verify the essential features of this figure.

It is noteworthy that the response curves have a local maximum if $2k^2 < \omega_0^2$, and that this maximum becomes increasingly pronounced as k decreases, indicating that the system responds more and more sensitively to frequencies within a narrow interval.

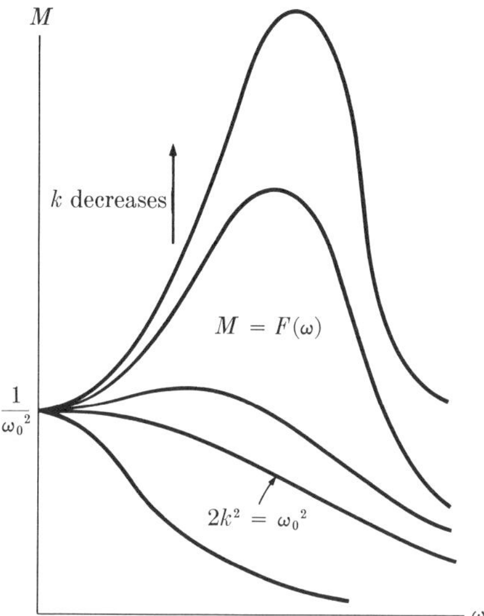

FIG. 13.   Response curves.

We indicate briefly the application of these results to the case of an arbitrary periodic function $b(t)$. Given a function $b(t)$ of period $p = 2\pi/\omega$, the *Fourier series* of $b(t)$ is defined to be

$$\frac{a_0}{2} + \sum_{j=1}^{\infty} a_j \cos j\omega t + b_j \sin j\omega t,$$

where

$$a_j = \frac{\omega}{\pi} \int_0^p b(t) \cos j\omega t \, dt$$

$$b_j = \frac{\omega}{\pi} \int_0^p b(t) \sin j\omega t \, dt.$$

It is shown in texts on advanced calculus that if $b(t)$ has a continuous derivative, then the Fourier series of $b(t)$ converges uniformly to $b(t)$. This means that we may approximate $b(t)$ as closely as we wish by a partial sum of its Fourier series. The differential equation then becomes (assuming $a_0 = 0$)

$$x'' + 2kx' + \omega_0^2 x = \sum_{j=1}^{n} a_j \cos j\omega t + b_j \sin j\omega t$$

and by the superposition principle the periodic solution is found by adding the periodic solutions of

$$x'' + 2kx' + \omega_0^2 x = a_j \cos j\omega t + b_j \sin j\omega t$$

for $j = 1, \ldots, n$. Hence the response of the system to sinusoidal functions determines its response to all periodic functions. To the extent that $F(\omega)$ is not constant, the amplitudes $\sqrt{a_j^2 + b_j^2}$ of the sinusoidal components of $b(t)$ are selectively magnified or suppressed in the response of the system. In an acoustical system such as a radio or phonograph this results in an undesirable distortion, or lack of "fidelity."

The Undamped Oscillator.　　We now turn to the case $k = 0$. The solutions of the reduced equation

$$x'' + \omega_0^2 x = 0 \tag{5.6}$$

are the sinusoidal functions

$$c_1 \cos \omega_0 t + c_2 \sin \omega_0 t$$

of period $p_0 = 2\pi/\omega_0$. A solution $w(t)$ of the complete equation

$$x'' + \omega_0^2 x = b(t) \tag{5.7}$$

can be obtained from the variation-of-constants formula (4.13). Setting $u_1(t) = \cos \omega_0 t$ and $u_2(t) = \sin \omega_0 t$, we find that the Wronskian $\varphi(t)$ is equal to ω_0. Substituting in (4.13) yields

$$w(t) = \frac{1}{\omega_0} \int_0^t b(s) \sin \omega_0(t - s)\, ds. \tag{5.8}$$

The solutions of (5.7) are thus

$$u(t) = c_1 \cos \omega_0 t + c_2 \sin \omega_0 t + \frac{1}{\omega_0} \int_0^t b(s) \sin \omega_0(t - s)\, ds.$$

Note that

$$w'(t) = \int_0^t b(s) \cos \omega_0(t - s)\, ds. \tag{5.9}$$

The discussion of the periodic solutions of (5.7) is complicated by the fact that the solutions of the reduced equation are periodic. Henceforth let p be the least period of $b(t)$. Note that if u is a solution of period T of (5.7), then, since $u'' + \omega_0^2 u$ also has period T, it follows that $b(t)$ has period T. Hence $T = np$ for some positive integer n. The search for periodic solutions of (5.7) can therefore be narrowed to solutions having periods which are integer multiples of p.

Theorem 5.4.　　(1) *If p/p_0 is irrational, Eq. (5.7) has a unique periodic solution. The period of this solution is p.*

(2)   *If p/p_0 is rational, but not an integer, (5.7) has a unique solution of period p, and every solution has period Np, where $p/p_0 = M/N$.*

(3)   *If $p = Mp_0$, then either every solution has period p, or no solution is periodic, according as the integrals*

$$\int_0^p b(t)\, \sin \omega_0 t\, dt \qquad and \qquad \int_0^p b(t)\, \cos \omega_0 t\, dt \qquad (5.10)$$

do, or do not, both vanish.

Proof.   Let

$$u(t) = c_1 \cos \omega_0 t + c_2 \sin \omega_0 t + w(t),$$

where $w(t)$ is given by (5.8). u has period np for some n if and only if $u(0) = u(np)$ and $u'(0) = u'(np)$, that is,

$$\begin{aligned}
c_1 + w(0) &= c_1 \cos \omega_0 np + c_2 \sin \omega_0 np + w(np) \\
\omega_0 c_2 + w'(0) &= -\omega_0 c_1 \sin \omega_0 np + \omega_0 c_2 \cos \omega_0 np + w'(np).
\end{aligned} \qquad (5.11)$$

Collecting terms in c_1 and c_2, and noting that

$$w(0) = w'(0) = 0$$

$$w(np) = \frac{1}{\omega_0} \int_0^{np} b(s)\, \sin \omega_0 (np - s)\, ds$$

$$w'(np) = \int_0^{np} b(s)\, \cos \omega_0 (np - s)\, ds,$$

(5.11) becomes

$$(1 - \cos \omega_0 np)c_1 - (\sin \omega_0 np)c_2 = \frac{1}{\omega_0} \int_0^{np} b(s)\, \sin \omega_0 (np - s)\, ds$$

$$(5.12)$$

$$(\omega_0 \sin \omega_0 np)c_1 + (\omega_0 - \omega_0 \cos \omega_0 np)c_2 = \int_0^{np} b(s)\, \cos \omega_0 (np - s)\, ds.$$

The periodic solutions of (5.7) thus correspond to the constants c_1, c_2 which satisfy (5.12). The determinant of the coefficients is readily found to be $2\omega_0(1 - \cos \omega_0 np)$. It vanishes only if $\cos \omega_0 np = 1$, i.e., if $\omega_0 np$ is an integer multiple of 2π, $\omega_0 np = 2\pi m$. Since $2\pi/\omega_0 = p_0$, this relation can be written $p/p_0 = m/n$. Thus the determinant of (5.12) vanishes only if p/p_0 is rational and n is chosen so that $p/p_0 = m/n$ for some m.

If p/p_0 is irrational, the determinant does not vanish for any n. Setting $n = 1$ therefore yields a unique solution c_1, c_2 and therefore a unique solution u of (5.7) having period p. Since u also has period np for every n, the solutions of (5.12) corresponding to values of n other than 1 do not yield any further solutions of (5.7). This completes the proof of part (1).

Next, let p/p_0 be a rational number, but not an integer. The determinant of (5.12) is again different from zero for $n = 1$, and consequently (5.7) has a unique solution of period p. Denote this solution by v,

$$v(t) = c_1' \cos \omega_0 t + c_2' \sin \omega_0 t + w(t).$$

Since

$$w(t) = v(t) - (c_1' \cos \omega_0 t + c_2' \sin \omega_0 t),$$

w is the difference of a function of period p and a function of period p_0. Since $p/p_0 = M/N$, w has period $Np = Mp_0$. An arbitrary solution u of (5.7),

$$u(t) = c_1 \cos \omega_0 t + c_2 \sin \omega_0 t + w(t),$$

is therefore a sum of a function of period p_0 and a function of period Mp_0, and therefore has period $Mp_0 = Np$. This completes the proof of part (2).

Finally, let $p = Mp_0$. Then for any integer n, $\omega_0 np = \omega_0 nMp_0 = 2\pi nM$, so that

$$\cos \omega_0 np = 1 \qquad \text{and} \qquad \sin \omega_0 np = 0.$$

Hence the coefficients of c_1 and c_2 in (5.12) vanish for every n. There are now two cases, according as the terms on the right of (5.12) do or do not both vanish. For $n = 1$ these terms are

$$-\frac{1}{\omega_0} \int_0^p b(s) \sin \omega_0 s \, ds \qquad \text{and} \qquad \int_0^p b(s) \cos \omega_0 s \, ds.$$

If both of these integrals vanish, (5.12) is satisfied by all c_1, c_2. Hence all solutions of (5.7) have period p.

Since $b(s) \sin \omega_0 s$ and $b(s) \cos \omega_0 s$ have period p, the terms on the right of (5.12) for arbitrary n satisfy

$$\int_0^{np} b(s) \sin \omega_0 s \, ds = n \int_0^p b(s) \sin \omega_0 s \, ds$$
$$\int_0^{np} b(s) \cos \omega_0 s \, ds = n \int_0^p b(s) \sin \omega_0 \, ds.$$

Hence if either of the right sides of (5.12) fails to vanish for $n = 1$, it fails to vanish for all n. Thus (5.12) has no solution for any n, and the differential equation has no periodic solution. This completes the proof of the theorem. $\square$

Let us consider the case $b(t) = A \sin \omega t$, i.e., the equation

$$x'' + \omega^2 x = A \sin \omega t. \tag{5.13}$$

If $\omega \neq \omega_0$, the solutions are given by

$$c_1 \cos \omega_0 t + c_2 \sin \omega_0 t + \frac{A}{\omega_0^2 - \omega^2} \sin \omega t.$$

If ω_0/ω is irrational, the only periodic solution is $\dfrac{A}{\omega_0^2 - \omega^2} \sin \omega t.$ If $\omega_0/\omega = M/N$, all solutions have period $2\pi N/\omega$.

If $\omega = \omega_0$, the solutions of (5.13) are given by

$$c_1 \cos \omega_0 t + c_2 \sin \omega_0 t - \frac{A}{2\omega_0} t \cos \omega_0 t. \tag{5.14}$$

Because $t \cos \omega_0 t$ is unbounded, there is no periodic solution.

The case $b(t) = A \sin \omega_0 t$, or in general the case of no periodic solution of (5.7), is called *resonance*.

The phenomenon of resonance in the general case is best understood in terms of the Fourier series of $b(t)$. Recall that the Fourier coefficients of $b(t)$ are

$$a_j = \frac{\omega}{\pi} \int_0^p b(t) \cos j\omega t$$

$$b_j = \frac{\omega}{\pi} \int_0^p b(t) \sin j\omega t,$$

where $p = \dfrac{2\pi}{\omega}$. If $p = Mp_0$, that is, if $\omega = \dfrac{1}{M} \omega_0$, we have

$$a_j = \frac{\omega}{\pi} \int_0^p b(t) \cos \frac{j}{M} \omega_0 t \, dt$$

$$b_j = \frac{\omega}{\pi} \int_0^p b(t) \sin \frac{j}{M} \omega_0 t \, dt.$$

If either a_M or b_M fails to vanish, the Fourier series contains a sinusoidal function of period $2\pi/\omega_0$, and this contributes an unbounded term as in (5.14) to the solution of the differential equation. The condition that the Fourier series of $b(t)$ does not contain such a term is precisely that the integrals (5.10) both vanish. If p is not a multiple of p_0 such a term is automatically excluded, and resonance cannot occur.

The resonance case

$$x'' + \omega_0^2 x = A \sin \omega_0 t$$

may be regarded as the limiting case $\omega \to \omega_0$ of

$$x'' + \omega_0^2 x = A \sin \omega t,$$

or even better, as the limiting case $k \to 0$ of

$$x'' + 2kx' + \omega_0^2 x = A \sin \omega_0 t.$$

Since $F(\omega_0) \to \infty$ as $k \to 0$, resonance arises physically whenever the damping is weak compared to the magnitude of the external force, and the force contains a sinusoidal term whose frequency is close to the natural frequency $\omega_0/2\pi$ of the oscillator. When resonance occurs the oscillator is likely to be destroyed. A familiar example is the shattering of a wine glass by the voice of a trained singer. It should be pointed out that non-mathematicians use the term "resonance" in a still wider sence. In common usage resonance is said to occur whenever an oscillator vibrates perceptibly with the frequency of the external force, and what we have called response curves are also called resonance curves.

EXERCISES

1. (a) Show that the response function

$$F(\omega) = [4k^2\omega^2 + (\omega_0^2 - \omega^2)^2]^{-\frac{1}{2}}, \qquad \omega > 0,$$

is a decreasing function if $2k^2 \geq \omega^2$.

(b) Show that $F(\omega)$ has a local maximum if $2k^2 < \omega_0^2$. Let $\bar{\omega}$ be the value of ω at which the maximum occurs. Compute $\bar{\omega}$ and $\lim_{k\to 0} \bar{\omega}$.

2. (a) Find the solution of $x'' + 2x' + x = \sin t$ satisfying $u(0) = u'(0) = 0$.

(b) What are $u(0)$ and $u'(0)$ if $u(t)$ is periodic? What is the phase of the periodic solution?

3. Let $u(t,x_0,y_0,\alpha)$ be the solution of $x'' + \alpha x = 0$ satisfying $u(0,x_0,y_0,\alpha) = x_0$, $u'(0,x_0,y_0,\alpha) = y_0$, where $\alpha \geq 0$. Show that $\lim_{\alpha\to 0} u(t,x_0,y_0,\alpha) = u(t,x_0,y_0,0)$.

4. Find the least period, p, of each of the following functions $b(t)$. Decide in each case whether the equation $x'' + x = b(t)$ has any periodic solutions, and if it does, compute them and find their least periods.

(a) $b(t) = \sin \pi t$ (b) $b(t) = \sin 2t + \sin 3t$
(c) $b(t) = 4 \sin^3 t$ (d) $b(t) = 2 \sin^2 t$

5. Let $b(t)$ have period p, and let u be a solution of period p of $x'' + 2kx' + \omega_0^2 x = b(t)$. On physical grounds we expect that the work done by the external force on the system during one period is equal to the work done by the system against friction, since the state of the system, and therefore its energy, is the same at the end of a cycle as at the beginning. Work being the integral of force with respect to distance, the work done by $b(t)$ is $\int_0^p b(t)u'(t)\, dt$, and the work done against friction is $\int_0^p 2k[u'(t)]^2\, dt$.

Prove that

$$2k \int_0^p [u'(t)]^2\, dt = \int_0^p b(t)u'(t)\, dt.$$

(Note that the case $k = 0$ is not excluded.)

6.   Show that if p/p_0 is irrational, every non-periodic solution of (5.7) is the sum of a function of least period p_0 and a function of least period p.

7.   (a)   Show that in the resonance case, (5.7) has a solution of the form $w(t) = at \sin \omega_0 t + bt \cos \omega_0 t + g(t)$, where $a^2 + b^2 \neq 0$ and $g(t)$ has period p. [Hint: Use the variation-of-constants formula in the form (4.12), note that the integrands have period p, and apply the method by which we proved Theorem 3.5, page 42.]

(b)   Hence conclude that in the resonance case all solutions are unbounded for $t > 0$ as well as for $t < 0$.

6.   Equations of Order n

In Sections 2 to 4 we discussed the most important properties of linear differential equations of the second order, as well as methods for obtaining solutions of equations with constant coefficients. All of this material generalizes in a simple and straightforward way to linear equations of order greater than 2. In fact, most of the theorems and proofs require only such minor adjustments as the occasional replacement of "2" by "n" to apply to the general case. The generalization of the material of Section 2 involves the use of determinants of order n, and the theorem that a system of n simultaneous linear algebraic equations in n unknowns has a unique solution if and only if the determinant of the coefficients is not zero. The reader who is not familiar with determinants of order greater than 2 will find a discussion in the next chapter.

Since the present section contains nothing essentially new, and since furthermore the subject matter of this chapter is merely a special case of the subject matter of Chapter 5, the discussion will be kept rather brief. Some matters are deferred altogether to Chapter 5, partly to avoid undue repetition, and partly because certain topics are more easily dealt with by matrix methods. This includes in particular the variation-of-constants formula for $n > 2$.

The general linear differential equation of order n has the form

$$x^{(n)} + a_1(t)x^{(n-1)} + \cdots + a_n(t)x = b(t), \tag{6.1}$$

where $x^{(k)} = \dfrac{d^k x}{dt^k}$. We shall sometimes write $x^{(0)} = x$. The reduced equation corresponding to (6.1) is the homogeneous equation

$$x^{(n)} + a_1(t)x^{(n-1)} + \cdots + a_n(t)x = 0. \tag{6.2}$$

We suppose that the functions $a_k(t)$ and $b(t)$ are continuous for all t. The left side of (6.1) will be denoted by Lx, so that a solution is a function u for which $Lu(t) = b(t)$.

We state the fundamental existence and uniqueness theorem for (6.1).

Theorem 6.1.   *Given a real number t_0 and complex numbers η_1, η_2, . . . , η_n, there exists a unique solution u of (6.1) defined for all t such that*

$$u(t_0) = \eta_1, \; u'(t_0) = \eta_2, \; \ldots \; , \; u^{(n-1)}(t_0) = \eta_n.$$

If u_1, u_2, . . . , u_m are any n-times-differentiable functions and c_1, . . . , c_m are constants, then obviously

$$L\left(\sum_{k=1}^{m} c_k u_k(t) \right) = \sum_{k=1}^{m} c_k L u_k(t). \tag{6.3}$$

From this relation it follows that Theorems 2.3–2.5 remain valid for nth-order equations. As before, to find all solutions of the complete equation we need one solution of the complete equation and all solutions of the reduced equation.

The theory of the homogeneous equation proceeds along the same lines as in the case $n = 2$. To begin with, we note that by virtue of the relation (6.3) any linear combination of solutions of (6.2) is a solution. Secondly, if u is a solution of (6.2) such that for some t_0, $u(t_0) = u'(t_0) = \cdots = u^{(n-1)}(t_0) = 0$, then it follows from Theorem 6.1 that $u(t) \equiv 0$.

Given n solutions u_1, . . . , u_n of (6.2), their Wronskian is the determinant

$$\varphi(t) = \begin{vmatrix} u_1(t) & u_2(t) & \cdots & u_n(t) \\ u_1'(t) & u_2'(t) & \cdots & u_n'(t) \\ \cdots & \cdots & \cdots & \cdots \\ u_1^{(n-1)}(t) & u_2^{(n-1)}(t) & \cdots & u_n^{(n-1)}(t) \end{vmatrix}.$$

Except that we must deal with n equations in n unknowns, we find exactly as before that $\varphi(t)$ is zero for all t or for no t according as u_1, . . . , u_n are linearly dependent or independent.

To show that (6.2) has n linearly independent solutions, let u_1, . . . , u_n be the solutions satisfying the initial conditions

$$u_k^{(j)}(0) = \begin{cases} 1 & \text{if } j = k - 1 \\ 0 & \text{if } j \neq k - 1 \end{cases}$$

where $k = 1, \ldots, n$ and $j = 0, 1, \ldots, n - 1$. The Wronskian of $u_1, \ldots, u_n$ evaluated at $t = 0$ is

$$\varphi(0) = \begin{vmatrix} 1 & 0 & 0 & \cdots & 0 \\ 0 & 1 & 0 & \cdots & 0 \\ 0 & 0 & 1 & \cdots & 0 \\ \cdot & \cdot & \cdot & & \cdot \\ 0 & 0 & 0 & \cdots & 1 \end{vmatrix} = 1.$$

It follows that $u_1, \ldots, u_n$ are linearly independent.

We carry through the proof that every solution of (6.2) is a linear combination of n linearly independent solutions. Let u be a solution, let $u_1, \ldots, u_n$ be n linearly independent solutions, and consider the system of n linear equations in the n unknowns $x_1, \ldots, x_n$

$$\sum_{k=1}^{n} u_k^{(j)}(0)x_k = u^{(j)}(0), \qquad j = 0, 1, \ldots, n - 1.$$

The determinant of the coefficients is the Wronskian of $u_1, \ldots, u_n$ evaluated at $t = 0$, and therefore does not vanish. Hence the system has a unique solution $x_k = c_k, \quad k = 1, \ldots, n$. Let $v(t) = \sum_{k=1}^{n} c_k u_k(t)$. Then v is a solution of (6.2) and $v^{(j)}(0) = u^{(j)}(0), \quad j = 0, 1, \ldots, n - 1$. It follows that $v = u$, that is, $u(t) = \sum_{k=1}^{n} c_k u_k(t)$.

The preceding discussion is summarized by the following theorem.

Theorem 6.2.   *The solutions of the complete equation* (6.1) *are given by*

$$w + c_1 u_1 + c_2 u_2 + \cdots + c_n u_n,$$

where w is a solution of the complete equation, $u_1, \ldots, u_n$ are n linearly independent solutions of the reduced equation, and $c_1, \ldots, c_n$ are arbitrary constants.

A set of n linearly independent solutions of (6.2) is called a *basis* for the solutions of (6.2).

We shall now construct a basis for the solutions of

$$x^{(n)} + a_1 x^{(n-1)} + \cdots + a_n x = 0, \tag{6.4}$$

where the coefficients a_k are constants. The polynomial

$$p(\lambda) = \lambda^n + a_1 \lambda^{n-1} + \cdots + a_n$$

is called the *characteristic polynomial* of (6.4). Let its distinct roots be $\lambda_1, \ldots, \lambda_s$, with respective multiplicities $m_1, \ldots, m_s$, so that

$$p(\lambda) = (\lambda - \lambda_1)^{m_1} \cdots (\lambda - \lambda_s)^{m_s}.$$

We define the operator L by

$$L = D^n + a_1 D^{n-1} + \cdots + a_n.$$

It follows from the algebra of operators that L has the factorization

$$L = (D - \lambda_1)^{m_1} \cdots (D - \lambda_s)^{m_s}.$$

Equation (6.4) may thus be written

$$Lx = 0$$

or

$$(D - \lambda_1)^{m_1}(D - \lambda_2)^{m_2} \cdots (D - \lambda_s)^{m_s} x = 0.$$

Since the factors $(D - \lambda_k)^{m_k}$ commute, a function u which satisfies

$$(D - \lambda_k)^{m_k} u(t) = 0$$

for some k obviously also satisfies

$$Lu(t) = 0.$$

In other words, a solution of any one of the s equations

$$(D - \lambda_k)^{m_k} x = 0 \tag{6.5}$$

is a solution of (6.4). Not surprisingly we obtain a basis for (6.4) by combining the bases of the equations (6.5).

To solve an equation of the form

$$(D - \lambda)^m x = 0, \tag{6.6}$$

we use the method of reduction to a system of first-order equations. Let $x_1 = x$, $x_2 = (D - \lambda)x_1$, and in general

$$x_{k+1} = (D - \lambda)x_k, \qquad k = 1, \ldots, m - 1 \tag{6.7}$$

so that

$$(D - \lambda)x_m = 0. \tag{6.8}$$

Equations (6.7) and (6.8) constitute a system of m first-order equations

$$\frac{dx_1}{dt} = \lambda x_1 + x_2$$

$$\frac{dx_2}{dt} = \lambda x_2 + x_3$$

$$\cdot \ \cdot \ \cdot \ \cdot \ \cdot \ \cdot \ \cdot \ \cdot \qquad\qquad (6.9)$$

$$\frac{dx_{m-1}}{dt} = \lambda x_{m-1} + x_m$$

$$\frac{dx_m}{dt} = \lambda x_m.$$

This system can be solved starting with the last equation and working backwards. Since $x_1 = x$, we thus obtain the solutions of (6.6). We carry out this computation.

The solutions of the last equation of (6.9) are

$$x_m = c_m e^{\lambda t},$$

where c_m is an arbitrary constant. Substituting into the preceding equation and solving yields

$$x_{m-1} = e^{\lambda t}(c_m t + c_{m-1}),$$

where c_{m-1} is an arbitrary constant. Continuing to substitute back and solving yields

$$x_{m-2} = e^{\lambda t}\left(\frac{c_m}{2}\, t^2 + c_{m-1}t + c_{m-2}\right)$$

and in general

$$x_{m-j} = e^{\lambda t}\left(\frac{c_m}{j!}\, t^j + \frac{c_{m-1}}{(j-1)!}\, t^{j-1} + \cdot \ \cdot \ \cdot + c_{m-j}\right).$$

Thus

$$x_1 = e^{\lambda t}\left(\frac{c_m}{(m-1)!}\, t^{m-1} + \frac{c_{m-1}}{(m-2)!}\, t^{m-2} + \cdot \ \cdot \ \cdot + c_1\right).$$

Writing c_j for the arbitrary constant $\dfrac{c_j}{(j-1)!}$ yields the solutions of (6.6) in the form

$$x = c_1 e^{\lambda t} + c_2 t e^{\lambda t} + c_3 t^2 e^{\lambda t} + \cdot \ \cdot \ \cdot + c_m t^{m-1} e^{\lambda t}.$$

Applying this result to each of the equations (6.5) gives us the following theorem.

Theorem 6.3. *The n functions*

$$t^j e^{\lambda_k t}, \qquad j = 0, 1, \ldots, m_k - 1, \quad k = 1, \ldots, s$$

are solutions of Eq. (6.4).

It remains to show that these functions are linearly independent.

Theorem 6.4. *If $\lambda_1, \ldots, \lambda_s$ are s distinct numbers and $p_1(t),$ $\ldots, p_s(t)$ are polynomials, then*

$$\sum_{k=1}^{s} p_k(t) e^{\lambda_k t} = 0$$

holds for all t only if all the $p_k(t)$ are identically zero.

Proof. For any polynomial $p(t)$ and any number λ,

$$Dp(t)e^{\lambda t} = (\lambda p(t) + p'(t))e^{\lambda t}.$$

If $\lambda \neq 0$ then $\lambda p(t) + p'(t)$ is a polynomial of the same degree as $p(t)$. It follows that if $\lambda \neq 0$ and $p(t)$ has degree d, then for any k,

$$D^k p(t)e^{\lambda t} = q(t)e^{\lambda t},$$

where $q(t)$ is a polynomial of degree d. In particular, if $p(t)$ is not identically zero, neither is $q(t)$.

We prove the theorem by induction on s. It is obviously true for $s = 1$. Suppose it is true for some integer $s - 1$, and let

$$\sum_{k=1}^{s} p_k(t) e^{\lambda_k t} = 0. \tag{6.10}$$

Divide by $e^{\lambda_1 t}$:

$$p_1(t) + \sum_{k=2}^{s} p_k(t) e^{(\lambda_k - \lambda_1)t} = 0.$$

Differentiate $d_1 + 1$ times, where d_1 is the degree of $p_1(t)$. This yields

$$\sum_{k=2}^{s} q_k(t) e^{(\lambda_k - \lambda_1)t} = 0,$$

where, since $\lambda_k - \lambda_1 \neq 0$, $k = 2, \ldots, s$, $q_k(t)$ is a polynomial of the same degree as $p_k(t)$. By the inductive hypothesis, $q_k(t) = 0$, so that $p_k(t) = 0$ also, $k = 2, \ldots, s$. Thus (6.10) becomes

$$p_1(t)e^{\lambda_1 t} = 0,$$

whence $p_1(t) = 0$ also. $\qquad \square$

It is now obvious that the n solutions

$$t^j e^{\lambda_k t}, \qquad j = 0, 1, \ldots, m_k - 1, \quad k = 1, \ldots, s$$

are linearly independent. A linear combination of them is of the form

$$\sum_{k=1}^{s} p_k(t) e^{\lambda_k t}$$

and can vanish identically only if all the polynomials $p_k(t)$ vanish, i.e., only if all the coefficients are zero.

EXAMPLE. $\qquad\qquad x''' + x'' - x' - x = 0.$

The characteristic polynomial is

$$p(\lambda) = \lambda^3 + \lambda^2 - \lambda - 1.$$

We find by inspection that $\lambda_1 = 1$ is a root. Dividing by $\lambda - 1$ yields $p(\lambda)/(\lambda - 1) = \lambda^2 + 2\lambda + 1 = (\lambda + 1)^2$. Hence $\lambda_2 = -1$ is a root of multiplicity 2. The solutions are therefore

$$c_1 e^t + (c_2 + c_3 t)e^{-t}. \qquad \square$$

It remains to construct a real basis in case that the coefficients a_k in (6.4) are real, but the characteristic roots are not all real. Since obviously the real and imaginary parts of a solution are themselves solutions, a complex root $\lambda = \alpha + i\beta$ of multiplicity m contributes the $2m$ real solutions

$$t^j(\cos \beta t)e^{\alpha t}, \quad t^j(\sin \beta t)e^{\alpha t}, \qquad j = 0, 1, \ldots, m - 1. \quad (6.11)$$

The same $2m$ solutions are contributed by the conjugate root $\bar{\lambda}$. We therefore replace the $2m$ complex solutions $t^j e^{\lambda t}$, $t^j e^{\bar{\lambda} t}$, $j = 0, 1, \ldots,$ $m - 1$ by the $2m$ real solutions (6.11), for each pair of complex conjugate roots λ, $\bar{\lambda}$. It follows from Exercise 8, Section 2, that the resulting set of n real solutions is linearly independent.

EXAMPLE. $\qquad\qquad x^{(4)} + 2x'' + x = 0.$

Since $p(\lambda) = \lambda^4 + 2\lambda^2 + 1 = (\lambda^2 + 1)^2$, the roots are i and $-i$, each of multiplicity 2. The solutions are therefore given in complex form by

$$(c_1 + c_2 t)e^{it} + (c_3 + c_4 t)e^{-it}$$

and in real form by

$$(c_1 + c_2 t)\cos t + (c_3 + c_4 t)\sin t. \qquad \square$$

A solution of the non-homogeneous equation

$$x^{(n)} + a_1 x^{(n-1)} + \cdots + a_n x = b(t) \tag{6.12}$$

can be found by the method of reduction to a system of first-order equations. We now denote the characteristic roots, not necessarily distinct, by $\lambda_1, \lambda_2, \ldots, \lambda_n$, each distinct root being listed a number of times equal to its multiplicity. Then (6.12) can be written

$$(D - \lambda_n)(D - \lambda_{n-1}) \cdots (D - \lambda_1)x = b(t).$$

Let $x_1 = x$, $x_2 = (D - \lambda_1)x_1$, and in general,

$$x_{k+1} = (D - \lambda_k)x_k, \qquad k = 1, \ldots, n - 1,$$

so that

$$(D - \lambda_n)x_n = b(t).$$

Equation (6.12) is thus equivalent to the system of n first-order equations

$$\frac{dx_1}{dt} = \lambda_1 x_1 + x_2$$
$$\frac{dx_2}{dt} = \lambda_2 x_2 + x_3$$
$$\cdots \cdots \cdots \cdots \tag{6.13}$$
$$\frac{dx_{n-1}}{dt} = \lambda_{n-1}x_{n-1} + x_n$$
$$\frac{dx_n}{dt} = \lambda_n x_n + b(t).$$

A solution $x_n = w_n(t)$ of the last equation is given by

$$w_n(t) = e^{\lambda_n t}\int e^{-\lambda_n t}b(t)\,dt.$$

Substituting into the preceding equation and solving yields a solution

$$w_{n-1}(t) = e^{\lambda_{n-1} t}\int e^{(\lambda_n - \lambda_{n-1})t}[\int e^{-\lambda_n t}b(t)\,dt]\,dt.$$

Continuing to substitute back and solving leads finally to a solution $w(t) = w_1(t)$ of (6.12). The process obviously involves n integrations.

If $b(t) = g(t)e^{\mu t}$, where $g(t)$ is a polynomial of degree d, it follows exactly as in the case $n = 2$ that $w_k(t) = r_k(t)e^{\mu t}$, where $r_k(t)$ is also a polynomial. The degree of $r_k(t)$ is equal to $d + l_k$, where l_k is the number of the roots $\lambda_k, \lambda_{k+1}, \ldots, \lambda_n$ which are equal to μ. Hence $w(t) = r(t)e^{\mu t}$, where the degree of $r(t)$ exceeds that of $g(t)$ by m, the multiplicity of μ as a characteristic root of (6.12). If all constants of integration are set equal to zero, $r(t)$ has no terms of degree lower than m. The method of undetermined coefficients thus remains valid without any change.

EXAMPLE. $$x''' + x' = \cos t.$$

The characteristic roots are 0, i, $-i$, so that the solutions of the reduced equation are

$$c_1 + c_2 \cos t + c_3 \sin t.$$

To find a solution of the complete equation we replace it by

$$x''' + x' = e^{it}.$$

The real part of a solution will be a solution of the original equation. Since i is a simple characteristic root, we let

$$w(t) = cte^{it}.$$

Differentiation yields

$$w'''(t) + w'(t) = -2ce^{it}.$$

Hence $c = -\frac{1}{2}$. The real part of $w(t)$ is $-\frac{1}{2}t \cos t$, and the solutions of the given equation are therefore

$$c_1 + c_2 \cos t + c_3 \sin t - \tfrac{1}{2}t \cos t. \qquad \square$$

In conclusion we prove that the zero solution of (6.4) is asymptotically stable if and only if all characteristic roots have negative real part.

Lemma 6.5.   *If $k, \lambda > 0$, then* $\lim\limits_{t \to \infty} t^k e^{-\lambda t} = 0$.

Proof.   Let $f(t) = t^k e^{-\mu t}$, where $\mu > 0$. Since $f'(t) = (kt^{k-1} - \mu t^k)e^{-\mu t}$, $f'(t) > 0$ for $0 < t < k/\mu$ and $f'(t) < 0$ for $t > k/\mu$. Hence $f(t) \leq f(k/\mu)$ for $t \geq 0$. Setting $c = f(k/\mu)$, we have $t^k e^{-\mu t} \leq c$ for $t \geq 0$. Now let $\mu = \lambda/2$, so $t^k e^{-(\lambda/2)t} \leq c$. Multiply both sides of this inequality by $e^{-(\lambda/2)t}$ to obtain $t^k e^{-\lambda t} \leq ce^{-(\lambda/2)t}$ for $t \geq 0$. The conclusion is now obvious. $\square$

Theorem 6.6.   *If all characteristic roots of (6.4) have negative real part, all solutions tend to zero as $t \to \infty$.*

If some root has positive real part, or if there is a repeated root with zero real part, some solutions are unbounded for $t \geq 0$.

In all other cases, all solutions of (6.4) are bounded for $t \geq 0$, but do not all tend to zero as $t \to \infty$.

The proof merely involves an inspection of the basis $t^j e^{\lambda_k t}$, $j = 0$, $1, \ldots, m_k - 1$, $k = 1, \ldots, s$. The details are left as an exercise.

EXERCISES

1. Find all solutions of

   (a) $x''' + 2x'' + x' = 0$ $\qquad$ (b) $x''' + 3x'' + 3x' + x = 0$
   (c) $x''' - x = 0$ $\qquad$ (d) $x^{(4)} - x = 0$
   (e) $(D - 1)(D^2 + 1)^2 x = 0$
   (f) $(D + 1)^5 x = 0$

2. Find all solutions of

   (a) $x''' - x = e^{-t}$ $\qquad$ (b) $x''' + x = \sin t$
   (c) $x''' + x = e^{-t}$ $\qquad$ (d) $x^{(4)} - x = \sin t$

3. Prove Theorem 6.6.

MISCELLANEOUS EXERCISES FOR CHAPTER 3

The following exercises deal in one way or another with the equation

$$x'' + a_1(t)x' + a_2(t)x = 0. \tag{1}$$

We assume that $a_1(t)$ and $a_2(t)$ are real functions continuous for all t. Exercises 4 to 7 are concerned with the connection between (1) and the non-linear first-order equation

$$\frac{dx}{dt} = -(x^2 + a_1(t)x + a_2(t)), \tag{2}$$

the so-called *Riccati equation* of (1). [A Riccati equation is an equation of the form $dx/dt = A(t)x^2 + B(t)x + C(t)$.] We shall consider only real solutions of (1) and (2).

1. Show that if $a_1(t)$ is differentiable, the substitution

$$x = e^{-\frac{1}{2}\int_0^t a_1(s)\, ds} y$$

transforms (1) into

$$y'' + a(t)y = 0,$$

where

$$a(t) = a_2(t) - \tfrac{1}{2}a_1'(t) - \tfrac{1}{4}a_1^2(t).$$

2. Let $f(t,x,y)$ be such that for any (t_0,x_0,y_0) the differential equation

$$\frac{d^2x}{dt^2} = f\left(t,x,\frac{dx}{dt}\right) \tag{3}$$

has a solution u satisfying $u(t_0) = x_0$, $u'(t_0) = y_0$. Show that if any (real) linear combination of two solutions is itself a solution, then (3) is a homogeneous linear differential equation. (See the proof of Theorem 3.2, Chapter 2.)

3. (Sturm Separation Theorem) Let u_1, u_2 be linearly independent (real) solutions of (1). Suppose that $u_1(t) \neq 0$ for $a < t < b$, and that $u_1(a) = u_1(b) = 0$. Prove that there is a number c, $a < c < b$, such that $u_2(c) = 0$. [Hint: If not, $u_2(t) \neq 0$ for $a \leq t \leq b$; apply Rolle's theorem to the function $u_1(t)/u_2(t)$.]

4. We showed in Section 3 that if $\lambda_1(t)$, $\lambda_2(t)$ are the solutions of $\lambda^2 + a_1(t)\lambda + a_2(t) = 0$, so that

$$\lambda^2 + a_1(t)\lambda + a_2(t) = [\lambda - \lambda_1(t)][\lambda - \lambda_2(t)],$$

then

$$D^2 + a_1(t)D + a_2(t) = [D - \lambda_1(t)][D - \lambda_2(t)]$$

holds only if λ_2 is constant. Is it possible to factor the operator even if λ_2 is not constant? Prove:

$$D^2 + a_1(t)D + a_2(t) = [D - \mu_1(t)][D - \mu_2(t)]$$

holds on an interval I if and only if $\mu_2(t)$ is a solution of (2) defined on I, and $\mu_1(t) = -\mu_2(t) - a_1(t)$.

5. Prove: If u is a solution of (1) and $u(t) \neq 0$ for t in an interval I, then $v(t) = u'(t)/u(t)$, t in I, is a solution of (2). Conversely, if v is a solution of (2) defined on an interval I, then

$$u(t) = e^{\int_{t_0}^{t} v(s)\, ds}, \qquad t_0,\, t \text{ in } I$$

in a solution of (1).

6. Show that if v_1, v_2 are any two solutions of (2) defined on an interval I, then (1) has a basis u_1, u_2 such that for t in I,

$$u_k(t) = e^{\int_{t_0}^{t} v_k(s)\, ds}, \qquad k = 1,\, 2, \quad t_0 \text{ in } I.$$

7. Let $\lambda_1(t)$, $\lambda_2(t)$ be defined as in Exercise 4. Suppose that $\lambda_1(t)$, $\lambda_2(t)$ are real, and that there are numbers α and β such that

$$\lambda_1(t) < \alpha < \lambda_2(t) < \beta$$

for $t \geq 0$.

(a)   Prove that (2) has a family of solutions v such that $\alpha \leqq v(t) \leqq \beta$ for $t \geqq 0$. (See Exercises 5 and 6 of the Miscellaneous Exercises for Chapter 2, page 69.)

(b)   Hence show that (1) has a basis u_1, u_2 such that for $t \geqq 0$,

$$u_k(t) = e^{\int_0^t v_k(s)\,ds},$$

where

$$\alpha \leqq v_k(t) \leqq \beta, \quad t \geqq 0, \quad k = 1, 2.$$

(c)   Suppose that it is possible to choose $\beta < 0$. Prove that all solutions of (1) tend to zero as $t \to \infty$.

4

VECTORS, MATRICES AND DETERMINANTS

This chapter contains a resumé of those portions of linear algebra which will be needed in the sequel. It is assumed that the reader is familiar with the subject matter of Chapter 3, Section 1. Certain basic facts about determinants and the invertibility of matrices are stated without proof. These facts are collected in Theorems 4.1 to 4.3.

1. Vectors

An ordered n-tuple $(x_1, x_2, \ldots, x_n)$ of complex numbers is called a *complex n-vector*, or simply a *vector*. The number x_i is called the ith coordinate of the vector. Vectors will usually be denoted by lower case letters without subscripts. If x is a vector, x_i will denote the ith coordinate of x, so that $x = (x_1, x_2, \ldots, x_n)$. Superscripts will be used to distinguish vectors. Thus if $x^1, x^2, \ldots, x^k$ are vectors, x^i_j denotes the jth coordinate of the vector x^i. Equality of vectors means equality of their coordinates, that is, $x = y$ means $x_i = y_i$, $i = 1, 2, \ldots, n$. Addition and subtraction of vectors is defined by

$$x \pm y = (x_1 \pm y_1, x_2 \pm y_2, \ldots, x_n \pm y_n).$$

If c is a complex number and x is a vector, the product cx denotes the vector $(cx_1, cx_2, \ldots, cx_n)$. The negative of a vector x is the vector $-x = (-x_1, -x_2, \ldots, -x_n)$, so that $-x = (-1)x$ and $x - x = 0$, where 0 denotes the vector all of whose coordinates are zero.

If $c_1, c_2, \ldots, c_k$ are numbers and $x^1, x^2, \ldots, x^k$ vectors, and if

$$x = \sum_{i=1}^{k} c_i x^i,$$ we say that x is a *linear combination* of the vectors

$x^1, \ldots, x^k$. The vectors $x^1, \ldots, x^k$ are said to be *linearly depen-dent* if there are complex numbers $c_1, \ldots, c_k$, not all zero, such that $\sum_{i=1}^{k} c_i x^i = 0$. If, on the other hand, $\sum_{i=1}^{k} c_i x^i = 0$ implies that $c_1 = c_2 = \cdots = c_k = 0$, we say the vectors are *linearly independent*. It is easy to see that $x^1, \ldots, x^k$ are linearly dependent if and only if at least one of them is a linear combination of the others (Exercise 1). Obviously, if one of the vectors is zero, the vectors are linearly dependent.

The set of all n-vectors is denoted by V_n. A set of n linearly inde-pendent n-vectors is called a *basis* for V_n. Of particular importance is the basis formed by the vectors $e^1, e^2, \ldots, e^n$ defined by $e_j^i = 0$ if $i \neq j$, $e_i^i = 1$, or

$$e^1 = (1,0,0, \ldots ,0)$$
$$e^2 = (0,1,0, \ldots ,0)$$
$$\cdots \cdots \cdots \cdots$$
$$e^n = (0,0,0, \ldots ,1).$$

To see that these vectors are linearly independent, note that

$$\sum_{i=1}^{n} c_i e^i = (c_1, c_2, \ldots ,c_n). \tag{1.1}$$

Hence $\sum_{i=1}^{n} c_i e^i = 0$ only if all the c_i are zero. We shall refer to e^1, $e^2, \ldots, e^n$ as the *standard basis*.

Equation (1.1) implies that for any vector x,

$$x = \sum_{i=1}^{n} x_i e^i,$$

and that if

$$x = \sum_{i=1}^{n} c_i e^i,$$

then

$$x_i = c_i, \qquad i = 1, \ldots, n.$$

We anticipate a generalization to be proved later: If $x^1, \ldots, x^n$ is a basis, then for every vector x there is a unique set of numbers $c_1, \ldots, c_n$ such that $x = \sum_{i=1}^{n} c_i x^i$. The numbers $c_1, \ldots, c_n$ are called the *components* of x with respect to the basis $x^1, \ldots, x^n$. Note that the components of a vector with respect to the standard basis are just the coordinates of the vector.

EXERCISES

1.   Prove that $x^1, \ldots, x^k$ are linearly dependent if and only if at least one of them can be expressed as a linear combination of the rest.

2.   Prove that any subset of a linearly independent set of vectors is linearly independent.

3.   Prove that if two of the vectors $x^1, \ldots, x^k$ are equal, the vectors $x^1, \ldots, x^k$ are linearly dependent.

4.   (a)   Let $x^1, \ldots, x^k$, $k \leq n$, be vectors such that $x_j^i = 0$ if $i < j$ while $x_i^i \neq 0$. Prove that $x^1, \ldots, x^k$ are linearly independent.

   (b)   Let $x^i = \displaystyle\sum_{k=1}^{i} e^k$, $i = 1, \ldots, n$. Prove that $x^1, \ldots, x^n$

is a basis.

2.   Matrices

A *matrix*, or more precisely, an $n \times n$ *matrix*, is an array of n^2 complex numbers a_{ij}, $i, j = 1, \ldots, n$:

$$\begin{bmatrix} a_{11} & a_{12} & \cdots & a_{1n} \\ a_{21} & a_{22} & \cdots & a_{2n} \\ \cdot & \cdot & \cdots & \cdot \\ a_{n1} & a_{n2} & \cdots & a_{nn} \end{bmatrix}.$$

Equivalently, a matrix is a function which assigns a complex number to each ordered pair of integers (i,j), $i, j = 1, \ldots, n$, the value of the function at (i,j) being denoted by a_{ij}. The number a_{ij} is called the (i,j)th entry of the matrix. A matrix will usually be denoted by a capital letter, and its (i,j)th entry by the same letter in lower case with subscripts. Thus if A is a matrix, its (i,j)th entry is a_{ij}. Whenever a matrix is displayed as a tableau, the tableau is enclosed in brackets. (The same tableau enclosed in vertical lines denotes the determinant of the matrix, which is merely a number. Determinants are discussed in Section 4.) Equality of matrices means equality of their entries. Thus $A = B$ means $a_{ij} = b_{ij}$, $i, j = 1, \ldots, n$.

If A is a matrix, the vector $(a_{i1}, a_{i2}, \ldots, a_{in})$ is called the ith *row* of A, and the vector $(a_{1j}, a_{2j}, \ldots, a_{nj})$ the jth *column* of A. The matrix all of whose entries are zero is called the *zero matrix* and is denoted by 0. The matrix whose (i,j)th entry is 1 or 0 according as $i = j$ or $i \neq j$ is called the *identity matrix*, and is denoted by I. Thus

$$I = \begin{bmatrix} 1 & 0 & 0 & \cdots & 0 \\ 0 & 1 & 0 & \cdots & 0 \\ \multicolumn{5}{c}{\cdots\cdots\cdots\cdots\cdots} \\ 0 & 0 & 0 & \cdots & 1 \end{bmatrix}.$$

Note that the ith row (and the ith column) of I is the vector e^i.

The vector $(a_{11}, a_{22}, \ldots, a_{nn})$ is called the *principal diagonal* of A. If all entries above (below) the principal diagonal vanish, A is said to be a *lower-* (*upper-*) *triangular* matrix. Thus A is lower-triangular if $i < j$ implies $a_{ij} = 0$, and A is upper-triangular if $i > j$ implies $a_{ij} = 0$. If A is either lower-triangular or upper-triangular, we say that A is *triangular*. If $a_{ij} = 0$ for all $i \neq j$, so that all entries off the principal diagonal vanish, A is called a *diagonal* matrix.

Addition and subtraction of matrices is defined entry-wise: $A + B$ is the matrix whose (i,j)th entry is $a_{ij} + b_{ij}$. In particular, $A - A = 0$. If c is a complex number and A is a matrix, cA is the matrix whose (i,j)th entry is ca_{ij}. The negative of A is defined by $-A = (-1)A$. The product Ac, where c is a number, is defined to be equal to cA.

If A is an $n \times n$ matrix and x is an n-vector, the product Ax is defined to be the n-vector whose ith coordinate is $\displaystyle\sum_{j=1}^{n} a_{ij}x_j$. Thus

$$Ax = \left(\sum_{j=1}^{n} a_{1j}x_j, \sum_{j=1}^{n} a_{2j}x_j, \ldots, \sum_{j=1}^{n} a_{nj}x_j \right).$$

This rule is most easily remembered as follows: The ith coordinate of Ax is the dot product of the ith row of A with x. (The dot product of two vectors x and y is the number $x \cdot y = x_1 y_1 + x_2 y_2 + \cdots + x_n y_n$.) We do not define the product xA.

Consider a system of n simultaneous linear algebraic equations in n unknowns $x_1, \ldots, x_n$:

$$\begin{aligned} a_{11}x_1 + a_{12}x_2 + \cdots + a_{1n}x_n &= b_1 \\ a_{21}x_1 + a_{22}x_2 + \cdots + a_{2n}x_n &= b_2 \\ &\cdots \\ a_{n1}x_1 + a_{n2}x_2 + \cdots + a_{nn}x_n &= b_n. \end{aligned}$$

In terms of the matrix-vector product defined above, this system assumes the simple form

$$Ax = b,$$

where A is the matrix formed from the coefficients, $x = (x_1, \ldots, x_n)$ is the unknown vector, and $b = (b_1, \ldots, b_n)$. This is one of the reasons for the definition of the matrix-vector product.

We shall sometimes denote the jth column of A by $A^{(j)}$. A first property of this notation is given in Exercise 1. Note that the ith coordinate of $A^{(j)}$ is a_{ij}.

A simple but important application of the matrix-vector product is its use in expressing linear combinations of vectors (Exercise 2).

EXERCISES

1. (a) Prove that $Ax = \displaystyle\sum_{j=1}^{n} x_j A^{(j)}$. (Hint: What is the ith coordinate of the vector on the right?)

    (b) Show that $Ae^i = A^{(i)}$.

2. Let $x^1, \ldots, x^n$ be vectors and $c_1, \ldots, c_n$ numbers. Show that $\displaystyle\sum_{i=1}^{n} c_i x^i = Ac$, where $c = (c_1, \ldots, c_n)$ and A is the matrix whose ith column is x^i.

3. Prove the following rules of matrix-vector multiplication, where A and B are matrices, x and y are vectors, and c is a number:

    (a) $A(x + y) = Ax + Ay$
    (b) $A(cx) = c(Ax)$
    (c) $(A + B)x = Ax + Bx$
    (d) $Ix = x$

3. Matrix Multiplication

Given two $n \times n$ matrices A and B, the product AB is defined to be the $n \times n$ matrix whose (i,j)th entry is $\displaystyle\sum_{k=1}^{n} a_{ik}b_{kj}$. The rule for forming this product can also be expressed this way: The (i,j)th entry of AB is the dot product of the ith row of A and the jth column of B.

In general $AB \neq BA$, that is, matrix multiplication is not commutative. This may be seen from the example

$$A = \begin{bmatrix} 1 & 2 \\ 0 & 1 \end{bmatrix}, \qquad B = \begin{bmatrix} 2 & 1 \\ 1 & 1 \end{bmatrix}.$$

Here

$$AB = \begin{bmatrix} 4 & 3 \\ 1 & 1 \end{bmatrix}$$

while

$$BA = \begin{bmatrix} 2 & 5 \\ 1 & 3 \end{bmatrix}.$$

We say that AB is obtained through left-multiplication of B by A, or right-multiplication of A by B. If A and B are two matrices such that $AB = BA$, we say that A and B *commute* with each other. Thus, for example, the zero matrix commutes with every matrix, since obviously $A0 = 0A = 0$ for every A.

Theorem 3.1. $(AB)^{(j)} = AB^{(j)}, \qquad j = 1, \ldots, n.$

Proof. We show that the vectors $(AB)^{(j)}$ and $AB^{(j)}$ have equal coordinates. The ith coordinate of $(AB)^{(j)}$ is the (i,j)th entry of AB, and is thus by definition equal to $\sum_{k=1}^{n} a_{ik}b_{kj}$. If x is a vector, the ith coordinate of Ax is by definition $\sum_{k=1}^{n} a_{ik}x_k$. Setting $x = B^{(j)}$, we have that $x_k = b_{kj}$, so that the ith coordinate of $AB^{(j)}$ is $\sum_{k=1}^{n} a_{ik}b_{kj}$. $\square$

The following associative and distributive properties of matrix multiplication follow easily from Theorem 3.1 and Exercise 1(a), Section 2. The proofs are left to the reader (Exercise 1).

Theorem 3.2

    (a) $(AB)x = A(Bx)$, *where x is a vector*
    (b) $(AB)C = A(BC)$
    (c) $(A + B)C = AC + BC, \quad A(B + C) = AB + AC$

Note also that $AI = IA = A$ for any matrix A, where I is the identity matrix.

Let A be a matrix. If there exists a matrix B such that $AB = BA = I$, we say that A is *invertible*. Suppose that A is invertible and that $AB = BA = I$ as well as $AC = CA = I$. We show that then $B = C$. Multiplying both sides of the equation $AB = I$ by C on the left yields $C(AB) = CI = C$, or, using the associative law, $(CA)B = C$. Since $CA = I$, we have $B = C$, the desired result. It follows that if A is invertible, there exists a *unique* matrix B such that $AB = BA = I$; B is called the *inverse* of A and is denoted by A^{-1} (read "A inverse"). The defining property of A^{-1} is thus

$$AA^{-1} = A^{-1}A = I.$$

Not every matrix has an inverse, i.e., is invertible. Obviously 0 has no inverse. The fact that there are non-zero matrices which are not invertible is closely connected with the failure of matrix multiplication to obey the cancellation law (see Exercises 2 and 3). Note that since $II = I$, I is invertible and $I^{-1} = I$.

Suppose that we are given an invertible matrix A. To compute A^{-1} we reason as follows. For any vector y, the equation $Ax = y$ has the solution $x = A^{-1}y$, as is readily verified by substitution: $A(A^{-1}y) = (AA^{-1})y = Iy = y$. Furthermore, this is the only solution. For if x satisfies $Ax = y$, left-multiplication by A^{-1} yields $A^{-1}(Ax) = A^{-1}y$, or $x = A^{-1}y$. To compute A^{-1}, we write out the equation $Ax = y$ as a system of simultaneous linear equations and solve it. The system is

$$a_{11}x_1 + a_{12}x_2 + \cdots + a_{1n}x_n = y_1$$
$$a_{21}x_1 + a_{22}x_2 + \cdots + a_{2n}x_n = y_2$$
$$\cdots \cdots \cdots \cdots \cdots \cdots \cdots \cdots \cdots$$
$$a_{n1}x_1 + a_{n2}x_2 + \cdots + a_{nn}x_n = y_n$$

where, of course, the coefficients a_{ij} are given numbers. The y's on the right must be left arbitrary, i.e., they should not be replaced by numbers. This system may be solved for $x_1, \ldots, x_n$ by various methods, the most efficient of which, in most cases, is that of Gaussian elimination, i.e., the successive elimination of unknowns by the addition of a multiple of one equation to another equation. The method does not matter. Once the system has been solved, we have expressions for $x_1, \ldots, x_n$ of the form $x_i = \sum_{j=1}^{n} b_{ij}y_j$, $i = 1, \ldots, n$, that is, $x = By$. Since we know that $x = A^{-1}y$, it follows that $By = A^{-1}y$. Since y is arbitrary, we may now set $y = e^i$ to obtain $B^{(i)} = (A^{-1})^{(i)}$, $i = 1, \ldots, n$. It follows that $B = A^{-1}$. As a check on the computations one should verify that $AA^{-1} = I$.

EXAMPLE.   Let

$$A = \begin{bmatrix} 1 & 2 & -1 \\ 1 & 1 & 1 \\ 1 & 1 & 2 \end{bmatrix}.$$

We wish to find A^{-1}. To avoid subscripts, we temporarily change notation and write x, y, z instead of x_1, x_2, x_3 and a, b, c instead of y_1, y_2, y_3. To solve the system

$$\begin{aligned} x + 2y - z &= a \\ x + y + z &= b \\ x + y + 2z &= c \end{aligned}$$

we begin by subtracting the first equation from the second and third. This eliminates x from the last two equations, and the system becomes

$$\begin{aligned} x + 2y - z &= a \\ -y + 2z &= b - a \\ -y + 3z &= c - a. \end{aligned}$$

Next, we subtract the second equation from the third to obtain

$$z = c - b.$$

Substituting back, we can now solve successively for y and x. We thus find

$$\begin{aligned} x &= -a + 5b - 3c \\ y &= a - 3b + 2c \\ z &= -b + c. \end{aligned}$$

It follows that

$$A^{-1} = \begin{bmatrix} -1 & 5 & -3 \\ 1 & -3 & 2 \\ 0 & -1 & 1 \end{bmatrix}.$$

The reader should check that $A A^{-1} = A^{-1}A = I$.

EXERCISES

1. (a) Prove that $(AB)x = A(Bx)$. [Hint: Apply in succession Exercise 1(a) of Section 2 and Theorem 3.1 to rewrite the product $(AB)x$. Rewrite $A(Bx)$ analogously.]

(b) Prove that $(AB)C = A(BC)$. [Hint: Show that $(AB)C$ and $A(BC)$ have equal columns.]

(c) Prove that $(A + B)C = AC + BC$ and $A(B + C) = AB + AC$.

2. Matrix multiplication does not obey the cancellation law, i.e., $AB = 0$ does not imply $A = 0$ or $B = 0$. Prove this by finding a non-zero 2×2 matrix A for which $A^2 = 0$. (A^2 means AA.)

3. Let A be a matrix such that $AB = 0$ for some non-zero matrix B. Prove that A is not invertible. (Hint: Suppose it is and derive a contradition.)

4. Prove that if A is invertible so is A^{-1}, and $(A^{-1})^{-1} = A$.

5. Prove that if A and B are invertible so is AB, and $(AB)^{-1} = B^{-1}A^{-1}$.

6. Compute the inverses of the following matrices:

(a) $\begin{bmatrix} 2 & 1 \\ 1 & 1 \end{bmatrix}$ (b) $\begin{bmatrix} 0 & 1 \\ 1 & 0 \end{bmatrix}$

(c) $\begin{bmatrix} 1 & 1 & -1 \\ 0 & 1 & -1 \\ 2 & 0 & 1 \end{bmatrix}$

4. Determinants; Fundamental Algebraic Theorems

To each $n \times n$ matrix A we assign a number called the *determinant* of A, and denoted by det (A). It is defined as follows. Let $j_1, j_2, \ldots, j_n$ be the integers from 1 to n in some order. (Such an ordering is sometimes called a permutation of the integers 1 to n. There are $n!$ of these orderings.) Count the number of times that a larger integer precedes a smaller in the given ordering $j_1, \ldots, j_n$. Denote this number by $N(j_1, \ldots, j_n)$. The determinant of A is defined to be the number

$$\det (A) = \Sigma(-1)^{N(j_1,\ldots,j_n)} a_{1j_1} a_{2j_2} \cdots a_{nj_n}, \tag{4.1}$$

the sum being extended over all $n!$ orderings of the column subscripts.

To illustrate the definition of the number $N(j_1, \ldots, j_n)$, let $n = 5$ and consider the ordering 2, 3, 5, 4, 1. We note that 2 precedes 1, 3 precedes 1, 5 precedes 4, 5 precedes 1, and 4 precedes 1. Therefore, $N(2,3,5,4,1) = 5$. Similarly, $N(1,2,4,3,5) = 1$ and $N(1,2,3,4,5) = 0$.

Formula (4.1) can be described as follows. Choose exactly one entry from each row and each column of A and form the product of these entries. This can be done in $n!$ ways. Arrange the factors in each product so that the row subscripts are in their natural order, and multiply the product by 1 or -1 according as the resulting ordering of the column subscripts has an even or odd number of inverted pairs. The sum of these $n!$ products is the determinant of A.

EXAMPLES.   Let A be a 2×2 matrix. There are only two orderings of the integers 1 and 2, viz., 1, 2 and 2, 1. We thus obtain two products, $a_{11}a_{22}$ and $a_{12}a_{21}$. Since $N(1,2) = 0$ and $N(2,1) = 1$, we have

$$\det (A) = a_{11}a_{22} - a_{12}a_{21}.$$

The reader will readily verify that if A is a 3×3 matrix,

$$\det (A) = a_{11}a_{22}a_{33} - a_{12}a_{21}a_{33} + a_{12}a_{23}a_{31}$$
$$- a_{11}a_{23}a_{32} + a_{13}a_{21}a_{32} - a_{13}a_{22}a_{31}. \qquad \square$$

When a matrix is displayed as a tableau, it is often convenient to place the tableau between vertical lines to denote the determinant. Thus we write, for example,

$$\det \begin{bmatrix} a_{11} & a_{12} \\ a_{21} & a_{22} \end{bmatrix} = \begin{vmatrix} a_{11} & a_{12} \\ a_{21} & a_{22} \end{vmatrix}.$$

Obviously $\det (0) = 0$. It is also easy to see that $\det (I) = 1$. Other elementary cases are given in Exercise 1.

We shall need a number of facts about determinants whose proofs would take us too far afield. (These proofs are given in most texts on linear algebra.) To begin with, the definition of $\det (A)$ is not only tedious computationally, but is ill suited to analytic applications. We therefore state a formula which expresses the determinant of a matrix in a form better adapted to our purposes.

Theorem 4.1.   *Let A be an $n \times n$ matrix. Let A_{ij} denote the $(n - 1) \times (n - 1)$ matrix obtained from A by deleting the ith row and jth column of A. Then for any i, $1 \leqq i \leqq n$,*

$$\det (A) = \sum_{j=1}^{n} (-1)^{i+j} a_{ij} \det (A_{ij}). \qquad (4.2)$$

The determinant of A_{ij} is called the *minor* of a_{ij}; $(-1)^{i+j} \det (A_{ij})$ is called the *cofactor* of a_{ij}. Formula (4.2) results from the fact, not proved here, that if in formula (4.1) the terms containing a specified entry a_{ij} are collected, and the entry a_{ij} is factored out of them, the coefficient of a_{ij} is its cofactor. Since the cofactor of a_{ij} contains no entry from the ith row, all entries a_{ik} with $k \neq j$ must occur in the remaining terms. Thus if the process of collecting terms and factoring out a_{ij} is carried out for each j, $1 \leqq j \leqq n$, while i is held fixed, formula (4.1) is transformed into (4.2). Formula (4.2) is called the *expansion of* $\det (A)$ *by the ith row.* (The expansion of $\det (A)$ by the jth column is obtained if the summation in (4.2) is taken over i instead of j.)

We shall also have occasion to use the following property of determinants:

Theorem 4.2. $\det (AB) = \det (A) \det (B)$.

It is an immediate consequence of Theorem 4.2 that an invertible matrix has a non-zero determinant, since $A A^{-1} = I$ yields

$$\det (A) \det (A^{-1}) = \det (I) = 1.$$

The converse is also true: if $\det (A) \neq 0$, then A is invertible. These facts form part of a chain linking the various concepts introduced in this chapter. We shall now state in the form of a theorem the connections which are important for the theory of linear systems of differential equations. For the sake of easy reference we shall call this omnibus theorem the "fundamental theorem of linear algebra," or "FTLA" for short.

Theorem 4.3 (FTLA).   *Let A be a matrix. The following statements are equivalent, i.e., each implies all the others.*

(1)   *A is invertible.*
(2) $\det (A) \neq 0$.
(3)   *The columns of A are linearly independent.*
(4)   *For every vector b, the equation $Ax = b$ has a solution.*
(5)   *$x = 0$ is the only solution of $Ax = 0$.*

Remark.   Note that (5) is equivalent to the statement that for any b the equation $Ax = b$ has at most one solution, since $Ax = b$ and $Ay = b$ implies $Ax = Ay$, or $A(x - y) = 0$, whence by (5), $x = y$. It follows from the equivalence of (4) and (5) that either $Ax = b$ has a unique solution for every b, or else there are vectors b such that it has no solution, while for all other b the solutions are not unique.

It may be worthwhile to point out that parts of this theorem are trivial. For example, it is obvious that (1) implies (4) and (5), and that (3) and (5) are equivalent. The next theorem is an immediate consequence of the equivalence of (3), (4) and (5).

Theorem 4.4.   *A set of n vectors $x^1, \ldots, x^n$ is linearly independent (and thus a basis for V_n) if and only if every vector x has a unique representation as a linear combination $\sum_{i=1}^{n} c_i x^i$.*

Proof. Let $x^1, \ldots, x^n$ be linearly independent, and let x be a given vector. We seek numbers $c_1, \ldots, c_n$ such that

$$\sum_{i=1}^{n} c_i x^i = x,$$

or equivalently, such that

$$Ac = x,$$

where A is the matrix whose columns are $x^1, \ldots, x^n$ and $c = (c_1, \ldots, c_n)$. Because the columns of A are linearly independent, it follows from the FTLA that there exists a unique vector c such that $Ac = x$.

Conversely, suppose that every vector x has a representation $x = \sum_{i=1}^{n} c_i x^i$, that is, the equation $Ac = x$ has a solution c for every x. It follows from the FTLA that the columns of A, i.e., the vectors $x^1, \ldots, x^n$, are linearly independent. (Note that the uniqueness of the representation also follows from the FTLA, and need not be assumed.) $\square$

EXERCISES

1.   Using only the definition of determinants [formula (4.1)] show that

    (a) $\det (I) = 1$
    (b)   If one row or one column of A is zero, $\det (A) = 0$
    (c)   If A is diagonal, $\det (A) = a_{11} a_{22} \cdots a_{nn}$
    (d)   If A is triangular, $\det (A) = a_{11} a_{22} \cdots a_{nn}$

2.   By evaluating a suitable determinant, decide whether the following vectors are linearly dependent or independent:

$$(1,0,3,2), \ (i, 1 + i, 0, 1), \ (0,1,1,1), \ (2,-1,i,0).$$

3.   Let $A = \begin{bmatrix} 1 & 1 \\ 1 & 1 \end{bmatrix}$. Construct vectors b^1 and b^2 such that $Ax = b^1$ has a solution and $Ax = b^2$ does not. Find all solutions of $Ax = b^1$.

5.   Eigenvalues, Eigenvectors, and the Diagonalization of Matrices

Let A be a matrix. If λ is a number such that the equation

$$Ax = \lambda x \tag{5.1}$$

has a non-zero solution vector x, λ is called an *eigenvalue* of A. If λ is an eigenvalue of A and x is a non-zero vector which satisfies (5.1), x is called an *eigenvector* of A, or more precisely, an eigenvector of A belonging to the eigenvalue λ. It is clear that if x is an eigenvector belonging to λ and c is a non-zero number, then cx is also an eigenvector belonging to λ. Eigenvalues (eigenvectors) are also called *characteristic roots* or *characteristic values (characteristic vectors)*.

We shall now show that the eigenvalues of a matrix are the roots of a certain polynomial associated with the matrix. By definition, λ is an eigenvalue of A if and only if the equation $Ax = \lambda x$ has a non-zero solution. This equation may be written $Ax - \lambda x = 0$, or

$$(A - \lambda I)x = 0, \tag{5.2}$$

where, as usual, I denotes the identity matrix. By the FTLA, Eq. (5.2) has a non-zero solution if and only if

$$\det (A - \lambda I) = 0. \tag{5.3}$$

Equation (5.3) is therefore a necessary and sufficient condition that λ is an eigenvalue of A.

Considering λ as a variable, we are thus led to associate with A the polynomial $p(\lambda) = \det (A - \lambda I)$, that is,

$$p(\lambda) = \begin{vmatrix} a_{11} - \lambda & a_{12} & \cdots & a_{1n} \\ a_{21} & a_{22} - \lambda & \cdots & a_{2n} \\ \cdots\cdots\cdots\cdots\cdots\cdots\cdots\cdots\cdots \\ a_{n1} & a_{n2} & \cdots & a_{nn} - \lambda \end{vmatrix}. \tag{5.4}$$

Obviously $p(\lambda)$ is a polynomial of degree n. Note that the coefficient of λ^n is $(-1)^n$ and that the constant term in $p(\lambda)$ is $\det (A)$. $p(\lambda)$ is called the *characteristic polynomial* of A; we have proved that *the eigenvalues of A are the roots of its characteristic polynomial*. It follows from the fundamental theorem of algebra (Theorem 1.1, Chapter 3) that *every matrix has at least one eigenvalue*. That a matrix may possibly have no more than one eigenvalue is illustrated by the identity matrix, whose only eigenvalue is 1; note that every non-zero vector is an eigenvector of I belonging to 1. By the multiplicity of an eigenvalue we shall mean its multiplicity as a root of the characteristic polynomial; if its multiplicity is 1, we call it a *simple* eigenvalue. An eigenvalue which is not simple is called a *multiple* or *repeated* eigenvalue.

In the sequel it will be necessary to compute the eigenvalues of certain matrices, as well as an eigenvector belonging to each of these eigenvalues. Computing the eigenvalues reduces to computing the roots of $p(\lambda)$. Suppose we have computed one such root; let us simply denote it by λ. To find an eigenvector belonging to λ we must compute a non-zero solution of $(A - \lambda I)x = 0$. Written out in coordinate form this equation becomes

$$
\begin{aligned}
(a_{11} - \lambda)x_1 + a_{12}x_2 + \cdots + a_{1n}x_n &= 0 \\
a_{21}x_1 + (a_{22} - \lambda)x_2 + \cdots + a_{2n}x_n &= 0 \\
\cdots \cdots \cdots \cdots \cdots \cdots \cdots \cdots \cdots \cdots \\
a_{n1}x_1 + a_{n2}x_2 + \cdots + (a_{nn} - \lambda)x_n &= 0.
\end{aligned}
$$

That this system has a non-zero solution is guaranteed by the fact that λ is an eigenvalue of A. We solve the system by the method of elimination. Since any multiple of a solution is itself a solution, it will be found that one (or perhaps even several) of the x_i may be chosen arbitrarily. The values assigned to these x_i should be chosen with a view to computational convenience, but must not all be zero.

EXAMPLE. We compute the eigenvalues and corresponding eigenvectors of the matrix

$$
\begin{bmatrix} 2 & 1 \\ 2 & 3 \end{bmatrix}. \tag{5.5}
$$

The characteristic polynomial is

$$
p(\lambda) = \begin{vmatrix} 2 - \lambda & 1 \\ 2 & 3 - \lambda \end{vmatrix} = \lambda^2 - 5\lambda + 4 = (\lambda - 4)(\lambda - 1)
$$

and therefore the eigenvalues are 4 and 1. To find an eigenvector belonging to 4 we solve $(A - 4I)x = 0$, i.e., the system

$$
\begin{aligned}
-2x_1 + x_2 &= 0 \\
2x_1 - x_2 &= 0.
\end{aligned}
$$

Note that the second equation is redundant, as it should be. Since $x_2 = 2x_1$, it is convenient to choose $x_1 = 1$, whence $x_2 = 2$. Thus $(1,2)$ is an eigenvector belonging to 4. In the same way we find that $(1,-1)$ is an eigenvector belonging to 1.

Theorem 5.1. *If $\lambda_1, \ldots, \lambda_k$ are distinct eigenvalues of A, and $x^1, \ldots, x^k$ are corresponding eigenvectors, then $x^1, \ldots, x^k$ are linearly independent.*

Proof. We shall deduce a contradiction from the assumption that $x^1, \ldots, x^k$ are linearly dependent. We suppose the vectors labeled so that $x^1, \ldots, x^r$ are linearly independent, where $1 \leq r < k$, while x^k can be written as a linear combination of $x^1, \ldots, x^r$. Let

$$x^k = \sum_{i=1}^{r} c_i x^i. \tag{5.6}$$

Then

$$Ax^k = A \left(\sum_{i=1}^{r} c_i x^i \right)$$

$$= \sum_{i=1}^{r} c_i A x^i$$

$$= \sum_{i=1}^{r} c_i \lambda_i x^i.$$

Since $Ax^k = \lambda_k x^k$, we thus have

$$\lambda_k x^k = \sum_{i=1}^{r} c_i \lambda_i x^i.$$

Substituting for x^k from (5.6) yields

$$\lambda_k \sum_{i=1}^{r} c_i x^i = \sum_{i=1}^{r} c_i \lambda_i x^i$$

or

$$\sum_{i=1}^{r} c_i (\lambda_k - \lambda_i) x^i = 0. \tag{5.7}$$

Since $x^1, \ldots, x^r$ are linearly independent, (5.7) implies

$$c_i(\lambda_k - \lambda_i) = 0, \qquad i = 1, \ldots, r.$$

But $\lambda_k - \lambda_i \neq 0$, $i = 1, \ldots, r$, whence

$$c_i = 0, \qquad i = 1, \ldots, r.$$

It follows that $x^k = 0$, a contradiction. We conclude that $x^1, \ldots, x^k$ are linearly independent. $\square$

Theorem 5.2.   *Let A be an $n \times n$ matrix having n distinct eigenvalues $\lambda_1, \ldots, \lambda_n$. Let P be an $n \times n$ matrix whose jth column is an eigenvector of A belonging to λ_j. Then P is invertible, and*

$$P^{-1}AP = D,$$

where D is the diagonal matrix

$$\begin{bmatrix} \lambda_1 & 0 & 0 & \cdots & 0 \\ 0 & \lambda_2 & 0 & \cdots & 0 \\ & \cdot & \cdot & \cdot & \\ 0 & 0 & 0 & \cdots & \lambda_n \end{bmatrix}.$$

Proof.   By the preceding theorem the columns of P are linearly independent. It follows from the FTLA that P is invertible. We compute the jth column of $P^{-1}AP$:

$$\begin{aligned} (P^{-1}AP)^{(j)} &= P^{-1}AP^{(j)} \\ &= P^{-1}\lambda_j P^{(j)} \\ &= \lambda_j P^{-1}P^{(j)} \\ &= \lambda_j (P^{-1}P)^{(j)} \\ &= \lambda_j I^{(j)} \\ &= \lambda_j e^j. \end{aligned}$$

Hence $P^{-1}AP = D$. $\square$

A matrix A is said to be *similar* to a matrix B if there exists an invertible matrix P such that

$$P^{-1}AP = B.$$

We say that P *transforms* A *into* B. An important problem of linear algebra is to find, for any given matrix A, the simplest matrix to which it is similar. The theorem just proved asserts that any matrix having only simple eigenvalues is similar to a diagonal matrix, and it also tells us how to construct a suitable P. A matrix which is similar to a diagonal matrix is said to be *diagonable*. While it is by no means necessary for a matrix to have n distinct eigenvalues in order to be diagonable, most matrices which arise in practice do in fact have n distinct eigenvalues. (The necessary and sufficient condition that a matrix is diagonable is easily seen to be that to every eigenvalue of multiplicity m there belong m linearly independent eigenvectors.) In Section 7 of the next chapter we shall prove that every matrix is similar to a triangular matrix.

EXERCISES

1. Find the eigenvalues (together with their multiplicities) and corresponding eigenvectors of the following matrices:

(a) $\begin{bmatrix} 0 & -1 \\ 1 & 0 \end{bmatrix}$ (b) $\begin{bmatrix} 1 & 1 \\ 0 & 1 \end{bmatrix}$

(c) $\begin{bmatrix} 1 & 1 & 1 \\ 0 & 1 & 0 \\ 0 & 0 & 1 \end{bmatrix}$ (d) $\begin{bmatrix} 1 & i \\ -i & 1 \end{bmatrix}$

2. For each of the following matrices A find a matrix P such that $P^{-1}AP$ is diagonal. Compute the products AP, PA and $P^{-1}AP$.

(a) $\begin{bmatrix} 0 & -1 \\ 1 & 0 \end{bmatrix}$ (b) $\begin{bmatrix} 2 & 1 \\ 2 & 3 \end{bmatrix}$ (c) $\begin{bmatrix} 1 & 1 \\ 1 & 1 \end{bmatrix}$

(d) $\begin{bmatrix} 0 & 1 & 0 & 0 \\ -1 & 0 & 0 & 0 \\ 0 & 0 & 0 & 2 \\ 0 & 0 & -2 & 0 \end{bmatrix}$ (e) $\begin{bmatrix} 0 & 1 & 1 \\ 0 & 1 & 0 \\ 0 & 0 & 2 \end{bmatrix}$

3. (a) Prove that if A is a diagonal matrix, the diagonal entries of A are its eigenvalues.

(b) Prove that if A is a triangular matrix, the diagonal entries of A are its eigenvalues.

4. Prove that similarity is an equivalence relationship:

(a) Every matrix is similar to itself.

(b) If A is similar to B, then B is similar to A.

(c) If A is similar to B, and B is similar to C, then A is similar to C.

5. Let $P^{-1}AP = B$. Prove that if λ is an eigenvalue of A, and x a corresponding eigenvector, then λ is also an eigenvalue of B, and $P^{-1}x$ is a corresponding eigenvector. (Similar matrices have the same eigenvalues.)

6. Use Theorem 4.2 to prove that similar matrices have the same characteristic polynomial. (Hence not only the eigenvalues, but also their multiplicities, are the same for similar matrices.)

7. Let $\lambda_1, \ldots, \lambda_n$ be the eigenvalues of A, each distinct eigenvalue being listed a number of times equal to its multiplicity. Prove that $\lambda_1 \lambda_2 \cdots \lambda_n = \det (A)$. (Hint: The constant term of $p(\lambda)$ is $\det (A)$.)

8. Let $A^k = AA \cdots A$ (k factors) if k is a positive integer, and $A^0 = I$.

(a) Let k be a non-negative integer. Prove that if λ is an eigenvalue of A, then λ^k is an eigenvalue of A^k.

(b)　Let c_1, c_2 be complex numbers and k_1, k_2 non-negative integers. Prove that if λ is an eigenvalue of A, then $c_1\lambda^{k_1} + c_2\lambda^{k_2}$ is an eigenvalue of $c_1A^{k_1} + c_2A^{k_2}$.

(c)　Given a polynomial $f(\lambda) = \sum_{k=0}^{m} c_k\lambda^k$ and a matrix A, we define the matrix $f(A)$ by $f(A) = \sum_{k=0}^{m} c_kA^k$. Prove that if λ is an eigenvalue of A, then $f(\lambda)$ is an eigenvalue of $f(A)$.

(d)　Let $p(\lambda)$ be the characteristic polynomial of A. Prove that $p(A)$ is a non-invertible matrix. (Actually $p(A)$ is the zero matrix. This fact is known as the Cayley-Hamilton theorem.)

(e)　Calculate $p(A)$ for the matrices (a), (b), and (d) of problem 1.

5

LINEAR SYSTEMS

The prerequisites for this chapter are Section 3 of Chapter 2, Section 1 of Chapter 3, and Chapter 4. The reader who wishes to proceed as rapidly as possible to Chapter 6 may confine himself to reading Sections 1, 2, 3 and 5.

1. Introduction

In this chapter we study systems of n simultaneous first-order linear differential equations, i.e., systems of the form

$$\frac{dx_1}{dt} = a_{11}(t)x_1 + a_{12}(t)x_2 + \cdots + a_{1n}(t)x_n + b_1(t)$$

$$\frac{dx_2}{dt} = a_{21}(t)x_1 + a_{22}(t)x_2 + \cdots + a_{2n}(t)x_n + b_2(t)$$

$$\cdot\ \cdot$$

$$\frac{dx_n}{dt} = a_{n1}(t)x_1 + a_{n2}(t)x_2 + \cdots + a_{nn}(t)x_n + b_n(t)$$

or, more compactly written,

$$\frac{dx_i}{dt} = \sum_{j=1}^{n} a_{ij}(t)x_j + b_i(t), \qquad i = 1, \ldots, n. \tag{1.1}$$

It follows from the discussion in Chapter 1, Section 2, that any linear differential equation or system can be put into this form. Thus the equations studied in Chapter 3 are special cases of (1.1). The functions $a_{ij}(t)$ and $b_i(t)$ are assumed to be continuous for all t, and are allowed to assume complex values. If all the functions $b_i(t)$ are identically zero, the system is said to be *homogeneous*, and otherwise *non-homogeneous*. A solution of (1.1) is by definition an n-tuple of functions $u_1, \ldots, u_n$ defined on an interval I such that

$$u_i'(t) = \sum_{j=1}^{n} a_{ij}(t)u_j(t) + b_i(t), \qquad i = 1, \ldots, n$$

for all t in I.

By a *vector-valued function* we mean a function u which assigns to each number t in some interval I an n-vector $u(t)$. We denote the coordinates of $u(t)$ by $u_1(t), u_2(t), \ldots, u_n(t)$, so that

$$u(t) = (u_1(t), u_2(t), \ldots, u_n(t)) \tag{1.2}$$

for each t in I. In this manner a vector-valued function u gives rise to n complex-valued functions $u_i, \quad i = 1, \ldots, n$, which we call the *coordinate functions* of u. Conversely, any n complex-valued functions $u_i, \quad i = 1, \ldots, n$ defined on a common interval I give rise to a vector-valued function u via the definition (1.2). A vector-valued function is said to be continuous or differentiable if its coordinate functions are. Derivatives and integrals are defined coordinate-wise, that is,

$$u'(t) = (u_1'(t), u_2'(t), \ldots, u_n'(t))$$

and

$$\int_a^b u(t)\,dt = \left(\int_a^b u_1(t)\,dt, \int_a^b u_2(t)\,dt, \ldots, \int_a^b u_n(t)\,dt \right).$$

A *matrix-valued function* is a function A which assigns to each number t in an interval I a matrix $A(t)$. Denoting the entries of $A(t)$ by $a_{ij}(t)$, so that

$$A(t) = \begin{bmatrix} a_{11}(t) & a_{12}(t) & \cdots & a_{1n}(t) \\ a_{21}(t) & a_{22}(t) & \cdots & a_{2n}(t) \\ \cdots & \cdots & \cdots & \cdots \\ a_{n1}(t) & a_{n2}(t) & \cdots & a_{nn}(t) \end{bmatrix}$$

for t in I, we see that a matrix-valued function may be regarded as a matrix whose entries are complex-valued functions. We say that A is continuous if the entries are continuous.

In vector-matrix notation the system (1.1) becomes

$$\frac{dx}{dt} = A(t)x + b(t)$$

where

$$x = (x_1, \ldots, x_n)$$
$$\frac{dx}{dt} = \left(\frac{dx_1}{dt}, \ldots, \frac{dx_n}{dt} \right)$$
$$b(t) = (b_1(t), \ldots, b_n(t))$$

and $A(t)$ is the matrix of the coefficients $a_{ij}(t)$. A solution is a vector-valued function u defined on an interval I such that

$$u'(t) = A(t)u(t) + b(t), \qquad t \text{ in } I.$$

In order to make room for superscripts, we shall hereafter use a dot instead of a prime to denote derivatives:

$$\dot{x} = \frac{dx}{dt}$$
$$\dot{u}(t) = u'(t)$$

We now state the fundamental existence and uniqueness theorem for linear systems, the proof of which is given in Chapter 7.

Theorem 1.1. *Consider the system*

$$\dot{x} = A(t)x + b(t),$$

where A and b are continuous for all t. Given a real number t_0 and a vector η, the system has a unique solution u defined for all t such that $u(t_0) = \eta$.

EXERCISES

1. Deduce the existence and uniqueness theorem for nth-order linear differential equations (Theorem 6.1, page 110) from Theorem 1.1.

2. Use Theorem 1.1 to obtain an existence and uniqueness theorem for the system

$$\frac{d^2x}{dt^2} = A(t)x + B(t)\frac{dx}{dt} + c(t)$$

where $x = (x_1, \ldots, x_n)$, A and B are $n \times n$ matrices continuous for all t, and c is a vector-valued function continuous for all t. (Hint: Transform the system into an equivalent system of $2n$ first-order equations.)

2. Homogeneous Systems

We now consider an nth-order homogeneous system

$$\dot{x} = A(t)x. \qquad (2.1)$$

Note that $u(t) \equiv 0$ is a solution. Also, if u is a solution such that $u(t_0) = 0$ for some t_0, uniqueness implies that $u(t) \equiv 0$.

To describe the structure of the solution space of (2.1) we need some definitions. Let $u^1, \ldots, u^k$ be vector-valued functions defined for all t. We say that $u^1, \ldots, u^k$ are *linearly dependent* if there exist complex numbers $c_1, \ldots, c_k$ not all zero such that $\sum_{i=1}^{k} c_i u^i(t) \equiv 0$. If this equation is satisfied only if $c_1 = c_2 = \cdots = c_k = 0$, the functions are said to be *linearly independent*. By a *linear combination* of $u^1, \ldots, u^k$ we mean a function u of the form $u(t) = \sum_{i=1}^{k} c_i u^i(t)$.

To elucidate the notion of linear dependence of functions, suppose that $u^1, \ldots, u^k$ are linearly dependent. By definition there are numbers, $c_1, \ldots, c_k$ not all zero such that $\sum_{i=1}^{k} c_i u^i(t) = 0$ for all t. This implies that for every t the vectors $u^1(t), \ldots, u^k(t)$ are linearly dependent. The converse is false, i.e., given functions $u^1, \ldots, u^k$ such that for every t the vectors $u^1(t), \ldots, u^k(t)$ are linearly dependent, it does not follow that the functions are linearly dependent (see Exercise 1). If $u^1, \ldots, u^k$ are functions such that for some t the vectors $u^1(t), \ldots, u^k(t)$ are linearly independent, then obviously the functions are linearly independent.

Theorem 2.1.   *Any linear combination of solutions of (2.1) is itself a solution.*

Proof.   Let
$$u(t) = \sum_{i=1}^{k} c_i u^i(t),$$

where $u^1, \ldots, u^k$ are solutions of (2.1) and $c_1, \ldots, c_k$ are numbers. Differentiating gives

$$\dot{u}(t) = \sum_{i=1}^{k} c_i \dot{u}^i(t)$$
$$= \sum_{i=1}^{k} c_i A(t) u^i(t)$$
$$= A(t) \sum_{i=1}^{k} c_i u^i(t)$$
$$= A(t) u(t). \qquad \square$$

Theorem 2.2.   *Solutions $u^1, \ldots, u^k$ of (2.1) are linearly dependent if and only if, for some t, the vectors $u^1(t), \ldots, u^k(t)$ are linearly dependent.*

Proof. The "only if" part is trivial; as we pointed out above, it holds even if the functions are not solutions of (2.1). Now suppose that there is a t_0 such that the vectors $u^1(t_0), \ldots, u^k(t_0)$ are linearly dependent, say $\sum_{i=1}^{k} c_i u^i(t_0) = 0$, with not all c_i equal to zero. Let $v(t) = \sum_{i=1}^{k} c_i u^i(t)$. Then v is a solution of (2.1) by Theorem 2.1. But $v(t_0) = 0$, whence by uniqueness $v(t) \equiv 0$. Hence $\sum_{i=1}^{k} c_i u^i(t) \equiv 0$. $\square$

Theorem 2.2 may be paraphrased by saying that linearly independent initial-value vectors $\eta^1, \ldots, \eta^k$ produce linearly independent solutions, and linearly dependent initial-value vectors produce linearly dependent solutions.

Theorem 2.3. *The system (2.1) has a set of n linearly independent solutions.*

Proof. Let $u^1, \ldots, u^n$ be the solutions of (2.1) satisfying the initial conditions $u^i(0) = e^i$, $i = 1, \ldots, n$. Since the e^i are linearly independent, so are the u^i. $\square$

Theorem 2.4. *Let $u^1, \ldots, u^n$ be n linearly independent solutions of (2.1). Then for every solution u there exists a unique set of numbers $c_1, \ldots, c_n$ such that $u(t) = \sum_{i=1}^{n} c_i u^i(t)$ for all t.*

Proof. Let u be a solution and write $u(0) = \eta$, $u^i(0) = \eta^i$, $i = 1, \ldots, n$. The vectors $\eta^1, \ldots, \eta^n$ are linearly independent by Theorem 2.2. Hence by Theorem 4.4, Chapter 4 there are numbers $c_1, \ldots, c_n$ such that $\eta = \sum_{i=1}^{n} c_i \eta^i$. Let $v(t) = \sum_{i=1}^{n} c_i u^i(t)$, so that v is a solution. Since $v(0) = \eta$, uniqueness implies $v = u$. Hence $u(t) = \sum_{i=1}^{n} c_i u^i(t)$. It remains to show that the c_i are unique. Suppose $u(t) = \sum_{i=1}^{n} d_i u^i(t)$. Then $u(0) = \sum_{i=1}^{n} d_i \eta^i$, so $\eta = \sum_{i=1}^{n} d_i \eta^i$. It follows, again from Theorem 4.4, Chapter 4, that $c_i = d_i$, $i = 1, \ldots, n$. $\square$

The preceding discussion is summarized by the statement that the solutions of (2.1) form an n-dimensional linear space over the complex numbers.

A set of n linearly independent solutions is called a *basis* for the solutions of (2.1). The basis $u^1, \ldots, u^n$ defined by the initial conditions $u^i(0) = e^i$, $i = 1, \ldots, n$, will be called the *standard basis*. If u is a solution with $u(0) = \eta$ and if $u^1, \ldots, u^n$ is the standard basis, then obviously $u(t) = \sum_{i=1}^{n} \eta_i u^i(t)$.

Solution Matrices.   In Chapter 4, Section 2, Exercise 2 we pointed out that a linear combination $\sum_{i=1}^{n} c_i x^i$ of n n-vectors can be written as the product Ac, where A is the $n \times n$ matrix whose ith column is x^i and $c = (c_1, \ldots, c_n)$. This fact leads to an extremely compact and powerful notation for the solutions of (2.1). If $u^1, \ldots, u^n$ are any n solutions of (2.1), a linear combination $\sum_{i=1}^{n} c_i u^i(t)$ takes the simple form $W(t)c$, where $W(t)$ is the matrix whose ith column is $u^i(t)$, and $c = (c_1, \ldots, c_n)$.

Definition.   An $n \times n$ matrix whose columns are solutions of (2.1) is called a *solution matrix* of (2.1). A solution matrix whose columns are linearly independent is called a *fundamental matrix* of (2.1). Let $u^1, \ldots, u^n$ be the standard basis of (2.1). The fundamental matrix whose columns are $u^1(t), \ldots, u^n(t)$ is called the *standard fundamental matrix* of (2.1), or simply the *s.f.m.*

Theorem 2.1 can now be paraphrased: *If W is a solution matrix of (2.1) and c is a vector, then $W(t)c$ is a solution.*

Theorem 2.4 becomes: *If W is a fundamental matrix and u is a solution then there exists a unique vector c such that $u(t) = W(t)c$.*

Combining Theorem 2.2 with the FTLA we obtain a number of equivalent characterizations of fundamental matrices.

Theorem 2.5.   *Let W be a solution matrix. The following are equivalent:*
    (a)   *W is fundamental*
    (b)   *For some t, $W(t)$ is invertible*
    (c)   *For some t, $\det W(t) \neq 0$*
    (d)   *$W(t)$ is invertible for all t*
    (e)   *For all t, $\det W(t) \neq 0$*

Note that a solution matrix W is the s.f.m. if and only if $W(0) = I$. Hence if W is the s.f.m. and u is a solution with $u(0) = \eta$, then $u(t) = W(t)\eta$.

Let V be an $n \times n$ matrix (not necessarily a solution matrix) whose entries are differentiable functions. The *derivative* of V, denoted by $\dot{V}$, is defined to be the matrix whose entries are the derivatives of the entries of V. A matrix with differentiable entries will be called a *differentiable matrix*.

Theorem 2.6.   *W is a solution matrix if and only if*

$$\dot{W}(t) = A(t)W(t)$$

for all t.

Proof.   Denote the columns of W by $u^1, \ldots, u^n$. Suppose that W is a solution matrix, so $\dot{u}^i(t) = A(t)u^i(t)$, $i = 1, \ldots, n$. The ith column of $A(t)W(t)$ is $A(t)u^i(t)$. Hence the matrices $\dot{W}(t)$ and $A(t)W(t)$ have the same columns, and are therefore equal.

Conversely, if $\dot{W}(t) = A(t)W(t)$, equating columns yields $\dot{u}^i(t) = A(t)u^i(t)$, so that W is a solution matrix. $\square$

The preceding theorem may be stated thus: *W is a solution matrix of the system (or vector differential equation)*

$$\dot{x} = A(t)x$$

if and only if W is a solution of the matrix differential equation

$$\dot{X} = A(t)X.$$

Theorem 2.7.   *Let V and W be differentiable matrices. Then*

$$\frac{d}{dt}[V(t)W(t)] = \dot{V}(t)W(t) + V(t)\dot{W}(t).$$

Proof.   The (i,j)th entry of $\dfrac{d}{dt}[V(t)W(t)]$ is by definition the derivative of the (i,j)th entry of $V(t)W(t)$, which is, using the product rule,

$$\frac{d}{dt}\left[\sum_{k=1}^{n} v_{ik}(t)w_{kj}(t)\right] = \sum_{k=1}^{n}[\dot{v}_{ik}(t)w_{kj}(t) + v_{ik}(t)\dot{w}_{kj}(t)].$$

But this is obviously the (i,j)th entry of $\dot{V}(t)W(t) + V(t)\dot{W}(t)$. $\square$

Theorem 2.8.   *If V is differentiable and invertible, then*

$$\frac{d}{dt}(V^{-1}(t)) = -V^{-1}(t)\dot{V}(t)V^{-1}(t).$$

The proof is left to the reader (Exercise 2).

EXERCISES

1. Let $u^1(t) = (1,t)$, $u^2(t) = (t,t^2)$. Show that for every t the vectors $u^1(t)$ and $u^2(t)$ are linearly dependent. Show also that u^1 and u^2 are linearly independent.

2. Prove Theorem 2.8. Note that the existence of the derivative of V^{-1} follows from the fact that the entries of V^{-1} are rational functions of the entries of V, and are therefore differentiable. To obtain the formula for the derivative of V^{-1} differentiate the identity $V(t)V^{-1}(t) = I$ and solve for $\dfrac{d}{dt}V^{-1}(t)$.

3. Let $W(t)$ be a solution matrix and B a constant matrix. Prove that $W(t)B$ is a solution matrix by (a) using only the definition of a solution matrix; (b) using Theorems 2.6 and 2.7.

4. Let $W(t)$ be a fundamental matrix and B a constant matrix. Prove that $W(t)B$ is fundamental if and only if B is invertible.

5. (a) Use Theorem 1.1 to prove that the matrix equation $\dot{X} = A(t)X$ has a unique solution W satisfying a given initial condition $W(t_0) = C$.

(b) Let $W(t)$ be an invertible solution of $\dot{X} = A(t)X$ [i.e., a fundamental matrix of $x = A(t)x$]. Prove that for every solution $V(t)$ there exists a matrix B such that $V(t) = W(t)B$. [Hint: Let $B = W^{-1}(0)V(0)$.]

6. Let $W(t)$ and $V(t)$ be fundamental matrices. Prove that there exists a constant invertible matrix B such that $V(t) = W(t)B$, by (a) using Exercise 5(b) only; (b) showing that $W^{-1}(t)V(t)$ has derivative identically zero and is therefore constant.

7. Let $W(t)$ be a fundamental matrix. Show that $W(t)W^{-1}(0)$ is the s.f.m.

8. Let $W(t)$ be a fundamental matrix and u a solution of $\dot{x} = A(t)x$ with $u(0) = \eta$. Show that $u(t) = W(t)W^{-1}(0)\eta$.

9. Let V be a differentiable matrix and u a differentiable vector. Show that

$$\frac{d}{dt}[V(t)u(t)] = \dot{V}(t)u(t) + V(t)\dot{u}(t).$$

10. Let W be an invertible solution of $\dot{X} = A(t)X$ and let $V(t) = W^{-1}(t)$. Prove that V is a solution of $\dot{X} = -XA(t)$.

3. Real Homogeneous Systems

We say that a system

$$\dot{x} = A(t)x \tag{3.1}$$

is real if the matrix $A(t)$ is real, i.e., has real entries, for all t. Our reason for having admitted non-real systems to consideration is not a desire for generality (there is no gain in generality; see Exercise 1), but the fact, familiar from Chapter 3, that a complex change of variables often simplifies a differential equation, and that complex functions are sometimes easier to deal with than real functions. However, if we start with a real system, we usually want to end by expressing the solutions in real form: we want a real basis, or, what amounts to the same thing, a real fundamental matrix. It turns out that the s.f.m. of a real system is real, and since the s.f.m. can be computed from an arbitrary fundamental matrix (see Exercise 7 of the preceding section), this problem has a simple solution.

Theorem 3.1. *Let* (3.1) *be real and let* u *be a solution such that* $u(t_0)$ *is real for some* t_0. *Then* $u(t)$ *is real for all* t.

Proof. Let $u(t) = v(t) + iw(t)$, where $i = \sqrt{-1}$ and v, w are real functions. Since

$$\dot{u}(t) = A(t)u(t),$$

we have

$$\dot{v}(t) + i\dot{w}(t) = A(t)v(t) + iA(t)w(t),$$

whence, equating real and imaginary parts,

$$\dot{v}(t)\dot{} = A(t)v(t)$$
$$\dot{w}(t) = A(t)w(t).$$

Since $w(t_0) = 0$, $w(t) \equiv 0$. Hence $u = v$. $\square$

Corollary 3.2. *If* (3.1) *is real, the standard basis is real, and hence the s.f.m. is real.*

EXERCISES

1. Let $A(t) = B(t) + iC(t)$, $i = \sqrt{-1}$, B, C real. Prove that $u = v + iw$ is a solution of $\dot{x} = A(t)x$ if and only if $y = v(t)$, $z = w(t)$ is a (real) solution of the (real) system of order $2n$

$$\dot{y} = B(t)y - C(t)z$$
$$\dot{z} = C(t)y + B(t)z.$$

2.   Prove: If $A(t)$ is not real, then (3.1) does not have a real fundamental matrix.

3.   Suppose that (3.1) is real and that W is a non-real fundamental matrix. Explain why one cannot expect to obtain a fundamental matrix by simply choosing the real (or the imaginary) part of W.

4. Non-Homogeneous Systems; The Variation-of-Constants Formula

We now turn to the non-homogeneous system

$$\dot{x} = A(t)x + b(t). \tag{4.1}$$

The corresponding homogeneous system

$$\dot{x} = A(t)x \tag{4.2}$$

is called the *reduced* system of (4.1).

Theorem 4.1.   *Let W be a fundamental matrix of (4.2) and v a solution of (4.1). For every vector c, the function*

$$W(t)c + v(t)$$

is a solution of (4.1). Conversely, if u is a solution of (4.1), there exists a unique vector c such that

$$u(t) = W(t)c + v(t).$$

Proof.   Given c, let $u(t) = W(t)c + v(t)$. Then

$$\begin{aligned}
\dot{u}(t) &= \dot{W}(t)c + \dot{v}(t) \\
&= A(t)W(t)c + A(t)v(t) + b(t) \\
&= A(t)[W(t)c + v(t)] + b(t) \\
&= A(t)u(t) + b(t).
\end{aligned}$$

Hence u is a solution of (4.1).

Conversely, given a solution u, let $g = u - v$. Then

$$\begin{aligned}
\dot{g}(t) &= \dot{u}(t) - \dot{v}(t) \\
&= A(t)u(t) + b(t) - A(t)v(t) - b(t) \\
&= A(t)(u(t) - v(t)) \\
&= A(t)g(t).
\end{aligned}$$

Thus g is a solution of (4.2), and since W is fundamental, there exists a unique vector c such that

$$g(t) = W(t)c$$

or

$$u(t) = W(t)c + v(t). \qquad \square$$

Thus the problem of finding all solutions of a non-homogeneous system reduces to finding a fundamental matrix of the reduced system and a single solution of the complete system. We now show that a solution of the complete system can actually be obtained from a fundamental matrix of the reduced system.

Let $W(t)$ be a fundamental matrix of (4.2). We try to find a vector-valued function $c(t)$ such that $W(t)c(t)$ is a solution of (4.1). Assuming such a function exists, let

$$v(t) = W(t)c(t).$$

Differentiating (see Exercise 9, page 146),

$$\begin{aligned}
\dot{v}(t) &= \dot{W}(t)c(t) + W(t)\dot{c}(t) \\
&= A(t)W(t)c(t) + W(t)\dot{c}(t) \\
&= A(t)v(t) + W(t)\dot{c}(t).
\end{aligned}$$

Our assumption is that

$$\dot{v}(t) = A(t)v(t) + b(t).$$

Hence

$$A(t)v(t) + b(t) = A(t)v(t) + W(t)\dot{c}(t)$$

or

$$W(t)\dot{c}(t) = b(t).$$

Thus

$$\dot{c}(t) = W^{-1}(t)b(t)$$

and finally

$$c(t) = \int W^{-1}(t)b(t)\, dt$$

for some choice of the indefinite integral. It follows that

$$v(t) = W(t)\int W^{-1}(t)b(t)\, dt.$$

It only remains to check that this formula does indeed furnish a solution of (4.1).

Theorem 4.2 (Variation-of-Constants Formula). *Let $W(t)$ be a fundamental matrix of (4.2). For any choice of the indefinite integral, the function*

$$v(t) = W(t)\int W^{-1}(t)b(t)\, dt \tag{4.3}$$

is a solution of (4.1).

Proof.

$$\dot{v}(t) = \dot{W}(t)\int W^{-1}(t)b(t)\,dt + W(t)W^{-1}(t)b(t)$$
$$= A(t)W(t)\int W^{-1}(t)b(t)\,dt + b(t)$$
$$= A(t)v(t) + b(t). \qquad \square$$

Theorem 4.3. *The solution v of (4.1) satisfying $v(t_0) = \eta$ is given by*

$$v(t) = u(t) + W(t)\int_{t_0}^{t} W^{-1}(s)b(s)\,ds \qquad (4.4)$$

where u is the solution of (4.2) satisfying $u(t_0) = \eta$, and W is a fundamental matrix of (4.2).

Corollary 4.4. *If W is the s.f.m. of (4.2), then the solution v of (4.1) satisfying $v(0) = \eta$ is given by*

$$v(t) = W(t)\left[\eta + \int_{0}^{t} W^{-1}(s)b(s)\,ds\right]. \qquad (4.5)$$

The proofs of Theorem 4.3 and the corollary are simple verifications, and are left to the reader. Note that the variation-of-constants formula, in any of the three forms (4.3), (4.4), or (4.5), reduces to Leibniz's formula when $n = 1$ (Chapter 2, Section 3).

EXERCISES

1.   Prove Theorem 4.3
2.   Prove Corollary 4.4.
3.   Let $Q(t,s)$ be the fundamental matrix of $\dot{x} = A(t)x$ which assumes the value I when $t = s$, that is,

$$\frac{d}{dt}Q(t,s) = A(t)Q(t,s) \qquad \text{and} \qquad Q(s,s) = I.$$

[Thus $Q(t,0)$ is the s.f.m.] Prove:
   (a)   For any fundamental matrix W, $W(t)W^{-1}(s) = Q(t,s)$
   (b) $Q(a,b)Q(b,c) = Q(a,c)$
   (c) $Q^{-1}(t,s) = Q(s,t)$
   (d) $\dfrac{d}{ds}Q(t,s) = -Q(t,s)A(s)$

   (e)   The solution of (4.1) satisfying $v(0) = 0$ is given by

$$v(t) = \int_{0}^{t} Q(t,s)b(s)\,ds.$$

5.   Systems with Constant Coefficients;
The Case of Simple Eigenvalues

If A is a constant matrix, the problem of solving

$$\dot{x} = Ax \qquad (5.1)$$

is purely algebraic. Once a fundamental matrix has been found, the non-homogeneous system

$$\dot{x} = Ax + b(t)$$

may be considered as solved also, its solutions being given by formula (4.4) or (4.5). The method of solving (5.1) consists in finding a linear change of variables which transforms (5.1) into a system which can be solved one equation at a time—for example, a system in which the first equation involves only x_1, the second only x_1 and x_2, and so on. Such a method was used to solve linear equations with constant coefficients in Chapter 3. By a linear change of variables we mean a substitution $x = Py$, where P is an invertible matrix; $y = P^{-1}x$ is then the inverse substitution.

Theorem 5.1.   *Let P be an invertible matrix. The change of variables $x = Py$ transforms*

$$\dot{x} = Ax$$

into

$$\dot{y} = P^{-1}APy.$$

Proof. $y = P^{-1}x$. Differentiating gives

$$\begin{aligned}
\dot{y} &= P^{-1}\dot{x} \\
&= P^{-1}Ax \\
&= P^{-1}APy. \qquad \square
\end{aligned}$$

Theorem 5.2.   *If $V(t)$ is a fundamental matrix of $\dot{y} = P^{-1}APy$, then $PV(t)$ is a fundamental matrix of $\dot{x} = Ax$.*

Proof.   Let $W(t) = PV(t)$. Then

$$\begin{aligned}
\dot{W}(t) &= P\dot{V}(t) \\
&= PP^{-1}APV(t) \\
&= APV(t) \\
&= AW(t).
\end{aligned}$$

Hence W is a solution matrix of $\dot{x} = Ax$. Since P and V are invertible, so is W. $\square$

For the remainder of this section we assume that the eigenvalues of A are simple. [Exercise 5 is devoted to a number of important properties of (5.1) which do not involve this assumption.]

Theorem 5.3.   *Let A have n distinct eigenvalues $\lambda_1, \ldots, \lambda_n$. Let P be a matrix whose jth column is an eigenvector of A belonging to λ_j. Then the change of variables $x = Py$ transforms $\dot{x} = Ax$ into $\dot{y} = Dy$, where D is the diagonal matrix whose (i,i)th entry is λ_i.*

Proof.   We know (Theorem 5.2, Chapter 4) that P is invertible and that $P^{-1}AP = D$. Thus $\dot{y} = Dy$ by Theorem 5.1. $\square$

A system of the form

$$\dot{y} = Dy, \tag{5.2}$$

where D is a diagonal matrix, is the simplest conceivable. Setting $d_{ii} = \lambda_i$, the system is

$$\frac{dy_1}{dt} = \lambda_1 y_1$$

$$\frac{dy_2}{dt} = \lambda_2 y_2$$

$$\cdots \cdots$$

$$\frac{dy_n}{dt} = \lambda_n y_n$$

and thus has the solutions

$$y_i = c_i e^{\lambda_i t}, \qquad i = 1, \ldots, n$$

or

$$y = (c_1 e^{\lambda_1 t}, c_2 e^{\lambda_2 t}, \ldots, c_n e^{\lambda_n t})$$

where $c_1, \ldots, c_n$ are arbitrary constants.

Theorem 5.4.   *If D is a diagonal matrix, $d_{ii} = \lambda_i$, then the s.f.m. of $\dot{y} = Dy$ is the diagonal matrix $V(t)$, where $v_{ii}(t) = e^{\lambda_i t}$.*

Proof.   Obviously $\dot{V}(t) = DV(t)$ and $V(0) = I$. $\square$

Thus the problem of constructing a fundamental matrix of (5.1) is completely solved for the case that A has n distinct eigenvalues. We summarize the result:

Theorem 5.5.   *Let A have n distinct eigenvalues, and define the matrices P and $V(t)$ as in Theorems 5.3 and 5.4. Then*

$$W(t) = PV(t)$$

is a fundamental matrix of $\dot{x} = Ax$.

Note that the matrix $W(t) = PV(t)$ is the s.f.m. of (5.1) only in the trivial case that $P = I$, that is, the case that A itself is a diagonal matrix. The s.f.m. is always given by $W(t)W^{-1}(0)$ (see Exercise 7, Section 2). If A is real, then, as we know, the s.f.m. is real. Thus, to find a real fundamental matrix in case that A is real but some of its eigenvalues are not, we simply compute $W(t)W^{-1}(0) = W(t)P^{-1}$.

EXAMPLE 1.   We solve the system

$$\dot{x}_1 = x_1 + x_2$$
$$\dot{x}_2 = 3x_1 - x_2 + e^t. \tag{5.3}$$

Here

$$A = \begin{bmatrix} 1 & 1 \\ 3 & -1 \end{bmatrix}$$

and

$$b(t) = (0, e^t).$$

The characteristic polynomial of A is $\lambda^2 - 4$. The eigenvalues are $\lambda_1 = 2$, $\lambda_2 = -2$, and corresponding eigenvectors are $x^1 = (1,1)$, $x^2 = (1,-3)$. A suitable P is therefore

$$P = \begin{bmatrix} 1 & 1 \\ 1 & -3 \end{bmatrix}.$$

Since

$$V(t) = \begin{bmatrix} e^{2t} & 0 \\ 0 & e^{-2t} \end{bmatrix},$$

a fundamental matrix of the reduced system is given by

$$W(t) = PV(t) = \begin{bmatrix} e^{2t} & e^{-2t} \\ e^{2t} & -3e^{-2t} \end{bmatrix}.$$

The solutions of the reduced system are given by $x = W(t)c$, or

$$x_1 = c_1 e^{2t} + c_2 e^{-2t}$$
$$x_2 = c_1 e^{2t} - 3c_2 e^{-2t}.$$

To find a solution of the complete system we use the variation-of-constants formula, for which we need $W^{-1}(t)$. A short computation yields

$$W^{-1}(t) = \tfrac{1}{4}\begin{bmatrix} 3e^{-2t} & e^{-2t} \\ e^{2t} & -e^{2t} \end{bmatrix}.$$

Thus

$$W^{-1}(t)b(t) = \tfrac{1}{4}(e^{-t}, -e^{3t}),$$
$$\int W^{-1}(t)b(t)\, dt = -\tfrac{1}{4}(e^{-t}, \tfrac{1}{3}e^{3t}),$$

and

$$W(t)\int W^{-1}(t)b(t)\,dt = (-\tfrac{1}{3}e^t, 0).$$

Hence

$$x_1 = -\tfrac{1}{3}e^t$$
$$x_2 = 0$$

is a solution of (5.3). The general solution of (5.3) is

$$x_1 = c_1 e^{2t} + c_2 e^{-2t} - \tfrac{1}{3}e^t$$
$$x_2 = c_1 e^{2t} - 3c_2 e^{-2t}. \qquad \square$$

EXAMPLE 2.   We solve the system

$$\dot{x}_1 = x_1 + x_2$$
$$\dot{x}_2 = -x_1 + x_2. \tag{5.4}$$

The eigenvalues are $\lambda_1 = 1 + i$, $\lambda_2 = 1 - i$, where $i = \sqrt{-1}$; $x^1 = (1,i)$, $x^2 = (1,-i)$ are corresponding eigenvectors. Thus

$$P = \begin{bmatrix} 1 & 1 \\ i & -i \end{bmatrix},$$

$$V(t) = \begin{bmatrix} e^{(1+i)t} & 0 \\ 0 & e^{(1-i)t} \end{bmatrix},$$

and

$$W(t) = PV(t) = \begin{bmatrix} e^{(1+i)t} & e^{(1-i)t} \\ ie^{(1+i)t} & -ie^{(1-i)t} \end{bmatrix}.$$

We go a step further and compute a real fundamental matrix. We find

$$W^{-1}(0) = \tfrac{1}{2}\begin{bmatrix} 1 & -i \\ 1 & i \end{bmatrix},$$

so that the s.f.m. is

$$W(t)W^{-1}(0) = \begin{bmatrix} e^t \cos t & e^t \sin t \\ -e^t \sin t & e^t \cos t \end{bmatrix}.$$

The solutions of (5.4) are therefore given in real form by

$$x_1 = c_1 e^t \cos t + c_2 e^t \sin t$$
$$x_2 = -c_1 e^t \sin t + c_2 e^t \cos t. \qquad \square$$

Remarks on Numerical Methods.   The one difficulty associated with solving a system $\dot{x} = Ax$ lies in calculating the eigenvalues of A; usually these must be computed by the methods of numerical analysis. However, one may dispense with formulas for the solutions altogether and deal only with numerical tables or graphs for the solutions. This is possible by virtue of Theorem 2.4, according to which the graphs of n linearly independent solutions suffice for the description of all solutions.

Since formulas have to be interpreted numerically or graphically any-way, it would seem natural to dispense with them. In the vast majority of cases this is probably correct. Nevertheless, numerical methods are not infallible, and in exceptional cases may yield totally spurious results. These exceptional cases arise when initial values arbitrarily close to each other produce solutions which in time diverge ever farther from each other. In attempting to compute a solution which tends to zero, for example, the slightest error in the approximations may lead to the computation of a solution which tends to infinity. This situation arises whenever the system has eigenvalues with both positive and negative real part. (For a simple example, see Exercise 4.) Although eigenvalues with positive real part occur infrequently in practice, numerical methods should never be trusted implicitly. The proper approach to this, as to every problem in differential equations, is to make use of every technique and resource at our disposal. We should compute the eigenvalues as well as tables or curves for the solutions.

EXERCISES

1. Continue the computations of Example 1 and
    (a) Find the s.f.m. of the reduced system.
    (b) Find the solution of the complete system satisfying $u(0) = 0$.
2. Find all solutions of

$$\text{(a)} \quad \dot{x}_1 = x_1 + 2x_2 + e^{3t}, \quad \dot{x}_2 = 2x_1 + x_2 + e^t$$
$$\text{(b)} \quad \dot{x}_1 = x_1 + x_2, \quad \dot{x}_2 = x_1 + x_2$$
$$\text{(c)} \quad \dot{x}_1 = x_1 - x_2 + x_3, \quad \dot{x}_2 = x_1 + x_2 - x_3, \quad \dot{x}_3 = 2x_1 - x_2$$
$$\text{(d)} \quad \dot{x}_1 = x_1 + ix_2, \quad \dot{x}_2 = -ix_1 + x_2$$
$$\text{(e)} \quad \dot{x}_1 = 2x_1 + x_2, \quad \dot{x}_2 = 3x_1 + 4x_2$$
$$\text{(f)} \quad \dot{x}_1 = 2x_1 + x_2, \quad \dot{x}_2 = 3x_1 + 4x_2 + e^t$$
$$\text{(g)} \quad \dot{x}_1 = x_2, \quad \dot{x}_2 = -x_1 + x_2 + x_3, \quad \dot{x}_3 = x_1$$
$$\text{(h)} \quad \dot{x}_1 = x_1 + 3x_2, \quad \dot{x}_2 = x_1 - x_2$$

3. Find the s.f.m. of $\dot{x}_1 = \alpha x_1 + \beta x_2$, $\dot{x}_2 = -\beta x_1 + \alpha x_2$, where α and β are real.
4. (a) Find all solutions of $\dot{x}_1 = x_2$, $\dot{x}_2 = 10x_1 + 9x_2$.
    (b) Find the solution satisfying $u(0) = (1, -1)$. If the reader has access to a computer, he will find it instructive to attempt to compute this solution numerically.
5. Let $W(t)$ be the s.f.m. of $\dot{x} = Ax$. Prove:
    (a) $W(a)W(b) = W(a + b)$ for any real numbers a, b
    (b) $W^{-1}(t) = W(-t)$
    (c) $AW(t) = W(t)A$ for all t

[Hints: (a) Show that $V(t) = W(t + c)$ is a fundamental matrix for any constant c, whence $W(t) = V(t)V^{-1}(0)$; (c) show that $AW(t) - W(t)A$ is identically zero by differentiating and using the uniqueness theorem.] These algebraic properties of $W(t)$ are further explored in the Miscellaneous Exercises at the end of the chapter. They are closely connected with the idea of the *exponential* of a matrix.

6. Physical Examples

EXAMPLE 1.   Two vessels of volumes v_1, v_2 liters are filled with salt solutions (see Fig. 14). The vessels are connected by tubes of negligible

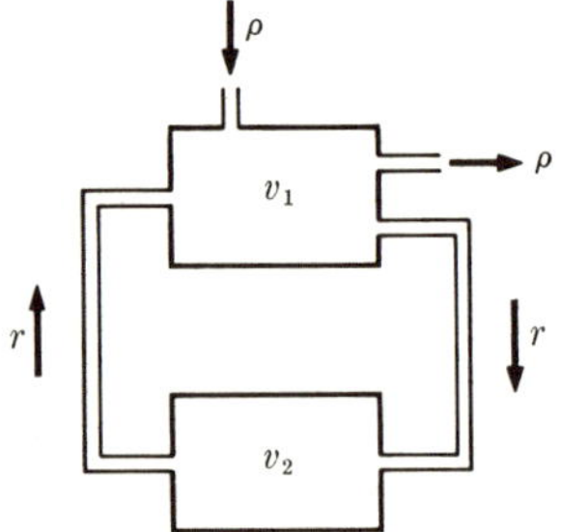

FIG. 14.   See Example 1.

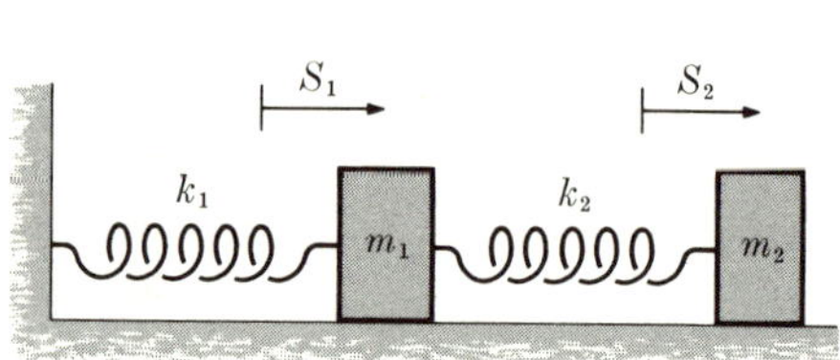

FIG. 15.   See Example 2.

length and volume through which the solutions pass from one vessel to the other at the rate of r l/sec. Water flows into the first vessel at the rate of ρ l/sec, the volume in the vessel being kept constant by an outflow valve through which the solution flows out at the rate of ρ l/sec. Within each vessel the concentration is uniform.

Let m_1, m_2 denote the mass of salt in each of the two vessels. Equating rate of change of mass to rate of entry minus rate of exit yields the equations

$$\dot{m}_1 = \frac{m_2}{v_2}r - \frac{m_1}{v_1}\rho - \frac{m_1}{v_1}r$$

$$\dot{m}_2 = \frac{m_1}{v_1}r - \frac{m_2}{v_2}r,$$

or, rearranging terms,

$$\dot{m}_1 = -\frac{\rho + r}{v_1}m_1 + \frac{r}{v_2}m_2$$

$$\dot{m}_2 = \frac{r}{v_1}m_1 - \frac{r}{v_2}m_2.$$

(6.1)

In practice (for example, in physiology) it is the concentrations $x_1 = \dfrac{m_1}{v_1}$, $x_2 = \dfrac{m_2}{v_2}$ which are of interest, rather than the masses. The differential system for x_1, x_2 is easily obtained from (6.1). It is

$$\dot{x}_1 = -\frac{\rho + r}{v_1}\, x_1 + \frac{r}{v_1}\, x_2$$

$$\dot{x}_2 = \frac{r}{v_2}\, x_1 - \frac{r}{v_2}\, x_2. \tag{6.2}$$

We simplify the notation by letting $\alpha = r/v_1$, $\beta = r/v_2$, $\gamma = \rho/v_1$. Then

$$\dot{x}_1 = -(\alpha + \gamma)x_1 + \alpha x_2$$
$$\dot{x}_2 = \beta x_1 - \beta x_2. \tag{6.3}$$

We leave it to the reader to compute the solutions of this system. It is understood that $\alpha > 0$, $\beta > 0$, $\gamma > 0$. $\square$

EXAMPLE 2.   A mass m_1 is attached by a spring of elastic constant k_1 to a rigid support (see Fig. 15). A second mass m_2 is attached to the first by a spring of elastic constant k_2. The masses rest on a smooth horizontal surface. Let s_1 and s_2 be the displacements of the masses from the positions they occupy when both springs are relaxed, i.e., neither stretched nor compressed. The force acting on m_2 is the tension in the second spring; the mass m_1 is subject to the tensions in both springs. The tension in the first spring is $k_1 s_1$, and that in the second is $k_2(s_2 - s_1)$. Applying Newton's second law to each mass in turn yields

$$m_1 \frac{d^2 s_1}{dt^2} = -k_1 s_1 + k_2(s_2 - s_1)$$

$$m_2 \frac{d^2 s_2}{dt^2} = -k_2(s_2 - s_1),$$

or

$$\ddot{s}_1 = -\frac{k_1 + k_2}{m_1}\, s_1 + \frac{k_2}{m_1}\, s_2$$

$$\ddot{s}_2 = \frac{k_2}{m_2}\, s_1 - \frac{k_2}{m_2}\, s_2, \tag{6.4}$$

where $\ddot{} = \dfrac{d^2}{dt^2}$. This is a system of two second-order linear equations. We write it as a system of first-order equations by setting $x_1 = s_1$,

$x_2 = \dot{s}_1,\ x_3 = s_2,\ x_4 = \dot{s}_2$ (see Chapter 1, Section 2). Then

$$\begin{aligned}
\dot{x}_1 &= x_2 \\
\dot{x}_2 &= -\frac{k_1 + k_2}{m_1}\, x_1 + \frac{k_2}{m_1}\, x_3 \\
\dot{x}_3 &= x_4 \\
\dot{x}_4 &= \frac{k_2}{m_2}\, x_1 - \frac{k_2}{m_2}\, x_3.
\end{aligned} \tag{6.5}$$

This system has the form $\dot{x} = Ax$, where $x = (x_1, x_2, x_3, x_4)$ and

$$A = \begin{bmatrix}
0 & 1 & 0 & 0 \\
-\dfrac{k_1 + k_2}{m_1} & 0 & \dfrac{k_2}{m_1} & 0 \\
0 & 0 & 0 & 1 \\
\dfrac{k_2}{m_2} & 0 & -\dfrac{k_2}{m_2} & 0
\end{bmatrix}.$$

Although there is no difficulty in solving this system, we simplify the computations by setting $k_1 = k_2 = 1$, $m_1 = m_2 = 1$. The system thus becomes

$$\begin{aligned}
\dot{x}_1 &= x_2 \\
\dot{x}_2 &= -2x_1 + x_3 \\
\dot{x}_3 &= x_4 \\
\dot{x}_4 &= x_1 - x_3.
\end{aligned} \tag{6.6}$$

Owing to the assumption that there is no friction, the eigenvalues are purely imaginary, and the solutions are linear combinations of sinusoidal functions. However, the periods of these sinusoidal functions are incommensurable. The solutions are therefore not periodic, except for certain sinusoidal solutions corresponding to very special initial conditions. Physicists call these sinusoidal solutions the *normal oscillations,* or *normal modes of oscillation,* of the spring-mass system. The computations are left as an exercise. $\square$

EXERCISES

EXAMPLE 1

1. (a) Show that the eigenvalues of (6.3) are given by

$$\lambda_1 = -\tfrac{1}{2}(\alpha + \beta + \gamma + \sqrt{(\alpha + \beta + \gamma)^2 - 4\beta\gamma})$$
$$\lambda_2 = -\tfrac{1}{2}(\alpha + \beta + \gamma - \sqrt{(\alpha + \beta + \gamma)^2 - 4\beta\gamma}).$$

(b)   Show that $\lambda_1 < \lambda_2 < 0$.

(c)   Show that the solutions of (6.3) are

$$x_1 = c_1 \alpha e^{\lambda_1 t} + c_2 \alpha e^{\lambda_2 t}$$
$$x_2 = c_1(\alpha + \gamma + \lambda_1)e^{\lambda_1 t} + c_2(\alpha + \gamma + \lambda_2)e^{\lambda_2 t} \qquad (6.7)$$

where c_1, c_2 are arbitrary constants.

2.   Let us ignore the restriction $x_1 \geq 0$, $x_2 \geq 0$ imposed by the physical interpretation of (6.3), and consider all (real) solutions (6.7). A real solution u_1, u_2 defines a parametrized curve $x_1 = u_1(t)$, $x_2 = u_2(t)$ in the (x_1,x_2)-plane. Our object is to obtain a picture of these curves, the so-called *trajectories* of (6.3). Note that the trajectory corresponding to the zero solution is simply the origin.

(a)   Show that if $c_1 = 0$, $c_2 \neq 0$, then

$$\frac{u_2(t)}{u_1(t)} = \frac{\alpha + \gamma + \lambda_2}{\alpha}$$

for all t. We obtain two trajectories, viz., the ray $x_2 = \dfrac{(\alpha + \gamma + \lambda_2)}{\alpha} x_1$,

$x_1 > 0$, and the ray $x_2 = \dfrac{\alpha + \gamma + \lambda_2}{\alpha} x_1$, $x_1 < 0$. Show that $\alpha + \gamma +$

$\lambda_2 > 0$, so that these rays have positive slope.

(b)   Show that if $c_1 \neq 0$, $c_2 = 0$, then

$$\frac{u_2(t)}{u_1(t)} = \frac{\alpha + \gamma + \lambda_1}{\alpha}$$

for all t. As in case (a), we obtain two rays. Show that $\alpha + \gamma + \lambda_1 < 0$, so that these rays have negative slope. Since neither enters the first quadrant, they correspond to solutions without physical significance.

(c)   Obviously all solutions tend to zero as $t \to \infty$. Show also that if $c_1 \neq 0$, $c_2 \neq 0$, then

$$\lim_{t \to \infty} \frac{u_2'(t)}{u_1'(t)} = \frac{\alpha + \gamma + \lambda_2}{\alpha}.$$

It follows that every trajectory (excepting only the two rays of case (b), and the origin itself) approaches the origin tangentially to one of the two rays $x_2 = \dfrac{\alpha + \gamma + \lambda_2}{\alpha} x_1$.

(d)   Show that if $c_1 \neq 0$, $c_2 \neq 0$, then

$$\lim_{t \to -\infty} \frac{u_2'(t)}{u_1'(t)} = \frac{\alpha + \gamma + \lambda_1}{\alpha}.$$

(e)   Make a sketch of the trajectories.

EXAMPLE 2

3. (a) Show that the eigenvalues of (6.6) are $\pm i\beta_1$, $\pm i\beta_2$, where

$$\beta_1 = \sqrt{\frac{3 + \sqrt{5}}{2}}, \qquad \beta_2 = \sqrt{\frac{3 - \sqrt{5}}{2}}.$$

Note that $\beta_1 > \beta_2 > 0$.

(b) Show that the solutions of (6.6) are

$$x_1 = c_1 \sin (\beta_1 t + \varphi_1) + c_2 \sin (\beta_2 t + \varphi_2)$$
$$x_2 = c_1\beta_1 \cos (\beta_1 t + \varphi_1) + c_2\beta_2 \cos (\beta_2 t + \varphi_2)$$
$$x_3 = c_1 \left(\frac{1 + \sqrt{5}}{2}\right) \sin (\beta_1 t + \phi_1) + c_2 \left(\frac{1 - \sqrt{5}}{2}\right) \sin (\beta_2 t + \varphi_2)$$
$$x_4 = c_1\beta_1 \left(\frac{1 + \sqrt{5}}{2}\right) \cos (\beta_1 t + \phi_1) + c_2\beta_2 \left(\frac{1 - \sqrt{5}}{2}\right) \cos (\beta_2 t + \varphi_2)$$

$$(6.8)$$

where c_1, c_2, φ_1, φ_2 are arbitrary.

(c) What are the solutions of (6.4), assuming

$$k_1 = k_2 = m_1 = m_2 = 1?$$

(d) Show that β_1/β_2 is irrational.

4. Let β_1, β_2, c_1, c_2 be non-zero numbers, and let $f(t) = c_1 \sin \beta_1 t + c_2 \sin \beta_2 t$. Prove that if f is periodic, β_1/β_2 is rational. (Hint: Let $f(t + p) = f(t)$ and use the formula for $\sin (A + B)$. A similar proof applies to $c_1 \cos \beta_1 t + c_2 \cos \beta_2 t$. It follows, with minor changes in the argument, that none of the functions (6.8) is periodic if $c_1 \neq 0$ and $c_2 \neq 0$.)

7. Systems with Constant Coefficients; The General Case

We now study the solutions of $\dot{x} = Ax$ without assuming that the eigenvalues of A are simple. The discussion to follow can be interpreted as yielding a method for computing a fundamental matrix, subject to the algebraic difficulties inherent in transforming a matrix to triangular form apparent from the proof of Theorem 7.2. However, our concern here is not so much with devising a practical method for computing solutions as it is with showing that the solutions have a certain analytic form, and that this form and the qualitative behavior of the solutions is closely associated with properties of the eigenvalues of A.

We shall prove that for every matrix A there exists an invertible matrix P such that $P^{-1}AP$ is triangular. The substitution $x = Py$ thus transforms $\dot{x} = Ax$ into the triangular system $\dot{y} = P^{-1}APy$, which can be solved one equation at a time. We begin with triangular systems.

Theorem 7.1.   *Let A be a lower-triangular $n \times n$ matrix, and u a solution of*

$$\dot{x} = Ax. \tag{7.1}$$

Then

$$u_i(t) = \sum_{j=1}^{i} q_{ij}(t)e^{a_{ii}t}, \qquad i = 1, \ldots, n,$$

where the $q_{ij}(t)$ are polynomials.

Proof.   We use induction. For $n = 1$ the theorem is obvious, since $u(t) = ce^{at}$ for some constant c. Let $n \geq 2$ and suppose the theorem holds for $n - 1$. In coordinate form the system (7.1) is

$$\begin{aligned}
\dot{x}_1 &= a_{11}x_1 \\
\dot{x}_2 &= a_{21}x_1 + a_{22}x_2 \\
&\cdot \cdot \cdot \cdot \cdot \cdot \cdot \cdot \cdot \cdot \\
\dot{x}_n &= a_{n1}x_1 + a_{n2}x_2 + \cdots + a_{nn}x_n.
\end{aligned}$$

By the inductive hypothesis,

$$u_i(t) = \sum_{j=1}^{i} q_{ij}(t)e^{a_{ii}t}, \qquad i = 1, \ldots, n - 1.$$

Now

$$\begin{aligned}
\dot{u}_n(t) &= \sum_{j=1}^{n} a_{nj}u_j(t) \\
&= \sum_{j=1}^{n-1} a_{nj} \sum_{k=1}^{j} q_{jk}(t)e^{a_{kk}t} + a_{nn}u_n(t) \\
&= \sum_{j=1}^{n-1} \sum_{k=1}^{j} a_{nj}q_{jk}(t)e^{a_{kk}t} + a_{nn}u_n(t).
\end{aligned}$$

Hence

$$u_n(t) = e^{a_{nn}t}\left[\int e^{-a_{nn}t} \sum_{j=1}^{n-1} \sum_{k=1}^{j} a_{nj}q_{jk}(t)e^{a_{kk}t}\, dt + c \right]$$

$$= ce^{a_{nn}t} + e^{a_{nn}t} \int \sum_{j=1}^{n-1} \sum_{k=1}^{j} a_{nj}q_{jk}(t)e^{(a_{kk}-a_{nn})t}\, dt,$$

where c is a constant. Interchanging the order of summing and integrating yields

$$u_n(t) = ce^{a_{nn}t} + e^{a_{nn}t} \sum_{j=1}^{n-1} \sum_{k=1}^{j} a_{nj} \int q_{jk}(t) e^{(a_{kk}-a_{nn})t} \, dt.$$

Since the integral of the product of a polynomial and e^{at} is again such a product (see Chapter 3, Section 4), we have that

$$u_n(t) = ce^{a_{nn}t} + e^{a_{nn}t} \sum_{j=1}^{n-1} \sum_{k=0}^{j} a_{nj} r_{jk}(t) e^{(a_{kk}-a_{nn})t}$$
$$= ce^{a_{nn}t} + \sum_{j=1}^{n-1} \sum_{k=1}^{j} a_{nj} r_{jk}(t) e^{a_{kk}t},$$

where the $r_{jk}(t)$ are polynomials. We define the polynomials $q_{nk}(t)$, $k = 1, \ldots, n$ by

$$q_{nk}(t) = \sum_{j-1}^{n-1} a_{nj} r_{jk}(t) \qquad \text{for } k = 1, \ldots, n-1$$
$$= c \qquad \text{for } k = n.$$

Then

$$u_n(t) = \sum_{k=1}^{n} q_{nk}(t) e^{a_{kk}t}.$$

The index k may now be replaced by j. $\qquad \square$

Theorem 7.2. *Every matrix is similar to a lower-triangular matrix.*

Proof. We must show that, given an $n \times n$ matrix A, there exists an invertible matrix P such that $P^{-1}AP$ is lower-triangular. We shall use induction on n. Since the case $n = 1$ is excessively trivial, it will be instructive to consider the case $n = 2$ separately. Let A be a 2×2 matrix, λ an eigenvalue of A, and P an invertible matrix whose second column is an eigenvector of A belonging to λ (see Exercise 1). Let $P^{-1}AP = B$. Then

$$B^{(2)} = (P^{-1}A)P^{(2)}$$
$$= P^{-1}(AP^{(2)})$$
$$= P^{-1}\lambda P^{(2)}$$
$$= \lambda P^{-1}P^{(2)}$$
$$= \lambda(P^{-1}P)^{(2)}$$
$$= \lambda I^{(2)}.$$

Hence B is lower-triangular.

Now let $n \geq 3$ and suppose that every $(n-1) \times (n-1)$ matrix is similar to a lower-triangular matrix. Let A be an $n \times n$ matrix. Let λ be an eigenvalue of A and Q an invertible $n \times n$ matrix whose last column is an eigenvector of A belonging to λ (see Exercise 1). Let $Q^{-1}AQ = B$. It follows exactly as in the case $n = 2$ that $B^{(n)} = \lambda I^{(n)}$. Let C be the $(n-1) \times (n-1)$ matrix obtained from B by deleting the last row and last column of B. Using an obvious notation, we have

$$
B = \begin{bmatrix} & & & & 0 \\ & & C & & 0 \\ & & & & \vdots \\ & & & & 0 \\ b_{n1} & b_{n2} & \cdots & b_{n,n-1} & \lambda \end{bmatrix}.
$$

By the inductive hypothesis there exists an invertible $(n-1) \times (n-1)$ matrix S such that $S^{-1}CS$ is lower-triangular. Let

$$
R = \begin{bmatrix} & & & & 0 \\ & & S & & 0 \\ & & & & \vdots \\ & & & & 0 \\ 0 & 0 & \cdots & 0 & 1 \end{bmatrix}.
$$

It is readily verified that

$$
R^{-1} = \begin{bmatrix} & & & & 0 \\ & & S^{-1} & & 0 \\ & & & & \vdots \\ & & & & 0 \\ 0 & 0 & \cdots & 0 & 1 \end{bmatrix}.
$$

Let $D = R^{-1}BR$. It is again easily verified that

$$
D = \begin{bmatrix} & & & & 0 \\ & & S^{-1}CS & & 0 \\ & & & & \vdots \\ & & & & 0 \\ d_{n,1} & & \cdots & d_{n,n-1} & \lambda \end{bmatrix}.
$$

Because $S^{-1}CS$ is lower-triangular, D is also lower-triangular. Since $R^{-1}BR = D$ and $B = Q^{-1}AQ$, we have $R^{-1}Q^{-1}AQR = D$. Finally, setting $P = QR$, we have $P^{-1}AP = D$, and the theorem is proved. $\qquad\square$

A slight modification of this proof shows that every matrix is similar to an upper-triangular matrix (Exercise 7).

We now consider a homogeneous linear system with constant coefficients

$$\dot{x} = Ax, \qquad\qquad (7.2)$$

where A is an arbitrary $n \times n$ matrix. Let u be a solution of (7.2). If $\lim_{t\to\infty} u_i(t) = 0$, $i = 1, \ldots, n$, we say that $\lim_{t\to\infty} u(t) = 0$. If $u_i(t)$ is unbounded for at least one i we say that $u(t)$ is unbounded. If all $u_i(t)$ are bounded, we say that $u(t)$ is bounded.

Theorem 7.3. (a) *If all eigenvalues of A have negative real part, every solution of (7.2) tends to zero as $t \to \infty$.*

(b) *If some eigenvalue of A has positive real part, (7.2) has a solution which is unbounded for $t \geq 0$.*

(c) *If some eigenvalue of A has zero real part, (7.2) has a solution which is bounded for all t, but does not tend to zero as $t \to \infty$.*

Proof of (a). Let $P^{-1}AP = B$, where B is lower-triangular. Let u be a solution of $\dot{x} = Ax$, and let $v(t) = P^{-1}u(t)$, so v is a solution of $\dot{x} = Bx$. By Theorem 7.1,

$$v_i(t) = \sum_{j=1}^{i} q_{ij}(t)e^{b_{ii}t}, \qquad i = 1, \ldots, n$$

for some polynomials $q_{ij}(t)$. Since the diagonal entries of a triangular matrix are its eigenvalues, this formula may be written

$$v_i(t) = \sum_{j=1}^{i} q_{ij}(t)e^{\lambda_i t}, \qquad i = 1, \ldots, n,$$

where $\lambda_1, \ldots, \lambda_n$ are the eigenvalues of B, each distinct eigenvalue being listed a number of times equal to its multiplicity. Because similar matrices have the same eigenvalues, $\lambda_1, \ldots, \lambda_n$ are also the eigenvalues of A. By hypothesis, Re $(\lambda_j) < 0$, $j = 1, \ldots, n$. Thus

$$\lim_{t\to\infty} v_i(t) = 0, \qquad i = 1, \ldots, n.$$

Now, $u(t) = Pv(t)$, or

$$u_i(t) = \sum_{j=1}^{n} p_{ij}v_j(t), \qquad i = 1, \ldots, n.$$

It follows that

$$\lim_{t \to \infty} u_i(t) = 0, \qquad i = 1, \ldots, n.$$

Proof of (b) *and* (c). Let λ be an eigenvalue of A and η an eigenvector belonging to λ. Let $u(t) = e^{\lambda t}\eta$. Then u is a solution of $\dot{x} = Ax$ (Exercise 2). In coordinate form,

$$u_i(t) = \eta_i e^{\lambda t}, \qquad i = 1, \ldots, n.$$

If Re $(\lambda) > 0$, $e^{\lambda t}$ is unbounded for $t \geq 0$. Since $\eta \neq 0$, it follows that $\eta_i \neq 0$ for at least one i. Hence at least one of the coordinate functions of $u(t)$ is unbounded for $t \geq 0$. This proves part (b).

If Re $(\lambda) = 0$, say $\lambda = i\beta$, then

$$u_i(t) = \eta_i(\cos \beta t + i \sin \beta t), \qquad i = 1, \ldots, n$$

so

$$|u_i(t)| = |\eta_i| \quad \text{for all } t, \qquad i = 1, \ldots, n.$$

Therefore u is bounded for all t. However, since $\eta \neq 0$, $u(t)$ does not tend to zero as $t \to \infty$. This proves part (c). Note that u is constant if $\beta = 0$, and periodic with least period $2\pi/\beta$ if $\beta \neq 0$. $\square$

Corollary 7.4. *A necessary and sufficient condition that all solutions of $\dot{x} = Ax$ tend to zero as $t \to \infty$ is that all eigenvalues of A have negative real part.*

A necessary condition that all solutions are bounded for $t \geq 0$ is that no eigenvalue has positive real part.

By analyzing the solutions of a triangular system in a little more detail, we can obtain the further result that if A has no eigenvalue with positive real part, and no repeated eigenvalue with zero real part, then all solutions of $\dot{x} = Ax$ are bounded for $t \geq 0$. The case of a repeated eigenvalue with zero real part is delicate. Two simple examples show that in this case the solutions may, or may not, be bounded for $t \geq 0$. First, if A is the zero matrix, all solutions of $\dot{x} = Ax$ are constant, and therefore bounded for all t. Note that 0 is the only eigenvalue of A. Second, if

$$A = \begin{bmatrix} 0 & 0 \\ 1 & 0 \end{bmatrix},$$

the solutions of $\dot{x} = Ax$ are $x_1 = c_1$, $x_2 = c_1 t + c_2$. The solutions for which $c_1 \neq 0$ are unbounded for $t \geq 0$ (as well as for $t \leq 0$). Again, 0 is the only eigenvalue of A.

EXERCISES

1. Let η be a non-zero vector. Find an invertible matrix whose last column is η.

2. Let η be a non-zero vector. Prove that $u(t) = e^{\lambda t}\eta$ is a solution of $\dot{x} = Ax$ if and only if λ is an eigenvalue of A and η is an eigenvector belonging to λ.

3. Let B be a lower-triangular $n \times n$ matrix.

   (a) Let k be an integer, $1 \leq k \leq n$, and let u be a solution of $\dot{x} = Bx$ which satifies $u_i(0) = 0$ for $i < k$, and $u_k(0) = 1$. Prove that $u_i(t) \equiv 0$ for $i < k$, and $u_k(t) = e^{b_{kk} t}$. [Hint: If $k = 1$, the result is obvious. If $k > 1$, $u_1, \ldots, u_{k-1}$ is a solution of the system

$$\dot{x}_i = \sum_{j=1}^{i} b_{ij} x_j, \qquad i = 1, \ldots, k - 1.$$

Conclude that $\dot{u}_k(t) = b_{kk} u_k(t)$.]

   (b) Let V be the s.f.m. of $\dot{x} = Bx$. Prove that $V(t)$ is lower-triangular for all t, and that $v_{ii}(t) = e^{b_{ii} t}$, $i = 1, \ldots, n$.

   (c) Conclude that, for every t, the eigenvalues of $V(t)$ are $e^{b_{ii} t}$, $i = 1, \ldots, n$.

4. Use the result of Exercise 3 to prove: If A is an $n \times n$ matrix with eigenvalues $\lambda_1, \ldots, \lambda_n$ (not necessarily distinct), and W is the s.f.m. of $\dot{x} = Ax$, then for every t the eigenvalues of $W(t)$ are $e^{\lambda_1 t}, \ldots, e^{\lambda_n t}$.

5. (a) Let u be a solution of $\dot{x} = Ax + b(t)$, where $b(t)$ has period $p > 0$. Prove that $v(t) = u(t + p)$ is also a solution, and hence that u has period p if and only if $u(0) = u(p)$. (The case that $b(t) \equiv 0$ is not excluded.)

   (b) Prove that $\dot{x} = Ax$ has a non-zero solution of period p if and only if 1 is an eigenvalue of $W(p)$, where $W(t)$ is the s.f.m.

   (c) Prove that $\dot{x} = Ax$ has a non-zero solution of period p if and only if, for some integer n, $2\pi n i/p$ is an eigenvalue of A.

   (d) Prove that $\dot{x} = Ax + b(t)$ has a unique solution of period p if and only if $\dot{x} = Ax$ has no solution of period p except $x = 0$.

6. Let u and v be solutions of $\dot{x} = Ax + b(t)$, where A is a matrix all of whose eigenvalues have negative real part. Prove that

$\lim\limits_{t \to \infty} (u(t) - v(t)) = 0$. Assuming that $b(t)$ is periodic, what can you say about the long-term behavior of the solutions? (See Exercise 5.)

7. Prove that every matrix is similar to an upper-triangular matrix. (Alter the proof of Theorem 7.2 by using first columns and first rows instead of last columns and last rows.)

8. Liouville's Formula

Theorem 8.1.   *Let $W(t)$ be a differentiable $n \times n$ matrix (not necessarily a solution matrix of a linear system), and let $\phi(t) = \det W(t)$. Then $\dot{\phi}(t)$ is the sum of n determinants, the ith of which is the determinant of the matrix obtained from $W(t)$ by replacing the entries of the ith row by their derivatives.*

Proof.   The determinant of a matrix W is a polynomial in the n^2 entries w_{ij}. Denote this polynomial by D, so

$$\det W = D(w_{11}, w_{12}, \ldots, w_{nn}).$$

If the entries are functions of t, we have

$$\phi(t) = \det W(t) = D(w_{11}(t), w_{12}(t), \ldots, w_{nn}(t)).$$

By the chain rule,

$$\dot{\phi}(t) = \sum_{i,j=1}^{n} \frac{\partial D}{\partial w_{ij}} \dot{w}_{ij}(t), \tag{8.1}$$

where the partial derivatives are evaluated at $w_{11}(t)$, $w_{12}(t)$, ..., $w_{nn}(t)$. To compute the partial derivatives, we recall (Theorem 4.1, Chapter 4) that for any i,

$$D = \sum_{j=1}^{n} D_{ij} w_{ij},$$

where D_{ij} is the cofactor of w_{ij}. Hence, for any i and k,

$$\frac{\partial D}{\partial w_{ik}} = \sum_{j=1}^{n} \frac{\partial D_{ij}}{\partial w_{ik}} w_{ij} + D_{ij} \frac{\partial w_{ij}}{\partial w_{ik}}.$$

The cofactor of an entry of the ith row does not depend on any entry of the ith row. Thus

$$\frac{\partial D_{ij}}{\partial w_{ik}} = 0$$

for all i, j, k. Also $\dfrac{\partial w_{ij}}{\partial w_{ik}}$ is equal to 1 if $j = k$ and zero otherwise. Thus

$$\frac{\partial D}{\partial w_{ik}} = D_{ik},$$

a result which is also obvious from the discussion following Theorem 4.1, Chapter 4. Substituting in (8.1), we have

$$\dot{\phi}(t) = \sum_{i,j=1}^{n} D_{ij}\dot{w}_{ij}(t) = \sum_{i=1}^{n}\left[\sum_{j=1}^{n} D_{ij}\dot{w}_{ij}(t)\right]. \tag{8.2}$$

Now, the term in brackets is the determinant of the matrix obtained from $W(t)$ by replacing the entries of the ith row by their derivatives. This proves the theorem. $\qquad\square$

Theorem 8.1 leads to a generalization of Abel's formula (Chapter 3, Section 2, Exercise 10).

Theorem 8.2 (Liouville's Formula).   *Let $W(t)$ be a solution matrix of $\dot{x} = A(t)x$. Then*

$$\det W(t) = \det W(t_0)e^{\int_{t_0}^{t}\sum_{i=1}^{n} a_{ii}(s)\,ds} \tag{8.3}$$

Proof.   Let $\phi(t) = \det W(t)$. Written out, the formula for $\dot{\phi}(t)$ is

$$\dot{\phi}(t) = \begin{vmatrix} \dot{w}_{11} & \dot{w}_{12} & \cdots & \dot{w}_{1n} \\ w_{21} & w_{22} & \cdots & w_{2n} \\ \cdots & \cdots & \cdots & \cdots \\ w_{n1} & w_{n2} & \cdots & w_{nn} \end{vmatrix} + \begin{vmatrix} w_{11} & w_{12} & \cdots & w_{1n} \\ \dot{w}_{21} & \dot{w}_{22} & \cdots & \dot{w}_{2n} \\ \cdots & \cdots & \cdots & \cdots \\ w_{n1} & w_{n2} & \cdots & w_{nn} \end{vmatrix} + \cdots$$

$$+ \begin{vmatrix} w_{11} & w_{12} & \cdots & w_{1n} \\ w_{21} & w_{22} & \cdots & w_{2n} \\ \cdots & \cdots & \cdots & \cdots \\ \dot{w}_{n1} & \dot{w}_{n2} & \cdots & \dot{w}_{nn} \end{vmatrix}. \tag{8.4}$$

Since $\dot{W}(t) = A(t)W(t)$, we have

$$\dot{w}_{ij}(t) = \sum_{k=1}^{n} a_{ik}(t)w_{kj}(t), \qquad i,j = 1, \ldots, n.$$

We substitute for $\dot{w}_{ij}$ in formula (8.4), thus eliminating all derivatives.

The first determinant in (8.4) is now

$$
\begin{vmatrix}
\displaystyle\sum_{k=1}^{n} a_{1k}w_{k1} & \displaystyle\sum_{k=1}^{n} a_{1k}w_{k2} & \cdots & \displaystyle\sum_{k=1}^{n} a_{1k}w_{kn} \\
w_{21} & w_{22} & \cdots & w_{2n} \\
\cdots & \cdots & \cdots & \cdots \\
w_{n1} & w_{n2} & \cdots & w_{nn}
\end{vmatrix}.
$$

Subtract from the first row a_{12} times the second, plus a_{13} times the third, and so on. This does not alter the determinant, and we thus obtain the result that the determinant is equal to

$$
\begin{vmatrix}
a_{11}w_{11} & a_{11}w_{12} & \cdots & a_{11}w_{1n} \\
w_{21} & w_{22} & \cdots & w_{2n} \\
\cdots & \cdots & \cdots & \cdots \\
w_{n1} & w_{n2} & \cdots & w_{nn}
\end{vmatrix} = a_{11}\phi.
$$

Simplifying the other determinants in (8.4) by the same method shows that they are respectively equal to $a_{22}\phi$, $a_{33}\phi$, $\ldots$, $a_{nn}\phi$. Thus

$$
\dot{\phi}(t) = [a_{11}(t) + a_{22}(t) + \cdots + a_{nn}(t)]\phi(t).
$$

But this means that ϕ is a solution of the linear first-order equation

$$
\frac{dx}{dt} = \left[\sum_{i=1}^{n} a_{ii}(t) \right] x
$$

and it follows that

$$
\phi(t) = \phi(t_0)e^{\int_{t_0}^{t} \sum_{i=1}^{n} a_{ii}(s)\,ds}. \qquad \square
$$

Liouville's formula expresses some remarkable facts about linear homogeneous systems $\dot{x} = A(t)x$. In the first place, it tells us that if $W(t)$ is the s.f.m., then

$$
\det W(t) = e^{\int_{0}^{t} \sum_{i=1}^{n} a_{ii}(s)\,ds}.
$$

Hence we can compute $\det W(t)$ even if we do not know $W(t)$—and we usually do not. It is even more remarkable that the determinant depends only on the diagonal entries of $A(t)$, and in fact, only on the integral of their sum. Liouville's formula has the following, very general, consequence:

Theorem 8.3.   *A necessary condition that all solutions of $\dot{x} = A(t)x$ are bounded for $t \geqq 0$ is that there exists a number K such that*

$$\mathrm{Re} \int_0^t \sum_{i=1}^n a_{ii}(s)\, ds \leqq K$$

for $t \geqq 0$.

The simple proof is left as an exercise.

Note, also, that Liouville's formula gives us a new proof of the fact that a solution matrix is invertible either for all t or for no t.

The sum of the diagonal entries of a matrix A is called the *trace* of A, and is denoted by $\mathrm{tr}\ A$:

$$\mathrm{tr}\ A = \sum_{i=1}^n a_{ii}.$$

The trace, like the determinant, is a numerical function of A. Like the determinant, it is invariant under similarity, i.e., similar matrices have the same trace. In fact (Exercise 3), the trace of A is the sum of the eigenvalues of A. In the trace notation, formula (8.3) becomes

$$\det W(t) = \det W(t_0) e^{\int_{t_0}^t \mathrm{tr}\, A(s)\, ds}. \tag{8.5}$$

EXERCISES

1.   Let $u_1, \ldots, u_n$ be solutions of the homogeneous linear equation

$$x^{(n)} + a_1(t)x^{(n-1)} + \cdots + a_n(t)x = 0,$$

and let $\phi(t)$ be their Wronskian. Prove that

$$\phi(t) = \phi(t_0) e^{-\int_{t_0}^t a_1(s)\, ds}.$$

Thus Abel's formula remains valid for $n > 2$.

2.   Prove Theorem 8.3.

3.   Let $p(\lambda)$ be the characteristic polynomial of the $n \times n$ matrix A.
   (a)   Show that the coefficient of λ^{n-1} in $p(\lambda)$ is equal to

$$(-1)^{n-1}(a_{11} + a_{22} + \cdots + a_{nn}).$$

(b)   Let $\lambda_1, \ldots, \lambda_n$ be the roots of $p(\lambda)$, each distinct root being listed a number of times equal to its multiplicity. Show that the coefficient of λ^{n-1} in $p(\lambda)$ is equal to

$$(-1)^{n-1}(\lambda_1 + \lambda_2 + \cdots + \lambda_n).$$

(c)   Conclude that $\operatorname{tr} A = \sum_{i=1}^{n} \lambda_i.$

4.   Apply Theorem 8.3 to systems $\dot{x} = Ax$ with constant coefficients. How good is the resulting theorem, compared with the second part of Corollary 7.4?

5.   Let A be a constant matrix with eigenvalues $\lambda_1, \ldots, \lambda_n$. Let $W(t)$ be the s.f.m. of $\dot{x} = Ax$. Without using Liouville's formula, prove that

$$\det W(t) = e^{\left(\sum_{i=1}^{n} \lambda_i\right)t}.$$

(Hint: Section 7, Exercise 4 and Chapter 4, Section 5, Exercise 7.)

6.   Show that the system

$$\dot{x}_1 = 2tx_1 + x_2$$
$$\dot{x}_2 = e^{-t}x_1 + (\cos t)x_2$$

has a solution which is unbounded for $t > 0$.

9.   Systems with Periodic Coefficient Matrix

The study of oscillations in numerous linear and nonlinear physical systems (electric and electronic circuits, machines, satellites, etc.) leads to linear homogeneous differential systems with periodic coefficient matrices. Such systems cannot be solved except in trivial cases, and to obtain information about the solutions we must investigate the systems directly. The basic results were obtained by Gaston Floquet in 1883, and are commonly called "Floquet theory." This theory yields a complete description of the possible analytic forms that the solutions may have, and shows that the long-term behavior of the solutions is determined by the eigenvalues of a certain matrix. We conclude our discussion of linear systems with an introduction to this theory.

We consider a system

$$\dot{x} = A(t)x, \tag{9.1}$$

where $A(t)$ is continuous for all t and of period $p > 0$, that is,

$A(t + p) = A(t)$ for all t. To clarify the exposition, we assume that p is the least period of $A(t)$. (The only consequence of this assumption is to exclude the trivial case that A is constant.) Throughout this discussion, $W(t)$ denotes the s.f.m. of (9.1).

Theorem 9.1. $W(t + p) = W(t)W(p)$ *for all* t.

Proof. Let

$$V(t) = W(t + p).$$

Then

$$\begin{aligned}
\dot{V}(t) &= \dot{W}(t + p) \\
&= A(t + p)W(t + p) \\
&= A(t)W(t + p) \\
&= A(t)V(t).
\end{aligned}$$

Hence $V(t)$ is a solution matrix, and because $W(t + p)$ is invertible, $V(t)$ is fundamental. It follows that

$$\begin{aligned}
W(t) &= V(t)V^{-1}(0) \\
&= W(t + p)W^{-1}(p)
\end{aligned}$$

or

$$W(t + p) = W(t)W(p). \qquad \square$$

A non-zero solution u of (9.1) is called *multiplicative* if there is a number α such that

$$u(t + p) = \alpha u(t) \tag{9.2}$$

for all t. By definition, α cannot be zero. Note that if $\alpha = 1$, then u has period p, and, in general, if α is an mth root of 1, then u has period mp.

Theorem 9.2.   *Equation* (9.1) *has a multiplicative solution.*

Proof. Let u be a non-zero solution of (9.1). Then

$$u(t) = W(t)\eta$$

for some $\eta \neq 0$. By Theorem 9.1,

$$u(t + p) = W(t)W(p)\eta.$$

Hence a necessary and sufficient condition that

$$u(t + p) = \alpha u(t)$$

is that

$$W(t)W(p)\eta = \alpha W(t)\eta,$$

or

$$W(p)\eta = \alpha\eta.$$

Thus, if α is an eigenvalue of $W(p)$ and η is an eigenvector belonging to α, then u satisfies (9.2). $\qquad \square$

We see, in fact, that u satisfies (9.2) if and only if α is an eigenvalue of $W(p)$ and $u(0)$ is an eigenvector belonging to α. The following corollaries are already included in the proof of Theorem 9.2.

Corollary 9.3. *Let $\alpha_1, \ldots , \alpha_k$ be distinct eigenvalues of $W(p)$, let η^j be an eigenvector belonging to α_j, and let*

$$u^j(t) \; = \; W(t)\eta^j, \qquad j = 1, \ldots , k.$$

Then $u^1, \ldots , u^k$ are linearly independent multiplicative solutions of (9.1).

Corollary 9.4. *If $W(p)$ has n distinct eigenvalues, (9.1) has a basis of multiplicative solutions.*

It is instructive to see a proof of the last corollary based on matrix similarity. Let $\alpha_1, \ldots , \alpha_n$ be distinct eigenvalues of $W(p)$. Let Q be a matrix whose jth column is an eigenvector of $W(p)$ belonging to α_j. Then

$$Q^{-1}W(p)Q \; = \; D,$$

where D is the diagonal matrix whose (j,j)th entry is α_j. Setting

$$V(t) \; = \; W(t)Q,$$

we have

$$\begin{aligned} V(t + p) \; &= \; W(t)W(p)Q \\ &= \; W(t)QD \\ &= \; V(t)D, \end{aligned}$$

whence

$$V^{(j)}(t + p) \; = \; \alpha_j V^{(j)}(t), \qquad j = 1, \ldots , n.$$

Thus $V(t)$ is a fundamental matrix whose columns are multiplicative solutions.

The eigenvalues of $W(p)$ are called the *characteristic multipliers*, or simply the *multipliers*, of (9.1). We shall see that they determine the asymptotic behavior of the solutions of (9.1).

Theorem 9.5. *Let $u(t)$ be a function (not necessarily a solution of (9.1)) which satisfies*

$$u(t + p) \; = \; \alpha u(t), \qquad u \neq 0.$$

Then there exists a function $\psi(t)$ of period p such that

$$u(t) = e^{\lambda t}\psi(t),$$

where

$$\lambda = \frac{1}{p}\left(\log |\alpha| + i \arg \alpha\right).$$

Proof. Let $r = |\alpha|$, and let θ be an argument of α. Then

$$\alpha = re^{i\theta}.$$

Because $\alpha \neq 0$, r is positive and therefore has a logarithm. Let

$$\gamma = \log r + i\theta.$$

Then

$$\begin{aligned}
e^{\gamma} &= e^{\log r + i\theta} \\
&= e^{\log r}e^{i\theta} \\
&= re^{i\theta} \\
&= \alpha.
\end{aligned}$$

Define the function $\psi(t)$ by

$$\psi(t) = e^{-\frac{\gamma}{p}t}u(t).$$

Then

$$\begin{aligned}
\psi(t + p) &= e^{-\frac{\gamma}{p}(t+p)}u(t + p) \\
&= e^{-\gamma}e^{-\frac{\gamma}{p}t}u(t + p) \\
&= \frac{1}{\alpha}e^{-\frac{\gamma}{p}t}\alpha u(t) \\
&= e^{-\frac{\gamma}{p}t}u(t) \\
&= \psi(t).
\end{aligned}$$

Hence ψ has period p. Since

$$u(t) = e^{\frac{\gamma}{p}t}\psi(t),$$

it only remains to set $\lambda = \dfrac{\gamma}{p}$, and the theorem is proved. $\qquad\square$

Theorem 9.6. *Let $u(t + p) = \alpha u(t)$, $u \neq 0$.*
  (a) *If $|\alpha| < 1$, then $\lim_{t \to \infty} u(t) = 0$.*
  (b) *If $|\alpha| > 1$, then $u(t)$ is unbounded for $t \geq 0$.*
  (c) *If $|\alpha| = 1$, then $u(t)$ is bounded for all t, but does not tend to zero
as $t \to \infty$.*

Proof.   Choose λ as in Theorem 9.5, so that

$$u(t) = e^{\lambda t}\psi(t),$$

where ψ is periodic. Because Re $(\lambda) = (1/p) \log |\alpha|$, the real part of λ is negative, positive, or zero, according as the absolute value of α is less than, greater than, or equal to 1. Note that

$$\begin{aligned}|u_i(t)| &= |e^{\lambda t}| \, |\psi_i(t)| \\ &= e^{\text{Re }(\lambda)t}|\psi_i(t)|, \qquad i = 1, \ldots, n.\end{aligned} \tag{9.3}$$

Because ψ is periodic, there is a number k so that $|\psi_i(t)| \leq k$ for all t and i. Hence

$$|u_i(t)| \leq k e^{\text{Re }(\lambda)t}.$$

It follows that, if Re $(\lambda) < 0$,

$$\lim_{t \to \infty} |u_i(t)| = 0, \qquad i = 1, \ldots, n.$$

If Re $(\lambda) > 0$, then $e^{\text{Re }(\lambda)t} \to \infty$ as $t \to \infty$. Since at least one of the periodic functions $\psi_i(t)$ is not zero, it follows from Eq. (9.3) that at least one of the functions $u_i(t)$ is unbounded for $t \geq 0$.

If Re $(\lambda) = 0$, then $|u_i(t)| = |\psi_i(t)|$; hence $|u_i(t)|$ is periodic, $i = 1, \ldots, n$. $\square$

We now return to the system (9.1). Let $\alpha_1, \ldots, \alpha_n$ be its characteristic multipliers, not necessarily distinct. The numbers

$$\lambda_j = \frac{1}{p} [\log |\alpha_j| + i \arg \alpha_j], \qquad j = 1, \ldots, n$$

are called the *characteristic exponents* of (9.1). Because arg α_j is determined only to within integral multiples of 2π, the imaginary part of λ_j is determined only to within integral multiples of $2\pi/p$. (We can eliminate this ambiguity by setting arg $\alpha_j =$ Arg α_j. In any case, only Re (λ_j) is of interest, and this is uniquely determined.) We summarize our results to this point.

Theorem 9.7.   (a)   *Corresponding to each characteristic exponent λ, there is a non-zero solution u such that*

$$u(t) = e^{\lambda t}\psi(t),$$

where ψ has period p.

(b)   *If (9.1) has a characteristic multiplier of absolute value less than 1 (greater than 1; equal to 1), then it has a non-zero solution which tends to zero as $t \to \infty$ (is unbounded for $t \geq 0$; is bounded for all t, but does not tend to zero as $t \to \infty$).*

(c) *If (9.1) has n distinct multipliers, then it has a fundamental matrix $V(t)$ of the form*

$$V(t) = \Psi(t)E(t),$$

where $\Psi(t)$ has period p and $E(t)$ is the diagonal matrix whose (j,j)th entry is $e^{\lambda_j t}$.

Thus a necessary condition that all solutions of (9.1) are bounded for $t \geq 0$ is that (9.1) has no multiplier of absolute value greater than 1. It also follows that if all the multipliers are simple and have absolute value less than 1, then all solutions tend to zero. This happens to be true even if (9.1) has repeated multipliers. Before we go on to consider the case of repeated multipliers, we pause to take stock.

We have established (at least if the multipliers are simple) that the solutions of (9.1) are linear combinations of functions which are products of exponentials and periodic functions, and that therefore their asymptotic behavior is determined by the absolute value of the multipliers. The difficulty is that the multipliers are the eigenvalues of $W(p)$, and the theory offers no way of computing them short of computing the solutions themselves. But if it were possible to compute the solutions (in the sense of finding formulas for them) there would be no need for the theory. This difficulty is very great indeed. It has been, and continues to be, a source of much research. Unfortunately, the numerous results obtained, and their applications to physical and mathematical problems, fall outside the scope of this book.

One simple result, however, follows from Liouville's formula,

$$\det W(t) = e^{\int_0^t \operatorname{tr} A(s)\, ds},$$

where setting $t = p$ gives

$$\det W(p) = e^{\int_0^p \operatorname{tr} A(s)\, ds}.$$

Let $\alpha_1, \ldots, \alpha_n$ be the multipliers of (9.1), not necessarily distinct. Because the determinant of a matrix is the product of its eigenvalues, we have that

$$\alpha_1 \alpha_2 \cdots \alpha_n = e^{\int_0^p \operatorname{tr} A(s)\, ds}.$$

Thus

$$|\alpha_1 \alpha_2 \cdots \alpha_n| = e^{\operatorname{Re} \int_0^p \operatorname{tr} A(s)\, ds}, \tag{9.4}$$

or, equivalently,

$$\sum_{i=1}^{n} \operatorname{Re}(\lambda_i) = \operatorname{Re} \frac{1}{p} \int_0^p \operatorname{tr} A(s)\, ds. \tag{9.5}$$

Theorem 9.8.   (a) *If* $\operatorname{Re} \int_0^p \operatorname{tr} A(s)\, ds < 0$, *then* (9.1) *has a multiplier of absolute value less than* 1.

(b)   *If* $\operatorname{Re} \int_0^p \operatorname{tr} A(s)\, ds > 0$, *then* (9.1) *has a multiplier of absolute value greater than* 1.

The proof is obvious from formula (9.4). For another elementary result, see Exercise 6.

EXAMPLE.   The system

$$\dot{x}_1 = (\cos^2 t)x_1 + (\sin t \cos t - 1)x_2$$
$$\dot{x}_2 = (\sin t \cos t + 1)x_1 + (\sin^2 t)x_2$$

is periodic with $p = \pi$. The trace is obviously equal to 1, so that $\int_0^\pi \operatorname{tr} A(s)\, ds = \pi > 0$. Hence some solutions are unbounded for $t \geq 0$. It is easy to verify that the s.f.m. is

$$W(t) = \begin{bmatrix} e^t \cos t & -\sin t \\ e^t \sin t & \cos t \end{bmatrix},$$

so that the solutions are

$$x_1 = c_1 e^t \cos t - c_2 \sin t$$
$$x_2 = c_1 e^t \sin t + c_2 \cos t.$$

The unbounded solutions correspond to $c_1 \neq 0$. Since

$$W(\pi) = \begin{bmatrix} -e^\pi & 0 \\ 0 & -1 \end{bmatrix},$$

the multipliers are $\alpha_1 = -e^\pi$ and $\alpha_2 = -1$, and one of them has absolute value greater than 1, as predicted. The fact that the other is equal to -1 corresponds to the existence of solutions of period $2p = 2\pi$; these are the solutions for which $c_1 = 0$. $\square$

By methods similar to those of Section 7, one can show that if all the multipliers (not necessarily simple) have absolute value less than 1, then all the solutions tend to zero. In order to simplify the presentation, we shall consider only the case $n = 2$.

Let $W(t)$ be the s.f.m. of

$$\dot{x} = A(t)x, \tag{9.6}$$

where $A(t)$ is a 2×2 matrix, and suppose that $W(p)$ has a repeated eigenvalue α. Let η be an eigenvector belonging to α, and let Q be an invertible matrix whose first column is η. Then

$$Q^{-1}W(p)Q = B,$$

where B is upper-triangular, i.e.,

$$B = \begin{bmatrix} \alpha & \beta \\ 0 & \alpha \end{bmatrix}.$$

Define the fundamental matrix $V(t)$ by

$$V(t) = W(t)Q.$$

Then

$$V(t + p) = W(t)W(p)Q = W(t)QB = V(t)B.$$

Hence

$$V^{(1)}(t + p) = V(t)B^{(1)} = \alpha V^{(1)}(t) \tag{9.7}$$

and

$$V^{(2)}(t + p) = V(t)B^{(2)} = \beta V^{(1)}(t) + \alpha V^{(2)}(t). \tag{9.8}$$

Thus $V^{(1)}(t)$ is a multiplicative solution, and we have

$$V^{(1)}(t) = e^{\lambda t}\psi^1(t),$$

where $\lambda = (1/p)[\log |\alpha| + i \arg \alpha]$ and ψ^1 is a function of period p. Substituting for $V^{(1)}$ in Eq. (9.8), we find that $V^{(2)}$ satisfies

$$V^{(2)}(t + p) = \beta e^{\lambda t}\psi^1(t) + \alpha V^{(2)}(t). \tag{9.9}$$

Define the function ψ^2 by

$$\psi^2(t) = -\frac{\beta}{\alpha p} t\psi^1(t) + e^{-\lambda t}V^{(2)}(t). \tag{9.10}$$

Using the periodicity of $\psi^1(t)$ and Eq. (9.9), we have

$$\begin{aligned}
\psi^2(t + p) &= -\frac{\beta}{\alpha p}(t + p)\psi^1(t) + e^{-\lambda(t+p)}V^{(2)}(t + p) \\
&= -\frac{\beta}{\alpha p} t\psi^1(t) - \frac{\beta}{\alpha}\psi^1(t) + e^{-\lambda t - \lambda p}[\beta e^{\lambda t}\psi^1(t) + \alpha V^{(2)}(t)] \\
&= -\frac{\beta}{\alpha p} t\psi^1(t) - \frac{\beta}{\alpha}\psi^1(t) + \frac{1}{\alpha}e^{-\lambda t}[\beta e^{\lambda t}\psi^1(t) + \alpha V^{(2)}(t)] \\
&= -\frac{\beta}{\alpha p} t\psi^1(t) + e^{-\lambda t}V^{(2)}(t) \\
&= \psi^2(t).
\end{aligned}$$

Thus ψ^2 has period p. Solving (9.10) for $V^{(2)}(t)$ yields

$$V^{(2)}(t) = \frac{\beta}{\alpha p} te^{\lambda t}\psi^1(t) + e^{\lambda t}\psi^2(t).$$

Lastly, we define ψ^3 by

$$\psi^3(t) = \frac{\beta}{\alpha p}\,\psi^1(t).$$

Note that $\psi^3 = 0$ if and only if $\beta = 0$. (In this case $W(p)$ is diagonal. See Exercise 5.) The columns of $V(t)$ therefore satisfy

$$\begin{aligned}
V^{(1)} &= e^{\lambda t}\psi^1(t) \\
V^{(2)}(t) &= e^{\lambda t}[\psi^2(t) + t\psi^3(t)],
\end{aligned} \tag{9.11}$$

where ψ^1, ψ^2, and ψ^3 have period p.

It follows that if Re $(\lambda) < 0$, then $V^{(j)}(t) \to 0$ as $t \to \infty$, $\quad j = 1, 2$.

EXERCISES

1. What effect does a change in the choice of arg α have on the representation $u(t) = e^{\lambda t}\psi(t)$ of Theorem 9.5?

2. Prove, without using Theorem 9.8 or any other result of this section: If Re $\int_0^p \operatorname{tr} A(s)\,ds > 0$, then (9.1) has a solution which is unbounded for $t \geq 0$.

3. Give a new proof of Theorem 3.6, Chapter 2, using the approach of this section.

4. Compute the product of the characteristic multipliers of the system

$$\begin{aligned}
\dot{x}_1 &= (\sin^2 t)x_1 + e^{\sin t}x_2 \\
\dot{x}_2 &= -x_1 + (\cos^2 t)x_2.
\end{aligned}$$

5. Let C be a 2×2 matrix with a double eigenvalue α. Prove that if C is similar to a diagonal matrix, then C is diagonal. Prove the corresponding statement for $n \times n$ matrices: If an $n \times n$ matrix with an n-fold eigenvalue α is similar to a diagonal matrix, then it is diagonal.

6. (a) Show that the s.f.m. of (9.1) satisfies $W(p)W(-p) = I$.

   (b) Let C be a 2×2 matrix such that $C^2 = I$ (C^2 means CC). Prove: (1) If α is an eigenvalue of C, then $\alpha = \pm 1$; (2) C is similar to a diagonal matrix.

   (c) Let $A(t)$ be a 2×2 matrix which has period p and is an odd function of t, i.e., $A(-t) = -A(t)$. Let $W(t)$ be the s.f.m. of $\dot{x} = A(t)x$. Show that $W(-t) = W(t)$, and conclude that $W^2(p) = I$.

   (d) Combine these results to prove that all solutions of $\dot{x} = A(t)x$ are periodic.

7. (a) Let C be a matrix with n distinct eigenvalues $\alpha_1, \ldots, \alpha_n$. Let $C^m = CC \cdots C$ (m factors). We say that $\lim\limits_{m \to \infty} C^m = 0$ if every entry of C^m tends to zero as $m \to \infty$. Prove that $\lim\limits_{m \to \infty} C^m = 0$ if and only if $|\alpha_i| < 1$, $i = 1, \ldots, n$. (Hint: Diagonalize C. The statement remains true if C has repeated eigenvalues, but is more difficult to prove.)

(b) Let $W(t)$ be the s.f.m. of (9.1). Prove that $W(t + mp) = W(t)W^m(p)$. Conclude from this that $\lim\limits_{t \to \infty} W(t) = 0$ if and only if $\lim\limits_{m \to \infty} W^m(p) = 0$. [Hint: Every $t \geq 0$ can be written $t = s + mp$, whence $W(t) = W(s)W^m(p)$, where $0 \leq s < p$.]

(c) Assume (9.1) has n distinct multipliers. Give a new proof that all solutions of (9.1) tend to zero if and only if all multipliers have absolute value less than 1.

MISCELLANEOUS EXERCISES FOR CHAPTER 5

The Exponential of a Matrix.   In Exercises 1 and 2 we complete the line of inquiry begun in Exercise 5, page 155, and Exercise 4, page 166. In Exercise 3 we apply the results to the exponential of a matrix.

1. Let $AB = BA$, let $W(t)$ be the s.f.m. of $\dot{x} = Ax$, and let $V(t)$ be the s.f.m. of $\dot{x} = Bx$. Prove:

(a) $W(t)B = BW(t)$ and $V(t)A = AV(t)$

(b) $W(t)V(t)$ is the s.f.m. of $\dot{x} = (A + B)x$

2. Let $W(t)$ be the s.f.m. of $\dot{x} = Ax$, let s be a real number, and let $V(t)$ be the s.f.m. of $\dot{x} = sAx$. Prove:

(a) $V(t) = W(st)$

(b) $V(1) = W(s)$

3. The *exponential* of a matrix A is defined to be the matrix $W(1)$, where $W(t)$ is the s.f.m. of $\dot{x} = Ax$. It is written e^A:

$$e^A = W(1).$$

Prove:

(a) If $AB = BA$, then $e^A e^B = e^{A+B}$

(b) e^A is invertible, and $(e^A)^{-1} = e^{-A}$

(c) The eigenvalues of e^A are $e^{\lambda_1}, \ldots, e^{\lambda_n}$, where $\lambda_1, \ldots, \lambda_n$ are the eigenvalues of A

(d) $\det e^A = e^{\operatorname{tr} A}$

(e) $Ae^A = e^A A$

(f) e^{tA} is the s.f.m. of $\dot{x} = Ax$

(g) $P^{-1}e^A P = e^{P^{-1}AP}$

The Adjoint of a Matrix

4. Given a matrix A, the matrix whose (i,j)th entry is $\bar{a}_{ji}$ is called the *adjoint* of A, and is denoted by A^*. Thus A^* is obtained by reflecting A in its principal diagonal and replacing all entries by their conjugates. [If A is real, the adjoint coincides with the *transpose*, whose (i,j)th entry is a_{ji}.] Obviously $(A^*)^* = A$. Prove that $(AB)^* = B^*A^*$.

5. (a) Let $W(t)$, $V(t)$ satisfy $\dot{W}(t) = A(t)W(t)$, $\dot{V}(t) = -V(t)A(t)$. Prove that $V(t)W(t)$ is constant.

(b) Let $W(t)$ satisfy $\dot{W}(t) = A(t)W(t)$, where $A^*(t) = -A(t)$. Show that $W^*(t)W(t)$ is constant. In particular, if $W(t)$ is the s.f.m. of $\dot{x} = A(t)x$, then $W^*(t)W(t) = I$. (If $A^* = -A$, we say that A is *skew-Hermitian*, and if $A^* = A^{-1}$, we say that A is *unitary*. Thus if A is skew-Hermitian, e^A is unitary.)

(c) Let $u(t)$ be a solution of $\dot{x} = A(t)x$, where $A^*(t) = -A(t)$. Prove that $\sum_{i=1}^{n} \bar{u}_i(t)u_i(t) = $ constant. Conclude that u is bounded for all t. (Recall that $\bar{u}_i u_i = |u_i|^2$.)

6

SECOND-ORDER AUTONOMOUS SYSTEMS

Except for the occurrence of some notions of matrix theory, this chapter is self-contained; Chapter 4 and Sections 1, 2, 3 and 5 of Chapter 5 constitute an adequate background. From the point of view of a systematic development of the theory, this chapter is a continuation of Section 6 of Chapter 2.

The subject matter of this chapter is the qualitative analysis of second-order autonomous systems, a topic which is important not only for its own sake, but because it introduces the reader to the general theory of non-linear differential equations. Here we should point out that linear systems play a central role in the study of non-linear systems; mathematicians frequently use the word "non-linear" to mean "not-necessarily-linear." This peculiar usage is made necessary by the historical circumstance that until relatively recently textbooks concerned themselves almost exclusively with linear systems.

1. Introduction

In this chapter we study systems of the form

$$\frac{dx}{dt} = f(x,y)$$

$$\frac{dy}{dt} = g(x,y). \tag{1.1}$$

Such systems include the important case of a single second-order
equation

$$\frac{d^2x}{dt^2} = g\left(x, \frac{dx}{dt}\right),\tag{1.2}$$

which, as we know, can be written

$$\begin{aligned}\frac{dx}{dt} &= y\\[2mm]\frac{dy}{dt} &= g(x,y).\end{aligned}\tag{1.3}$$

As in Chapter 5, we shall use dots to denote derivatives: $\dot{x} = dx/dt$,
$\dot{u}(t) = u'(t)$, etc. *We suppose that the functions f and g in Eq.* (1.1)
*are real-valued functions defined and continuous for all real values of
x and y, and that f and g have continuous partial derivatives with respect
to x and y.* Unless an exception is explicitly made, this hypothesis is
tacitly incorporated in all theorems.

A solution of (1.1) is by definition a pair of real-valued functions
u, v defined on a common interval I such that

$$\begin{aligned}\dot{u}(t) &= f(u(t),v(t))\\\dot{v}(t) &= g(u(t),v(t))\end{aligned}$$

for t in I.

We remind the reader of the geometric interpretation of (1.1) as a
lineal field in the three-dimensional (t,x,y)-space, and of a solution as
a function pair whose graph is tangent to the field (Chapter 1, Section
5). Owing to the autonomous character of (1.1), we shall be able to replace this interpretation by a two-dimensional one, and to treat (1.1)
as a vector field in the (x,y)-plane. First, however, we shall discuss the
three-dimensional geometry of the solution curves, i.e., the graphs of
the solutions.

The graph of a solution, or, for that matter, of any function pair
u, v defined on an interval I, is the set of points (t,x,y) which satisfy
$x = u(t)$, $y = v(t)$, t in I. If u, v is a solution of (1.1), its graph has a
tangent line at every point $(t,u(t),v(t))$ on it, and a set of direction numbers for this tangent line is given by 1, $f(u(t),v(t))$, $g(u(t),v(t))$. Conversely, this property of the graph implies that u, v is a solution. The
simplest and most important case is that of a constant solution. Clearly
$u(t) \equiv x_0$, $v(t) \equiv y_0$ is a solution if and only if $f(x_0,y_0) = g(x_0,y_0) = 0$.
The graph of this solution is the line $x = x_0$, $y = y_0$, i.e., the line
parallel to the t-axis which intersects the (x,y)-plane in the point (x_0,y_0).

EXAMPLE 1.
$$\dot{x} = y$$
$$\dot{y} = -x.$$
(1.4)

The solutions are $u(t) = A \sin(t + \varphi)$, $v(t) = A \cos(t + \varphi)$; we may suppose $A \geqq 0$ and $0 \leqq \varphi < 2\pi$. $A = 0$ yields the zero solution; its graph is the t-axis. If $A > 0$, the graph is a helix lying on the circular cylinder $x^2 + y^2 = A^2$ (Fig. 16). The effect of varying φ while A is

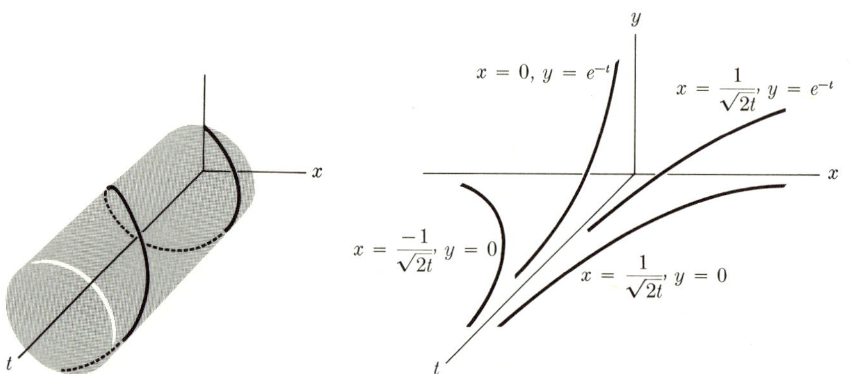

FIG. 16.   Solution curves of
$\dot{x} = y$, $\dot{y} = -x$.

FIG. 17.   Solution curves of
$\dot{x} = -x^3$, $\dot{y} = -y$.

held fixed is to translate the helix parallel to the t-axis. Thus every solution curve other than the t-axis is a circular helix. Obviously all solutions are defined for all t and have period 2π.   □

EXAMPLE 2.
$$\dot{x} = -x^3$$
$$\dot{y} = -y.$$
(1.5)

The solutions are $u(t) \equiv 0$, $v(t) = ce^{-t}$, defined for all t, and $u(t) = \pm 1/\sqrt{2t - c_1}$, $v(t) = c_2 e^{-t}$, defined for $t > c_1/2$. Some of the solution curves are shown in Fig. 17. Note that the only solutions defined for all t are those for which $u(t) \equiv 0$, and whose graphs therefore lie in the (t,y)-plane. All solutions are defined for large t, and tend to zero as $t \to \infty$.   □

As with first-order equations, we say that a solution is maximal if it has no continuation, i.e., if its graph is not contained in the graph of any other solution. The following theorem is proved in Chapter 7.

Theorem 1.1.   *Given numbers t_0, x_0, y_0, there exists a unique maximal solution u, v of Eq. (1.1) such that $u(t_0) = x_0$ and $v(t_0) = y_0$. The interval of definition of u, v is open, and if it has an endpoint α, then $u^2(t) + v^2(t) \to \infty$ as $t \to \alpha$.*

We usually suppress the adjective "maximal"; the context makes it clear whether it is implied.

A pair of functions u, v from an interval I to R gives rise to a function w from I to R^2 (the plane) via the definition $w(t) = (u(t),v(t))$. The derivative of w is defined, as in Chapter 5, by $\dot{w}(t) = (\dot{u}(t),\dot{v}(t))$. Similarly, a pair of functions f, g from R^2 to R gives rise to a function h from R^2 to R^2 via the definition $h(x,y) = (f(x,y),g(x,y))$. Thus we write $w = (u,v)$, $\dot{w} = (\dot{u},\dot{v})$, $h = (f,g)$. In this notation, w is a solution of (1.1) if $\dot{w}(t) = h(w(t))$.

Theorem 1.2. *Let* $w = (u,v)$ *be a solution of (1.1), and let* φ *be a number. Let* $w_1(t) = w(t + \varphi)$. *Then* w_1 *is also a solution.*

Proof. $\dot{w}_1(t) = \dot{w}(t + \varphi) = h(w(t + \varphi)) = h(w_1(t))$. $\square$

It follows that if a solution curve is translated parallel to the t-axis, another solution curve is obtained. The family of all such translates of a given solution curve C forms a cylinder, namely, the cylinder whose generators are the lines through C parallel to the t-axis. If the given solution happens to be constant, it coincides with all its translates, a property which in fact defines a constant function. In this case, and only in this case, the cylinder degenerates to a line.

Theorem 1.3. *Let* $w_1 = (u_1,v_1)$ *and* $w_2 = (u_2,v_2)$ *be solutions of (1.1). Suppose there are numbers* t_1 *and* t_2 *such that* $w_1(t_1) = w_2(t_1)$. *Then* $w_1(t) = w_2(t + t_2 - t_1)$.

Proof. Let $w(t) = w_2(t + t_2 - t_1)$. By Theorem 1.2, w is a solution. Now $w(t_1) = w_2(t_2)$, so by hypothesis $w(t_1) = w_1(t_1)$. Uniqueness implies that $w = w_1$. $\square$

In Chapter 2, we showed that a solution of an autonomous first-order equation having the uniqueness property is either constant, strictly increasing, or strictly decreasing. If a solution of such an equation assumes the same value twice it is therefore constant. A solution of (1.1) which assumes the same value twice need not be constant, but it must be periodic.

Theorem 1.4. *Let* w *be a maximal solution of (1.1). Suppose that there are numbers* $t_1 < t_2$ *such that* $w(t_1) = w(t_2)$. *Then* w *is defined for all* t *and has period* $p = t_2 - t_1$.

Proof. Let w be defined on I. Let I_1 be the interval obtained by translating I p units to the left, so that t is in I_1 if and only if $t + p$ is in I. Let $w_1(t) = w(t + p)$ for t in I_1. A translate of a maximal solution is obviously maximal; hence w_1 is maximal. Now $w_1(t_1) = w(t_2)$, so by hypothesis $w_1(t_1) = w(t_1)$. By uniqueness $w_1 = w$, and therefore also $I_1 = I$. But the only interval which coincides with a translate of itself is R. Hence w is defined for all t and $w(t + p) = w(t)$. $\square$

Trajectories.   Let u, v be a solution of (1.1), let C be its graph, and let Γ be the projection of C onto the (x,y)-plane. Note that (x,y) lies on Γ if and only if there is a number t such that the point (t,x,y) lies on C, or, in analytic terms, such that $x = u(t)$, $y = v(t)$. Thus the equations

$$x = u(t), \qquad y = v(t) \tag{1.6}$$

are parametric equations for Γ. The curve Γ is called the *trajectory* of u, v. Note the distinction between C and Γ, that is, between a solution curve and a trajectory: the former is the *graph* of a solution, the latter is its *range*.

Let u_1, v_1 be a translate of u, v,

$$u_1(t) = u(t + \varphi), \qquad v_1(t) = v(t + \varphi),$$

let C_1 be the graph of u_1, v_1, and let Γ_1 be its trajectory. Since C_1 is obtained by translating C parallel to the t-axis, obviously $\Gamma_1 = \Gamma$. Thus all translates of a given solution u, v have the same trajectory. Analytically, this means that for any φ, the equations

$$x = u(t + \varphi), \qquad y = v(t + \varphi) \tag{1.7}$$

are also parametric equations for Γ, a self-evident fact. If we think of Γ as the path of a point P whose position at time t is $(u(t),v(t))$, then the parametrization (1.7) corresponds to the path of a point P_1 whose motion is identical with that of P, except that P and P_1 reach a given position at different times. The constant time difference φ is called the difference in phase between u, v and u_1, v_1.

A trajectory could also be defined as a curve in the (x,y)-plane which is parametrized by a solution of (1.1), or as the intersection of the (x,y)-plane with the cylinder generated by the translates of a solution curve. All these definitions are clearly equivalent.

Theorem 1.5.   *Through each point of the (x,y)-plane there passes a unique trajectory, which is either a point, a simple closed curve, or a simple arc. The trajectory of a constant solution is a point. The trajectory of a non-constant periodic solution is a simple closed curve. The trajectory of a non-periodic solution is a simple arc.*

Proof. Given a point (x_0, y_0), let u, v be the solution of (1.1) which satisfies $u(0) = x_0$, $v(0) = y_0$. The trajectory of u, v is the curve Γ: $x = u(t)$, $y = v(t)$; obviously (x_0, y_0) lies on Γ. Suppose there is another trajectory Γ_1: $x = u_1(t)$, $y = v_1(t)$ which passes through (x_0, y_0), so that for some t_1, $x_0 = u_1(t_1)$, $y_0 = v_1(t_1)$. Then $u_1(t_1) = u(0)$, $v_1(t_1) = v(0)$, and, by Theorem 1.3, $u_1(t) = u(t - t_1)$, $v_1(t) = v(t - t_1)$. Thus u_1, v_1 is a translate of u, v, and $\Gamma_1 = \Gamma$.

If u, v is a constant solution, so that $u(t) \equiv x_0$, $v(t) \equiv y_0$, then obviously Γ reduces to the point (x_0, y_0) itself.

If u, v is not constant, Γ is a simple curve, i.e., it has no self-intersections. For if $u(t_1) = u(t_2)$ and $v(t_1) = v(t_2)$ for some $t_1 < t_2$, then, by Theorem 1.4, u, v is periodic with period $p = t_2 - t_1$. It follows that Γ coincides with the curve $x = u(t)$, $y = v(t)$, where t is restricted to the interval $[0, p]$, that is, Γ is a simple closed curve. Finally, if u, v is not periodic, $u(t_1) = u(t_2)$, $v(t_1) = v(t_2)$ implies $t_1 = t_2$, and consequently Γ is a simple arc. $\square$

Hereafter we reserve the term "periodic" for non-constant periodic functions.

The trajectory of a constant solution is called a *singular point*. Equivalently, (x_0, y_0) is a singular point if $f(x_0, y_0) = g(x_0, y_0) = 0$. The trajectory of a periodic solution is called an *orbit*. The (x, y)-plane itself is called the *phase plane*. The decomposition of the phase plane into trajectories of (1.1) is called the *phase portrait* of (1.1). The ultimate goal of the theory is to obtain this phase portrait, since all the important properties of the solutions may be inferred from it. Usually much less than the complete phase portrait is required, or of interest, for the applications. Both physically and mathematically, primary interest attaches to the behavior of the trajectories near the singular points. Also of great importance is the question of the existence of orbits, and of the behavior of the trajectories near a given orbit.

By the phase portrait of (1.2) we mean the phase portrait of the equivalent system (1.3). Since every trajectory Γ of (1.3) is of the form $x = u(t)$, $y = \dot{u}(t)$, where u is a solution of (1.2), Γ is an orbit if and only if u is a non-constant periodic solution. Note that all the singular points of (1.3) lie on the x-axis.

There is a more direct geometric connection between (1.1) and its trajectories than we have so far indicated. The connection arises if we use (1.1) to define a vector field in the (x, y)-plane. To a given point (x, y) we assign the vector whose horizontal and vertical components are $f(x, y)$ and $g(x, y)$. We shall denote this vector simply by $(f(x, y), g(x, y))$, thus making no notational distinction between points and vectors. In our notation the unit vectors $\vec{i}$ and $\vec{j}$ are given by $\vec{i} = (1,0)$, $\vec{j} = (0,1)$, so that $f(x,y)\vec{i} + g(x,y)\vec{j} = (f(x,y),g(x,y))$. We shall also write $\vec{V}(x,y) = (f(x,y),g(x,y))$. Note that if (x_0,y_0) is a singular point, $\vec{V}(x_0,y_0)$ is the zero vector. Figures 18 and 19 show the vector fields of Eqs. (1.4) and (1.5).

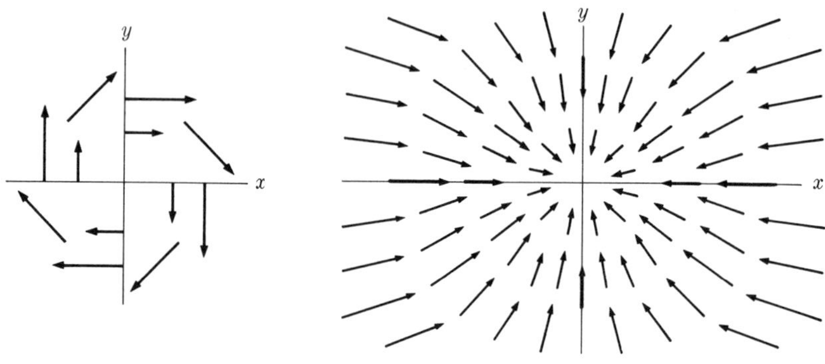

FIG. 18.   Vector field of
$\dot{x} = y$, $\dot{y} = -x$.

FIG. 19.   Vector field of
$\dot{x} = -x^3$, $\dot{y} = -y$.

Consider a point P which moves in the (x,y)-plane in such a way that, whatever its coordinates (x,y), its velocity vector is $\vec{V}(x,y)$. Let the coordinates of P at time t be $(u(t),v(t))$, so that its velocity vector is $(\dot{u}(t),\dot{v}(t))$. By assumption, $(\dot{u}(t),\dot{v}(t)) = \vec{V}(u(t),v(t))$, so $\dot{u}(t) = f(u(t),v(t))$, $\dot{v}(t) = g(u(t),v(t))$. Thus u, v is a solution of (1.1). Conversely, if the coordinates of a moving point P at time t are given by a solution of (1.1), then the velocity vector of P at any point in its path is equal to $\vec{V}(x,y)$. A trajectory Γ is therefore the path of a moving point whose velocity vector at (x,y) is $\vec{V}(x,y)$. We shall speak indiscriminately of Γ as being a trajectory of (1.1) or as being a trajectory of the vector field $\vec{V}(x,y)$. There is, in fact, no distinction between a second-order autonomous system and a plane velocity vector field. A point whose position at time t is given by a solution of (1.1) is called a *representative point*. The speed with which the point moves, the so-called *phase speed*, is $\sqrt{\dot{u}^2(t) + \dot{v}^2(t)}$. The phase speed at (x,y) is thus the non-negative number $\sqrt{f^2(x,y) + g^2(x,y)}$; it vanishes only at a singular point. It follows from the uniqueness theorem that a representative point is either always or never at rest. We now show that if a representative point approaches a point (x_0,y_0) as $t \to \infty$ (or, tracing its path backward in time, as $t \to -\infty$), then (x_0,y_0) is a singular point.

Theorem 1.6.   *Let u, v be a solution such that $(u(t),v(t)) \to (x_0,y_0)$ as $t \to \infty$ or $-\infty$. Then (x_0,y_0) is a singular point.*

Proof.   We consider first the case that $t \to \infty$. Because

$$\dot{u}(t) = f(u(t),v(t))$$

and f is continuous,

$$\lim_{t \to \infty} \dot{u}(t) = \lim_{t \to \infty} f(u(t),v(t)) = f(x_0,y_0).$$

Exactly as in the proof of Theorem 6.4, Chapter 2, either of the assumptions that $f(x_0,y_0)$ is positive or that it is negative leads to a contradiction. Hence $f(x_0,y_0) = 0$. The same argument applied to $\dot{v}(t)$ shows that $g(x_0,y_0) = 0$.

The case that $t \to -\infty$ may be dealt with in the same way, due attention being paid to the fact that integrating from right to left reverses inequalities. It is much more instructive, however, to obtain the result in this case as a corollary to the first case. We define functions u_1, v_1 by $u_1(t) = u(-t)$, $v_1(t) = v(-t)$, so that $\dot{u}_1(t) = -\dot{u}(-t)$, $\dot{v}_1(t) = -\dot{v}(-t)$, and therefore

$$\dot{u}_1(t) = -f(u_1(t),v_1(t))$$
$$\dot{v}_1(t) = -g(u_1(t),v_1(t)).$$

Thus u_1, v_1 is a solution of

$$\dot{x} = -f(x,y)$$
$$\dot{y} = -g(x,y). \tag{1.8}$$

The hypothesis implies that $(u_1(t),v_1(t)) \to (x_0,y_0)$ as $t \to \infty$, and it follows from the first part of the theorem that (x_0,y_0) is a singular point of (1.8). Obviously (1.1) and (1.8) have the same singular points, and the proof is finished. $\square$

According to our definition, a trajectory Γ is a curve in the sense of analytic geometry, i.e., a set of points. However, if Γ is not a singular point, it also has a natural orientation, namely the direction given to it by its tangent vectors $\vec{V}(x,y)$, or equivalently, the direction in which Γ is traversed by a representative point. When we draw a trajectory, we indicate this orientation by an arrow. (Figures 20 and

21 show the trajectories of Eqs. (1.4) and (1.5).) Finally, we may consider a non-singular trajectory not only as an oriented curve, but

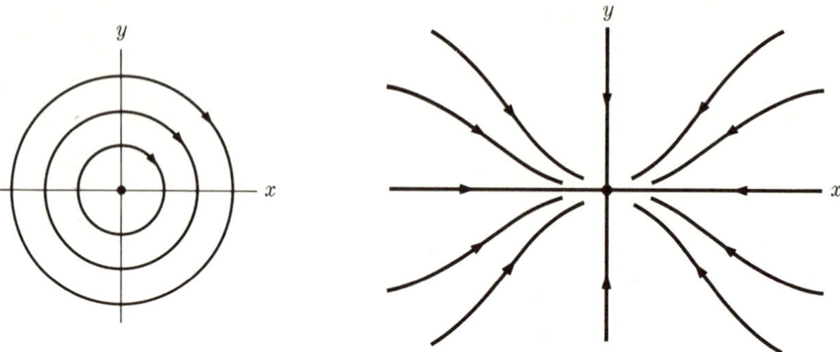

Fig. 20. Trajectories of
$\dot{x} = y$, $\dot{y} = -x$.

Fig. 21. Trajectories of
$\dot{x} = -x^3$, $\dot{y} = -y$.

as such a curve together with the solutions which parametrize it. It is sometimes necessary to distinguish between these three possible meanings of the work "trajectory." When verbal precision is needed to avoid misunderstanding, we use the term *characteristic* for the first meaning. A characteristic, therefore, has neither orientation nor preferred parametrization. Note that the characteristics of the vector field $-\vec{V}(x,y)$ are the same as those of $\vec{V}(x,y)$, whereas the trajectories differ in being oppositely oriented. In fact, as we saw in the proof of Theorem 1.6, if $x = u(t)$, $y = v(t)$ is a trajectory of $\vec{V}(x,y)$, then $x = u(-t)$, $y = v(-t)$ is a trajectory of $-\vec{V}(x,y)$. Usually these distinctions are not reflected in our terminology; we rely on the context to make it clear in what sense the word "trajectory" is being used.

We make one further hypothesis in addition to those stated at the beginning of the chapter. *We suppose that the singular points of (1.1) are isolated, in other words, that every singular point is the center of a disk which contains no other singular point.* This hypothesis excludes cases which are trivial or pathological, and devoid of physical or mathematical interest.

EXERCISES

1. Let f and g be bounded. Show that every solution of (1.1) is defined for all t.

2. Let $M(x,y)$ be positive for all (x,y), and M_x, M_y continuous. Show that the system $\dot{x} = M(x,y)f(x,y)$, $\dot{y} = M(x,y)g(x,y)$ has the same trajectories as (1.1).

3. (a) Show that the system $\dot{x} = \dfrac{f}{1 + f^2 + g^2}$, $\dot{y} = \dfrac{g}{1 + f^2 + g^2}$ has the same trajectories as (1.1).

(b) Given a system (1.1), show that there is a system having the same trajectories whose solutions are defined for all t.

4. Identify the singular points of

$$\dot{x} = [(x - 1)^2 + y^2]y, \qquad \dot{y} = -[(x - 1)^2 + y^2]x$$

and sketch the trajectories. [Hint: Compare with Eq. (1.4).]

5. Compute the solutions of the following systems and sketch the trajectories.

(a) $\dot{x} = x,\ \dot{y} = y$ (b) $\dot{x} = -x,\ \dot{y} = -y$
(c) $\dot{x} = x,\ \dot{y} = -y$ (d) $\dot{x} = -x,\ \dot{y} = 2y$

6. Show that a system of the form $\dot{x} = f(x)$, $\dot{y} = g(y)$ has no orbits. (See Corollary 6.3, page 58.) Can a system of the form $\dot{x} = f(x)$, $\dot{y} = g(x,y)$ have any orbits?

2. Singular Points of Linear Systems

We consider a linear (homogeneous) system

$$\begin{aligned} \dot{x} &= ax + by \\ \dot{y} &= cx + dy. \end{aligned} \tag{2.1}$$

The hypothesis that the singular points of (2.1) are isolated implies that the matrix

$$A = \begin{bmatrix} a & b \\ c & d \end{bmatrix}$$

is invertible, or $ad - bc \neq 0$. It follows that the origin is the only singular point of (2.1). Our object is to classify linear systems according to the geometry of their trajectories.

Linear systems have a property which it is helpful to keep in mind, although we shall make no explicit use of it. This is, that the entire phase portrait is determined by the trajectories in an arbitrary neighborhood Ω of the origin. Let γ be an arc of a trajectory lying in Ω, let

k be a number, and let $k\gamma$ be the arc obtained by multiplying the points of γ by k. Since a multiple of a solution of (2.1) is again a solution, $k\gamma$ is also an arc of a trajectory. Since multiples of arcs lying in Ω fill out the whole plane, the assertion is proved. Note that if some arc γ subtends an angle of at least 180° at the origin, then γ by itself generates the phase portrait in this way.

By using the notation $z = (x,y)$, we can write (2.1) in the form

$$\dot{z} = Az. \tag{2.2}$$

Let P be a real invertible matrix, let $B = P^{-1}AP$, and consider the system

$$\dot{z} = Bz. \tag{2.3}$$

If $z = w(t)$ is a trajectory of (2.3), then $z = Pw(t)$ is obviously a trajectory of (2.2). Thus to each trajectory Γ of (2.3) corresponds the trajectory $P\Gamma$ of (2.2) which is obtained by multiplying the points of Γ by P. The phase portrait of (2.3) is therefore transformed into the phase portrait of (2.2) if the points of the plane are subjected to the linear transformation which maps z into Pz. It can be shown that such a transformation amounts to (at most) a rotation about the origin, or a reflection in a line through the origin, followed by a stretching of the plane away from (or compression toward) some line through the origin, and possibly another such stretching in a direction perpendicular to the first. Rotations and reflections are congruences and do not alter the geometry of the phase portrait at all. The effect of the transformation P is therefore at worst a distortion resulting from two mutually perpendicular expansions or contractions. (For another way of looking at the effect of P, see Exercise 3.) The properties of the phase portrait in which we are interested are preserved by such a distortion, and we may therefore replace (2.2) by (2.3), having chosen P so that B is as simple as possible.

Our procedure will be to find for a given matrix A the simplest real matrix B which is similar to A, and then to analyze the phase portrait of $\dot{z} = Bz$, which we call the *canonical form* of $\dot{z} = Az$. In each case we shall indicate the effect of the linear transformation P on the phase portrait.

We distinguish the following cases. Note that by hypothesis A does not have a zero eigenvalue.

 I. A has real distinct eigenvalues λ, μ.
  (a) λ and μ have the same sign.
  (b) λ and μ have opposite signs.

II. A has a repeated (real) eigenvalue.
    (a) A is diagonal.
    (b) A is not diagonal.
III. A has complex eigenvalues $\alpha \pm i\beta$.
    (a) $\alpha = 0$.
    (b) $\alpha \neq 0$.

In cases I(a), II, and III(b), we shall further have to take into account whether the eigenvalues have positive or negative real part.

We shall occasionally make use of the fact that the trajectories of $\dot{z} = -Bz$ differ from those of $\dot{z} = Bz$ only in their orientation.

I.   *A has real distinct eigenvalues* λ, μ. In this case, A is similar to the diagonal matrix

$$B = \begin{bmatrix} \lambda & 0 \\ 0 & \mu \end{bmatrix}.$$

The canonical form is thus

$$\dot{x} = \lambda x$$
$$\dot{y} = \mu y$$

and the trajectories are

$$x = c_1 e^{\lambda t}, \qquad y = c_2 e^{\mu t}. \tag{2.4}$$

(i)   Suppose λ and μ are negative. To be definite, let $\mu < \lambda < 0$. Note that every trajectory tends to the origin as $t \to \infty$.

Setting $c_1 = c_2 = 0$ in (2.4) yields the singular point.

If $c_1 = 0$ and $c_2 \neq 0$, we obtain the two rays $x = 0$, $y > 0$ (for $c_2 > 0$) and $x = 0$, $y < 0$ (for $c_2 < 0$).

If $c_1 \neq 0$ and $c_2 = 0$, we again obtain two rays, the positive and negative parts of the x-axis.

To describe the trajectories for which $c_1 \neq 0$ and $c_2 \neq 0$, it suffices to consider the trajectories in the first quadrant, since the others are obtained from them by reflection in the coordinate axes. Setting $c_1 > 0$, $c_2 > 0$ yields

$$\log \frac{x}{c_1} = \lambda t, \qquad \log \frac{y}{c_2} = \mu t,$$

or, eliminating t,

$$\frac{1}{\lambda} \log \frac{x}{c_1} = \frac{1}{\mu} \log \frac{y}{c_2}.$$

Hence

$$y = k x^{\mu/\lambda},$$

where $k = c_2 c_1^{-\mu/\lambda}$. This is an approximately parabolic arc. Since $\mu/\lambda > 1$, both y and dy/dx tend to zero as $x \to 0$. Thus every trajectory with $c_1 c_2 \neq 0$ approaches the origin tangentially to the x-axis.

The phase portrait is shown in Fig. 22(a). The distorted phase portrait of $\dot{z} = Az$ is shown in Fig. 22(b). The essential features are

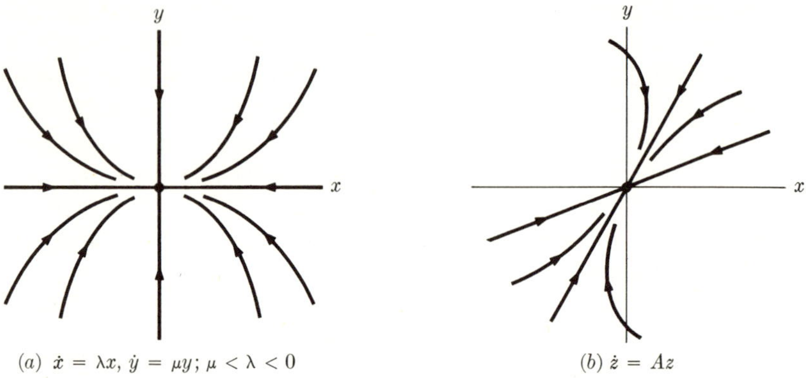

(a) $\dot{x} = \lambda x, \dot{y} = \mu y; \mu < \lambda < 0$ (b) $\dot{z} = Az$

Fig. 22.   Stable two-tangent node.

the same, but the special role of the coordinate axes may in general be assumed by any two lines through the origin.

The singular point in this case is called a *stable two-tangent node*.

(ii)   If λ and μ are positive, with $0 < \lambda < \mu$, the phase portrait looks exactly the same, except that now every trajectory tends to the origin as $t \to -\infty$, i.e., the orientation of all the trajectories is reversed (Fig. 23). The singular point is called an *unstable two-tangent node*.

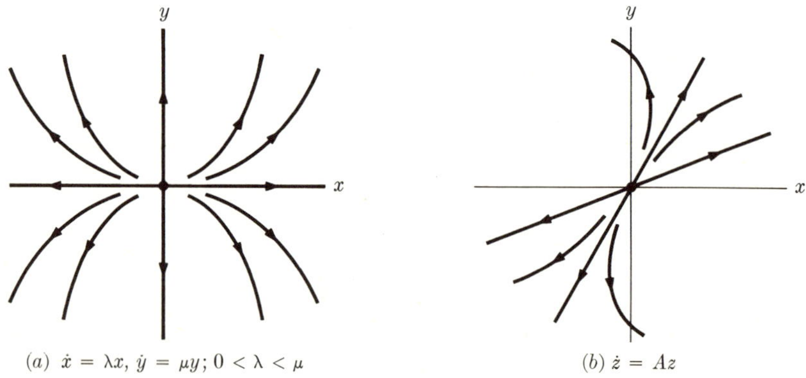

(a) $\dot{x} = \lambda x, \dot{y} = \mu y; 0 < \lambda < \mu$ (b) $\dot{z} = Az$

Fig. 23.   Unstable two-tangent node.

(iii)   We now consider the case that λ and μ have opposite signs. To be definite, let $\mu < 0 < \lambda$. We still have the four rays lying on the coordinate axes. However, the trajectories lying on the x-axis tend away from the origin, whereas the trajectories lying on the y-axis tend toward the origin. The remaining trajectories are again found by symmetry from the trajectories in the first quadrant. These are given as before by $y = kx^{\mu/\lambda}$. Now, however, $\mu/\lambda < 0$, so that $y \to \infty$ as $x \to 0$, and $y \to 0$ as $x \to \infty$. The trajectories are approximately hyperbolic, with the coordinate axes as asymptotes (Fig. 24(a)). Note that two trajectories approach the origin as $t \to \infty$, two approach the origin as $t \to -\infty$, whereas every other trajectory spends at most a finite length of time in any given neighborhood of the origin. The origin is called a *saddle point* in this case. The distorted phase portrait is shown in Fig. 24(b).

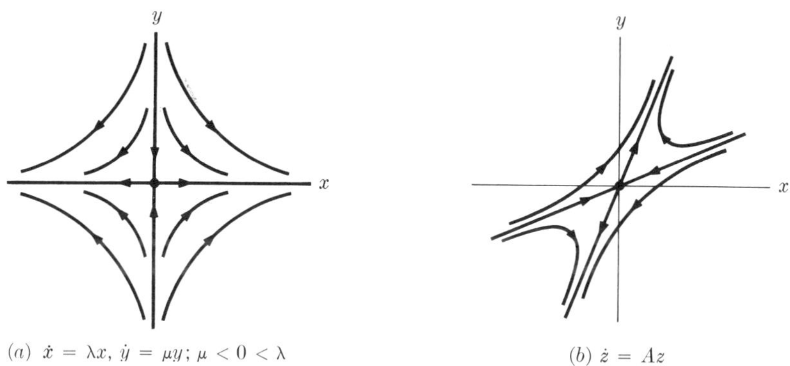

(a) $\dot{x} = \lambda x,\ \dot{y} = \mu y;\ \mu < 0 < \lambda$ $\qquad\qquad$ (b) $\dot{z} = Az$

FIG. 24.   Saddle point.

II.   *A has a repeated eigenvalue λ.*
(i)   If A is diagonal, the system $\dot{z} = Az$ is

$$\dot{x} = \lambda x$$
$$\dot{y} = \lambda y$$

and the trajectories are

$$x = c_1 e^{\lambda t}, \qquad y = c_2 e^{\lambda t}.$$

If $c_1 = 0$ and $c_2 \neq 0$, we obtain two rays on the y-axis. For $c_1 \neq 0$, the trajectories satisfy $y = (c_2/c_1)x$. Thus every trajectory other than the singular point is a ray. The rays tend toward or away from the origin according as $\lambda < 0$ or $\lambda > 0$. The origin is called a *stellar node*, stable if $\lambda < 0$ and unstable if $\lambda > 0$ (Fig. 25).

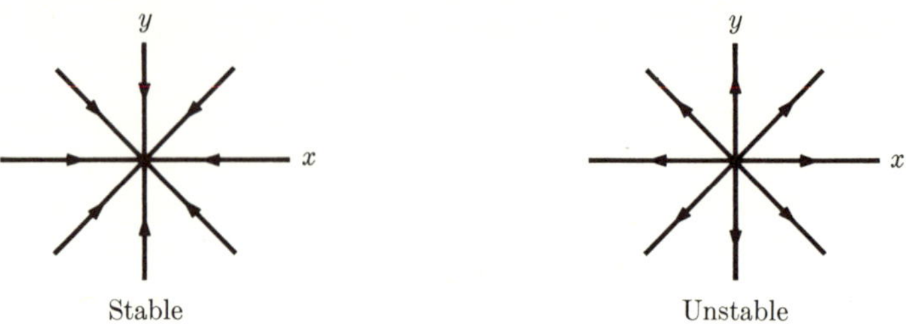

Fig. 25.   Stellar node.

(ii)   If A is not diagonal, then it is not similar to a diagonal matrix. For if $P^{-1}AP$ were diagonal, it would be equal to λI, whence $A = \lambda I$, contrary to hypothesis. We show that if A is not diagonal, then it is similar to the matrix

$$B = \begin{bmatrix} \lambda & \lambda \\ 0 & \lambda \end{bmatrix}.$$

Let Q be a real invertible matrix whose first column is an eigenvector of A belonging to λ. Then

$$Q^{-1}AQ = \begin{bmatrix} \lambda & c \\ 0 & \lambda \end{bmatrix},$$

where, by the preceding remark, $c \neq 0$. Let

$$R = \begin{bmatrix} 1 & 0 \\ 0 & \lambda/c \end{bmatrix}.$$

and let $P = QR$. Direct verification shows that $P^{-1}AP = B$.

The system $\dot{z} = Bz$ is now

$$\dot{x} = \lambda x + \lambda y$$
$$\dot{y} = \lambda y,$$

and therefore

$$x = c_1 e^{\lambda t} + c_2 \lambda t e^{\lambda t}$$
$$y = c_2 e^{\lambda t}.$$

Suppose first that $\lambda < 0$. Obviously all trajectories tend to the origin as $t \to \infty$. For $c_1 \neq 0$, $c_2 = 0$, we obtain two rays on the x-axis. If $c_2 \neq 0$,

$$\frac{y}{x} = \frac{c_2}{c_1 + c_2 \lambda t} \to 0$$

as $t \to \infty$. Hence every trajectory Γ approaches the origin tangentially to the x-axis. If $c_2 > 0$, Γ lies above the x-axis. In this case, $x < 0$ for t large, and Γ approaches the origin from the left. If $c_2 < 0$, Γ lies below the x-axis, $x > 0$ for t large, and Γ approaches the origin from the right. The origin is called a *stable one-tangent node* (Fig. 26).

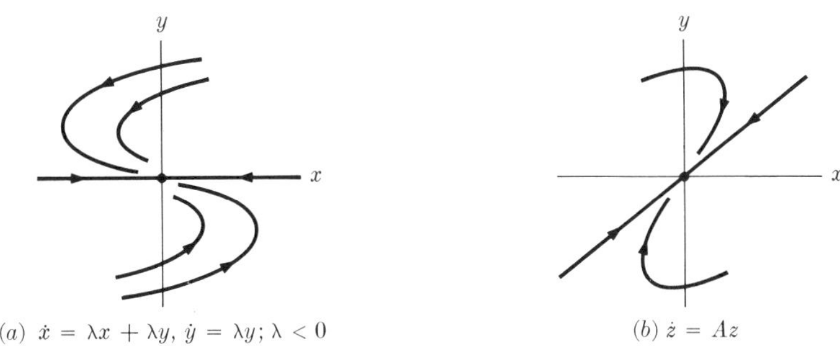

(a) $\dot{x} = \lambda x + \lambda y,\ \dot{y} = \lambda y;\ \lambda < 0$

(b) $\dot{z} = Az$

FIG. 26. Stable one-tangent node.

If $\lambda > 0$, the phase portrait differs only in the orientation of the trajectories, which now tend away from the origin. The origin is called an *unstable one-tangent node*.

The property common to all these types of nodes (one-tangent, two-tangent, and stellar) is that all trajectories approach the singular point, either as $t \to \infty$ or as $t \to -\infty$, and that every trajectory has a limiting direction at the singular point. This defining property of nodes can also be expressed by saying that any trajectory together with the singular point forms an arc which has a tangent at the singular point. The distinction between the three types of nodes arises when we consider the number of tangents which are obtained in this way. This number is one for a one-tangent node and two for a two-tangent node, whereas every line through the singular point is a tangent for a stellar node.

III. *A has complex eigenvalues* $\alpha \pm i\beta$. We begin by proving that A is similar to

$$B = \begin{bmatrix} \alpha & -\beta \\ \beta & \alpha \end{bmatrix}.$$

Let $\eta + i\mu$ be an eigenvector of A belonging to $\alpha + i\beta$, where η and μ are real vectors. It follows easily that $\eta - i\mu$ is an eigenvector belonging to $\alpha - i\beta$. Hence $\eta + i\mu$ and $\eta - i\mu$ are linearly independent, and therefore η and μ are also linearly independent. (The simple proof is left to the reader.) Let P be the matrix whose columns are μ and η. Clearly P is invertible. Equating real and imaginary parts in

$$A(\eta + i\mu) = (\alpha + i\beta)(\eta + i\mu)$$

yields

$$A\eta = \alpha\eta - \beta\mu$$
$$A\mu = \beta\eta + \alpha\mu.$$

Thus $(AP)^{(1)} = \beta\eta + \alpha\mu$ and $(AP)^{(2)} = \alpha\eta - \beta\mu$. A simple computation shows that these are also the columns of PB. Hence $AP = PB$, or $P^{-1}AP = B$.

We now consider the system $\dot{z} = Bz$,

$$\dot{x} = \alpha x - \beta y$$
$$\dot{y} = \beta x + \alpha y. \tag{2.5}$$

Instead of using the methods of Chapter 5 to solve this system, we take this opportunity to introduce a method of considerable importance for non-linear equations—the use of polar coordinates. Let

$$x = r \cos \theta, \qquad y = r \sin \theta, \tag{2.6}$$

where $r > 0$. (Polar coordinates are not defined at the origin, and r is always equal to $\sqrt{x^2 + y^2}$.) Differentiating yields

$$\dot{x} = \dot{r} \cos \theta - r \sin \theta \, \dot{\theta}$$
$$\dot{y} = \dot{r} \sin \theta + r \cos \theta \, \dot{\theta}.$$

Hence, solving for $\dot{r}$ and $\dot{\theta}$,

$$\dot{r} = \cos \theta \, \dot{x} + \sin \theta \, \dot{y}$$
$$\dot{\theta} = \frac{1}{r}(-\sin \theta \, \dot{x} + \cos \theta \, \dot{y}).$$

An arbitrary system

$$\dot{x} = f(x,y)$$
$$\dot{y} = g(x,y)$$

thus assumes, for $(x,y) \neq (0,0)$, the polar form

$$\dot{r} = f(r \cos \theta, r \sin \theta) \cos \theta + g(r \cos \theta, r \sin \theta) \sin \theta$$
$$\dot{\theta} = \frac{1}{r}[-f(r \cos \theta, r \sin \theta) \sin \theta + g(r \cos \theta, r \sin \theta) \cos \theta]. \tag{2.7}$$

The polar system expresses the velocity of a representative point at (r,θ) in terms of $\dot{r}$, the rate of change of its distance from the origin, and $\dot\theta$, its angular velocity about the origin. Note that $\dot r$ and $r\dot\theta$ are simply the components of $\vec V(x,y)$ along two mutually perpendicular lines through (x,y), of which the first is the line through the origin.

In the system (2.5),

$$f(x,y) = \alpha x - \beta y = \alpha r \cos\theta - \beta r \sin\theta$$
$$g(x,y) = \beta x + \alpha y = \beta r \cos\theta + \alpha r \sin\theta.$$

Substituting in (2.7) yields

$$\dot r = \alpha r$$
$$\dot\theta = \beta. \tag{2.8}$$

Thus

$$r = r_0 e^{\alpha t}, \qquad \theta = \theta_0 + \beta t. \tag{2.9}$$

By substituting in (2.6), we obtain the solutions of (2.5). There is no point in doing so, however, since the formulas (2.9) show the geometric character of the trajectories much more clearly.

The sign of β may obviously be chosen according to our convenience. (Changing the sign of β amounts to a reflection in a line through the origin.) If $\beta > 0$, $\theta \to \infty$ as $t \to \infty$, so that every trajectory winds infinitely many times around the origin, the direction of increasing t being counterclockwise. If $\beta < 0$, $\theta \to -\infty$ as $t \to \infty$, so that the direction of increasing t is clockwise. There are now several different cases.

(i)   If $\alpha = 0$, then r is constant. The trajectories are circles centered on the origin (Fig. 27(a)). Every trajectory of (2.5), other than the

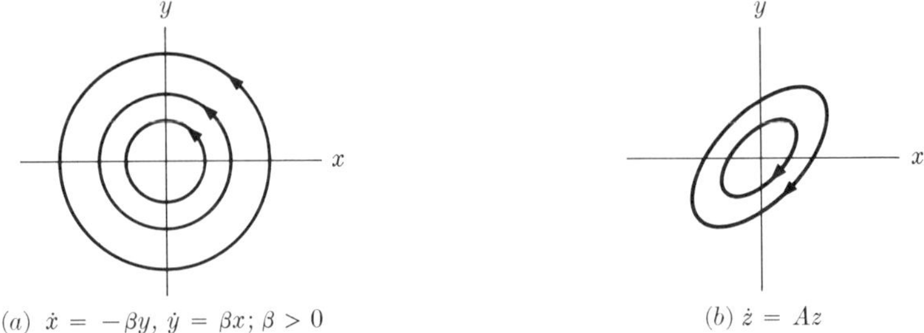

Fig. 27.   Center.

origin itself, is therefore an orbit. Note that all have period $2\pi/\beta$, the time for one revolution around the origin. The orbits of $\dot z = Az$ are ellipses centered on the origin (see Exercise 3). The phase portrait in the general case is indicated in Fig. 27(b).

A singular point surrounded by orbits is called a *center*. Note that a center is not approached by any other trajectory.

(ii) If $\alpha < 0$, then $r \to 0$ as $t \to \infty$. Hence the trajectories are spirals which approach the origin with increasing t. By eliminating t from (2.8), we find $dr/d\theta = (\alpha/\beta)r$, so that $r = ce^{(\alpha/\beta)\theta}$, the equation of a logarithmic spiral. The singular point is called a *stable focus*. It is distinguished from a stable node in that the trajectories wind onto the singular point, and therefore do not approach it with a limiting direction. The phase portraits are shown in Fig. 28.

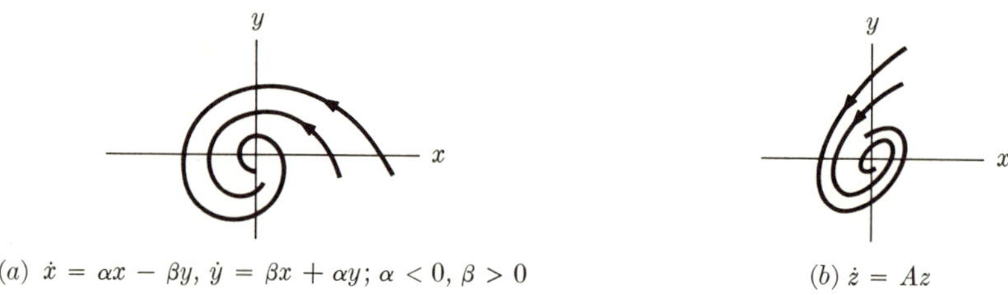

(a) $\dot{x} = \alpha x - \beta y,\ \dot{y} = \beta x + \alpha y;\ \alpha < 0,\ \beta > 0$ (b) $\dot{z} = Az$

FIG. 28. Stable focus.

(iii) If $\alpha > 0$, then $r \to 0$ as $t \to -\infty$. Hence the trajectories are spirals which approach the origin as $t \to -\infty$. The singular point is an *unstable focus*. In order to duplicate the phase portrait for the case $\alpha < 0$ in all but orientation, we must change the sign of β as well as of α (Fig. 29). If the sign of β is not changed, the sign of $dr/d\theta$ is

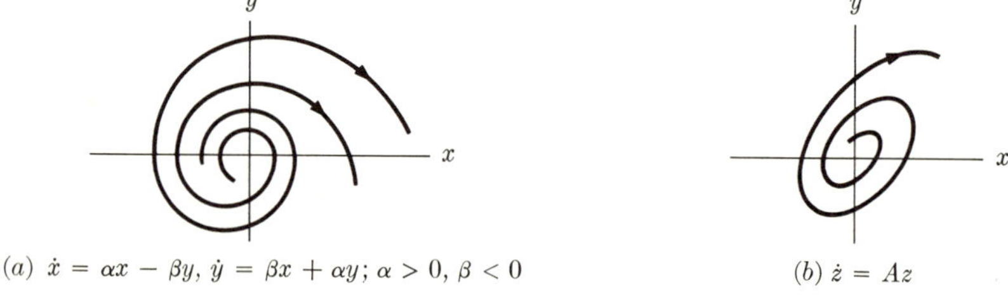

(a) $\dot{x} = \alpha x - \beta y,\ \dot{y} = \beta x + \alpha y;\ \alpha > 0,\ \beta < 0$ (b) $\dot{z} = Az$

FIG. 29. Unstable focus.

reversed, and this changes the sense in which the trajectories wind onto the origin.

EXERCISES

1.   Show that if A is invertible, a suitable translation of coordinates transforms the non-homogeneous system $\dot{z} = Az + b$ into a homogeneous system.

2.   The property of linear homogeneous systems that a multiple of a trajectory is a trajectory is shared by a larger class of systems, which are simply called homogeneous. Let m be a positive integer. A function $f(x,y)$ is said to be *homogeneous of degree* m if $f(kx,ky) = k^m f(x,y)$ for all (real) numbers k. We say that a system $\dot{x} = f(x,y)$, $\dot{y} = g(x,y)$ is homogeneous if f and g are homogeneous of the same degree. The degree of f and g is also called the degree of the system. Note that a linear homogeneous system has degree one. Prove:

(a)   A homogeneous system has a singular point at the origin, and (with the standard hypothesis that the singular points are isolated) no other singular point.

(b)   If u, v is a solution of a homogeneous system of degree m, so is $\bar{u}(t) = ku(k^{m-1}t)$, $\bar{v}(t) = kv(k^{m-1}t)$, for any number k.

(c)   If Γ is a trajectory of a homogeneous system, so is $k\Gamma$, for any number k.

3.   (a)   Let P be a real invertible matrix and C a circle with center at the origin. Show that PC is an ellipse with center at the origin. [Hint: Let $P^{-1} = \begin{bmatrix} \alpha & \beta \\ \gamma & \delta \end{bmatrix}$, so that $P^{-1}(x,y) = (\alpha x + \beta y,\ \gamma x + \delta y)$. Note that PC consists of all points (x,y) for which $P^{-1}(x,y)$ is a point of C.]

(b)   In the same way, show that if C is a line through the origin, so is PC.

4.   The vector $\vec{V}(x,y)$ assigned by Eq. (2.1) to the point (x,y) is $(ax + by,\ cx + dy)$. Use the invertibility of A to show that no two points are assigned the same vector, and that given a vector (v_1,v_2) there is a point (x,y) for which $\vec{V}(x,y) = (v_1,v_2)$.

5.   Show that the origin is a saddle point if $\det A < 0$, and a node, focus or center if $\det A > 0$.

6.   Sketch the trajectories of the following systems. Begin by identifying the nature of the singular point. Locate the rectilinear trajectories, if any, by setting $y = mx$ and solving for m. (Test $x \equiv 0$ separately.) Another method for computing the rectilinear trajectories is given in Exercise 8. An inspection of the vector field yields the orientation of the trajectories and is often helpful in other ways.

(a) $\dot{x} = x + y, \qquad \dot{y} = 3x + y$
(b) $\dot{x} = y, \qquad \dot{y} = x + y$

(c) $\dot{x} = y, \qquad \dot{y} = x$
(d) $\dot{x} = y, \qquad \dot{y} = -2x$
(e) $\dot{x} = 2x, \qquad \dot{y} = x + y$
(f) $\dot{x} = x - 2y, \qquad \dot{y} = 4x - 3y$
(g) $\dot{x} = x + 2y, \qquad \dot{y} = -4x - 3y$
(h) $\dot{x} = -3x + 2y, \qquad \dot{y} = -2x + y$
(i) $\dot{x} = x - y, \qquad \dot{y} = 2x - y$
(j) $\dot{x} = 2x + y, \qquad \dot{y} = 3x + 4y$
(k) $\dot{x} = -y, \qquad \dot{y} = x + y$

7.   Classify the singular points of the following equations and sketch the trajectories.

(a) $\ddot{x} - \dot{x} + x = 0$ (b) $\ddot{x} + \dot{x} + x = 0$
(c) $\ddot{x} + x = 0$ (d) $\ddot{x} + \dot{x} - x = 0$
(e) $\ddot{x} + 2\dot{x} + x = 0$ (f) $\ddot{x} - 3\dot{x} + 2x = 0$

8.   (a)   Let A have real eigenvalues, and let $\eta = (\eta_1, \eta_2)$ be a real eigenvector of A. Show that the ray through the point (η_1, η_2) is a trajectory of $\dot{z} = Az$. (It is understood that det $A \neq 0$.)

    (b)   Conversely, given that the ray through some point (η_1, η_2) is a trajectory, show that η is an eigenvector of A.

3. Elementary Singular Points; Stability

Introduction.   A common problem in differential equations is to determine the behavior of the solutions near a given constant solution. Physically, a constant solution represents an equilibrium, and many problems of science and technology can be reduced to the question of the stability or instability of an equilibrium. In the context of such problems, only those solutions are of interest whose initial values lie near the equilibrium, and the question arises whether these solutions remain near the equilibrium, and possibly even tend toward it in the course of time. Let us make these ideas precise for the case of a second-order autonomous system

$$\begin{aligned} \dot{x} &= f(x,y) \\ \dot{y} &= g(x,y). \end{aligned} \tag{3.1}$$

Definition.   Let (x_0, y_0) be a singular point of (3.1). We say that (x_0, y_0) is *stable* if for every $\epsilon > 0$ there is a $\delta > 0$ such that every solution u, v which satisfies

$$[u(t_0) - x_0]^2 + [v(t_0) - y_0]^2 < \delta^2 \qquad (3.2)$$

for some t_0, also satisfies

$$[u(t) - x_0]^2 + [v(t) - y_0]^2 < \epsilon^2$$

for all $t > t_0$. Geometrically, this means that, given a disk D_ϵ of radius ϵ about (x_0,y_0), there is a disk D_δ so that every trajectory which has a point in D_δ remains thereafter in D_ϵ. If (x_0,y_0) is not stable, we say it is *unstable*. If (x_0,y_0) is stable, and if δ can be chosen so that (3.2) implies that $(u(t),v(t)) \to (x_0,y_0)$ as $t \to \infty$, we say that (x_0,y_0) is *asymptotically stable*. A singular point which is stable but not asymptotically stable will be called *weakly stable*.

Note that the singular point of a linear system $\dot{z} = Az$, with $\det A \neq 0$, is stable if no eigenvalue of A has positive real part, asymptotically stable if both eigenvalues have negative real part, and weakly stable if the eigenvalues are purely imaginary. *Thus stable nodes and foci are asymptotically stable, saddle points and unstable nodes and foci are unstable, and centers are weakly stable.*

It should be mentioned that stability is not always desirable in an equilibrium. Clocks and certain electronic circuits, for example, are designed to oscillate, and it is frequently advantageous that they attain their oscillatory state spontaneously, i.e., that any equilibrium be unstable. In such a system it is the oscillatory state that must be asymptotically stable. (See the remarks on limit cycles later in this chapter.)

Another useful concept is that of an *attractor*. We say that a singular point is an attractor if all nearby trajectories tend to it as $t \to \infty$ or as $t \to -\infty$. In other words, a singular point of (3.1) is an attractor if either it is asymptotically stable, or if it is an asymptotically stable singular point of $\dot{x} = -f(x,y)$, $\dot{y} = -g(x,y)$. Thus all nodes and foci are attractors.

The behavior of the trajectories near a singular point is of considerable interest for the phase portrait as a whole. Indeed, to obtain the phase portrait, one usually begins with the neighborhoods of the singular points, much as in graphing a function one begins with its critical points. While the phase portrait near a singular point of a non-linear system may closely resemble one of the linear cases, it may also be quite different from any of these. Furthermore, even if the singular point is a focus, for example, it may be extremely difficult to establish this fact. Even the simpler question, whether a singular point is stable or unstable, may be very difficult to answer. These difficulties can be largely overcome in the case of an elementary singular point.

Definition.   Let (x_0, y_0) be a singular point of (3.1). Let

$$
\begin{aligned}
a &= f_x(x_0, y_0), \qquad b = f_y(x_0, y_0) \\
c &= g_x(x_0, y_0), \qquad d = g_y(x_0, y_0).
\end{aligned}
\tag{3.3}
$$

We say that (x_0, y_0) is an *elementary singular point* if the matrix

$$
A = \begin{bmatrix} a & b \\ c & d \end{bmatrix}
$$

is invertible, or $ad - bc \neq 0$.

We shall see that the trajectories near an elementary singular point behave in most cases like the trajectories of $\dot{z} = Az$ near $z = 0$. The equation $\dot{z} = Az$ is called the *linear approximation* to (3.1) at (x_0, y_0). We now investigate the behavior of the trajectories of (3.1) near a given elementary singular point (x_0, y_0).

Preliminary Transformations.   We begin by introducing rectangular coordinates with origin at the singular point. Let

$$
\chi = x - x_0, \qquad \eta = y - y_0
$$

so that (3.1) becomes

$$
\begin{aligned}
\dot{\chi} &= f(x_0 + \chi, \, y_0 + \eta) \\
\dot{\eta} &= g(x_0 + \chi, \, y_0 + \eta).
\end{aligned}
\tag{3.4}
$$

Using the fact that f and g have continuous partial derivatives at (x_0, y_0), we have

$$
\begin{aligned}
f(x_0 + \chi, \, y_0 + \eta) &= f(x_0, y_0) + f_x(x_0, y_0)\chi + f_y(x_0, y_0)\eta + R_1(\chi, \eta) \\
g(x_0 + \chi, \, y_0 + \eta) &= g(x_0, y_0) + g_x(x_0, y_0)\chi + g_y(x_0, y_0)\eta + R_2(\chi, \eta),
\end{aligned}
$$

where

$$
\frac{R_i(\chi, \eta)}{\sqrt{\chi^2 + \eta^2}} \to 0 \quad \text{as} \quad \sqrt{\chi^2 + \eta^2} \to 0, \qquad i = 1, 2.
\tag{3.5}
$$

Employing the notation (3.3), and noting that $f(x_0,y_0) = g(x_0,y_0) = 0$, we can therefore write (3.4) in the form

$$\dot{\chi} = a\chi + b\eta + R_1(\chi,\eta)$$
$$\dot{\eta} = c\chi + d\eta + R_2(\chi,\eta).$$

In the new coordinates the singular point is at the origin.

Following convention, we now change notation and write x in place of χ and y in place of η:

$$\dot{x} = ax + by + R_1(x,y)$$
$$\dot{y} = cx + dy + R_2(x,y). \tag{3.6}$$

In the new notation, (3.5) becomes

$$\lim_{r \to 0} \frac{R_i(x,y)}{r} = 0, \qquad i = 1,\, 2, \tag{3.7}$$

where $r = \sqrt{x^2 + y^2}$. The significance of (3.7) is that in the neighborhood of the origin the non-linear terms R_1, R_2 are small compared to r. In conjunction with the hypothesis that A is invertible, this implies that near the origin the vector $(R_1(x,y), R_2(x,y))$ is small in comparison to the vector $(ax + by,\, cx + dy)$, so that the vector

$$(ax + by + R_1,\, cx + dy + R_2)$$

has nearly the same magnitude and direction as the vector obtained by dropping the non-linear terms. It is therefore reasonable to expect that the non-linear terms do not greatly affect the trajectories near the origin. It is easy to prove that elementary singular points are isolated (Exercise 1).

In vector notation, (3.6) assumes the form

$$\dot{z} = Az + R(z), \tag{3.8}$$

where $R(z) = (R_1(z), R_2(z)) = (R_1(x,y), R_2(x,y))$. Instead of investigating the trajectories near $z = 0$ directly, we proceed as in Section 2 and replace (3.8) by a simpler system whose trajectories are related to those of (3.8) by a linear transformation P. An easy verification shows that if $w(t)$ is a solution of

$$\dot{z} = P^{-1}APz + P^{-1}R(Pz), \tag{3.9}$$

then $Pw(t)$ is a solution of (3.8). We will choose P so that $P^{-1}AP$ assumes one of the standard forms of Section 2, and then investigate the trajectories of (3.9) near $z = 0$. For a given P, we write

$$B = P^{-1}AP, \qquad Q(z) = P^{-1}R(Pz)$$

so that (3.9) becomes

$$\dot{z} = Bz + Q(z). \tag{3.10}$$

It is essential to know that the functions $Q_i(x,y)$ satisfy

$$\lim_{r \to 0} \frac{Q_i(x,y)}{r} = 0, \qquad i = 1, 2. \tag{3.11}$$

The proof of this fact is left to the reader (Exercise 2).

We shall refer to (3.10) as the canonical form of (3.1) relative to the singular point (x_0, y_0).

I.   *A has complex eigenvalues $\alpha \pm i\beta$, $\alpha \neq 0$.*   The canonical form for this case is

$$\begin{aligned}
\dot{x} &= \alpha x - \beta y + Q_1(x,y) \\
\dot{y} &= \beta x + \alpha y + Q_2(x,y),
\end{aligned} \tag{3.12}$$

where we suppose $\beta > 0$. For $(x,y) \neq (0,0)$, we transform this system to polar coordinates. Using Eq. (2.7), we readily find that

$$\begin{aligned}
\dot{r} &= \alpha r + F_1(r,\theta) \\
\dot{\theta} &= \beta + \frac{1}{r} F_2(r,\theta),
\end{aligned} \tag{3.13}$$

where

$$\begin{aligned}
F_1(r,\theta) &= Q_1(r \cos\theta, r \sin\theta) \cos\theta + Q_2(r \cos\theta, r \sin\theta) \sin\theta \\
F_2(r,\theta) &= -Q_1(r \cos\theta, r \sin\theta) \sin\theta + Q_2(r \cos\theta, r \sin\theta) \cos\theta.
\end{aligned} \tag{3.14}$$

Since $\displaystyle\lim_{r \to 0} \frac{Q_i(r \cos\theta, r \sin\theta)}{r} = 0$, it follows that $\displaystyle\lim_{r \to 0} \frac{F_i(r,\theta)}{r} = 0$. We may therefore choose a $\delta > 0$ so that $r \leq \delta$ implies that $|F_1(r,\theta)| \leq (|\alpha|/2)r$ and $|F_2(r,\theta)| \leq (\beta/2)r$. We consider first the case that $\alpha < 0$. Then

$$\begin{aligned}
\dot{r} &\leq \alpha r - \frac{\alpha}{2} r = \frac{\alpha}{2} r \\
\dot{\theta} &\geq \beta - \frac{\beta}{2} = \frac{\beta}{2},
\end{aligned} \tag{3.15}$$

provided $r \leq \delta$. If $r = r(t)$, $\theta = \theta(t)$ is a trajectory which satisfies $r(0) \leq \delta$, it follows that $r(t) \leq \delta$ for $t \geq 0$. Integrating the inequalities

$\dot{r}(t) \leq (\alpha/2)r(t)$, $\dot{\theta}(t) \geq \beta/2$ yields

$$r(t) \leq r(0)e^{(\alpha/2)t}$$

$$\theta(t) \geq \theta(0) + \frac{\beta}{2}t \tag{3.16}$$

for $t > 0$. Hence $r(t) \to 0$ and $\theta(t) \to \infty$ as $t \to \infty$. We conclude that every trajectory in the disk $r \leq \delta$ spirals toward the origin, which is thus a stable focus.

If $\alpha > 0$, the inequality for $F_1(r,\theta)$ yields

$$\dot{r} \geq \alpha r - \frac{\alpha}{2}r = \frac{\alpha}{2}r \tag{3.17}$$

so that

$$r(t) \geq r(0)e^{(\alpha/2)t}$$

for $t > 0$ and $r(t) \leq \delta$. Hence $r(t)$ increases with t and every trajectory (the singular point excepted) eventually leaves the disk $r \leq \delta$. The origin is therefore unstable. Integrating (3.17) from 0 to $t < 0$ yields

$$r(t) \leq r(0)e^{(\alpha/2)t}, \qquad t < 0$$

so that $r(t) \to 0$ as $t \to -\infty$. Similarly, the inequality $\dot{\theta} \geq \beta/2$ yields

$$\theta(t) \leq \theta(0) + \frac{\beta}{2}t, \qquad t < 0$$

so that $\theta(t) \to -\infty$ as $t \to -\infty$. The origin is therefore a focus in this case also. We summarize the results.

Theorem 3.1. *If A has complex eigenvalues $\alpha \pm i\beta$ with $\alpha \neq 0$, the singular point is a focus, stable if $\alpha < 0$ and unstable if $\alpha > 0$.*

The case $\alpha = 0$ will be taken up later.

II. *A has real, distinct, eigenvalues λ, μ of the same sign.* The canonical form is now

$$\dot{x} = \lambda x + Q_1(x,y)$$
$$\dot{y} = \mu y + Q_2(x,y),$$

or, in polar coordinates,

$$\dot{r} = (\lambda \cos^2 \theta + \mu \sin^2 \theta)r + F_1(r,\theta)$$

$$\dot{\theta} = \frac{\mu - \lambda}{2} \sin 2\theta + \frac{1}{r} F_2(r,\theta), \tag{3.18}$$

where F_1, F_2 are given by (3.14).

We consider first the equation for $\dot{r}$. Let λ and μ be negative, say $\mu < \lambda < 0$. Then

$$\lambda \cos^2 \theta + \mu \sin^2 \theta \leq \lambda \cos^2 \theta + \lambda \sin^2 \theta = \lambda,$$

so

$$\dot{r} \leq \lambda r + F_1(r,\theta).$$

Choose $\delta > 0$ so that $r \leq \delta$ implies $|F_1(r,\theta)| \leq (|\lambda|/2)r$. Then

$$\dot{r} \leq \frac{\lambda}{2} r$$

and

$$r(t) \leq r(0)e^{(\lambda/2)t}$$

for $t > 0$ and $r(t) \leq \delta$. Since $\lambda < 0$, it follows as in case I that $r(t) \to 0$ as $t \to \infty$. We conclude that if A has distinct negative eigenvalues, the singular point is asymptotically stable.

If λ and μ are positive, say $0 < \lambda < \mu$, then, choosing δ as before, and noting that

$$\lambda \cos^2 \theta + \mu \sin^2 \theta \geq \lambda \cos^2 \theta + \lambda \sin^2 \theta = \lambda,$$

we have

$$\dot{r} \geq \lambda r + F_1(r,\theta) \geq \lambda r - \frac{\lambda}{2} r,$$

or

$$\dot{r} \geq \frac{\lambda}{2} r$$

for $r \leq \delta$. It follows exactly as with (3.17) that the origin is unstable, and that every trajectory in the disk $r \leq \delta$ tends to the origin as $t \to -\infty$.

In order to investigate the manner in which the trajectories approach the origin as $t \to \infty$ (or $-\infty$, in the unstable case), we use the equation for $\dot{\theta}$. We now choose $\delta > 0$ so that $|F_2(r,\theta)| \leq (|\mu - \lambda|/4)r$ for $r \leq \delta$.

Suppose that $\mu < \lambda < 0$, so that $\mu - \lambda < 0$. Then for $r \leq \delta$,

$$\dot{\theta} \leq \frac{\mu - \lambda}{2} \sin 2\theta - \frac{\mu - \lambda}{4},$$

or

$$\dot{\theta} \leq \frac{\mu - \lambda}{2} (\sin 2\theta - \tfrac{1}{2}). \tag{3.19}$$

Similarly,

$$\dot{\theta} \geq \frac{\mu - \lambda}{2} (\sin 2\theta + \tfrac{1}{2}). \tag{3.20}$$

If $\theta = \pi/4$ or $\theta = 5\pi/4$, $\sin 2\theta = 1$, and (3.19) yields $\dot\theta < 0$. If $\theta = 3\pi/4$ or $\theta = 7\pi/4$, $\sin 2\theta = -1$, and (3.20) yields $\dot\theta > 0$. Hence a trajectory which intersects the line $y = x$ does so in the direction of decreasing θ, whereas a trajectory which intersects the line $y = -x$ does so in the direction of increasing θ (see Fig. 30). Therefore, if a trajectory enters either the sector $-\pi/4 < \theta < \pi/4$ or the sector $5\pi/4 < \theta < 3\pi/4$, then it must thereafter remain in that sector. It follows that no trajectory winds around the origin, and we are justified in calling the origin a node. More detailed analysis shows that all but two trajectories approach the origin tangentially to the x-axis, while the exceptional trajectories approach the origin tangentially to the y-axis. The origin is thus a stable two-tangent node.

Finally, if $0 < \lambda < \mu$, the direction of the trajectories is reversed, and the origin is an unstable node. We therefore have the following result.

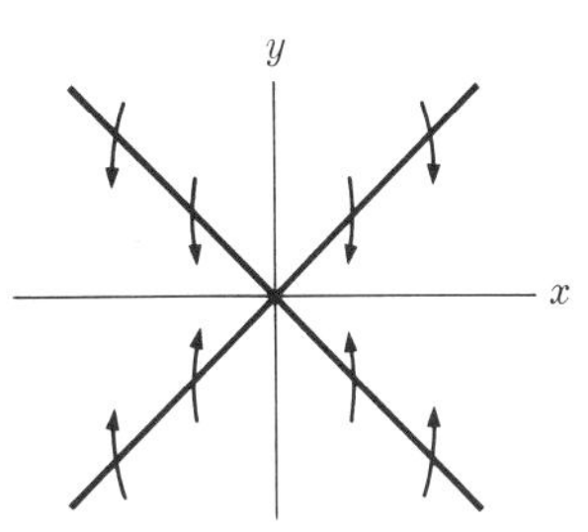

Fig. 30. See proof of Theorem 3.2.

Theorem 3.2. *If A has real distinct eigenvalues of the same sign, the singular point is a node, stable or unstable according as the eigenvalues are negative or positive.*

III. *A has a double eigenvalue λ.* The canonical form is

$$\begin{aligned} \dot x &= \lambda x + Q_1(x,y) \\ \dot y &= \lambda y + Q_2(x,y), \end{aligned} \tag{3.21}$$

or

$$\begin{aligned} \dot x &= \lambda x + \lambda y + Q_1(x,y) \\ \dot y &= \lambda y + Q_2(x,y), \end{aligned} \tag{3.22}$$

depending on whether A is, or is not, diagonal. The polar equations are

$$\begin{aligned} \dot r &= \lambda r + F_1(r,\theta) \\ \dot\theta &= \frac{1}{r} F_2(r,\theta) \end{aligned} \tag{3.23}$$

in the first case, and

$$\begin{aligned} \dot r &= \lambda(1 + \tfrac{1}{2}\sin 2\theta)r + F_1(r,\theta) \\ \dot\theta &= -\lambda \sin^2\theta + \frac{1}{r} F_2(r,\theta) \end{aligned} \tag{3.24}$$

in the second.

It is obvious from the first of Eqs. (3.23) that the origin is an attractor, stable or unstable according as λ is negative or positive. The same follows also from the first of Eqs. (3.24). Notice that

$$1 + \tfrac{1}{2} \sin 2\theta \geqq \tfrac{1}{2}.$$

Thus, if $\lambda < 0$,

$$\dot{r} \leqq \frac{\lambda}{2} r + F_1(r,\theta),$$

and if $\lambda > 0$,

$$\dot{r} \geqq \frac{\lambda}{2} r + F_1(r,\theta).$$

A familiar argument therefore leads to the desired conclusion.

Neither of the $\dot{\theta}$ equations allows us to decide whether the origin is a focus or a node. It is possible to show by examples that it may indeed be either, depending on F_1 and F_2, and thus, ultimately, on R_1 and R_2. In the case of a double eigenvalue, the linear approximation therefore does not decide whether the singular point is a focus or a node. However, one can prove that if $f(x,y)$ and $g(x,y)$ have continuous partial derivatives of the second order, then the origin is a stellar node for (3.21) and a one-tangent node for (3.22).

We summarize these facts as follows.

Theorem 3.3.   *If A has a double eigenvalue λ, the singular point is an attractor, stable or unstable according as λ is negative or positive. If, furthermore, f and g have continuous second partial derivatives, the singular point is a node.*

IV.   *A has real eigenvalues λ, μ of opposite sign.*   The canonical form is

$$\begin{aligned}
\dot{x} &= \lambda x + Q_1(x,y) \\
\dot{y} &= \mu y + Q_2(x,y),
\end{aligned} \tag{3.25}$$

where we assume $\mu < 0 < \lambda$. Recall that the origin is a saddle point for the linear approximation. As we shall see, it is also a saddle point for (3.25). Instead of using the polar equations, we investigate (3.25) directly.

Let γ be the smaller of the two positive numbers λ and $-\mu$. Let $m > 1$, and choose $\delta > 0$ so that $x^2 + y^2 \leqq \delta^2$ implies

$$|Q_i(x,y)| \leqq \frac{\gamma}{2\sqrt{1+m^2}} \sqrt{x^2 + y^2}, \qquad i = 1, 2.$$

Figure 31 shows the circle $x^2 + y^2 = \delta^2$ and the four lines $y = \pm mx$, $x = \pm my$, whose intersections with the circle are labeled A, B, C, D, ... We begin by determining the direction of the velocity vector

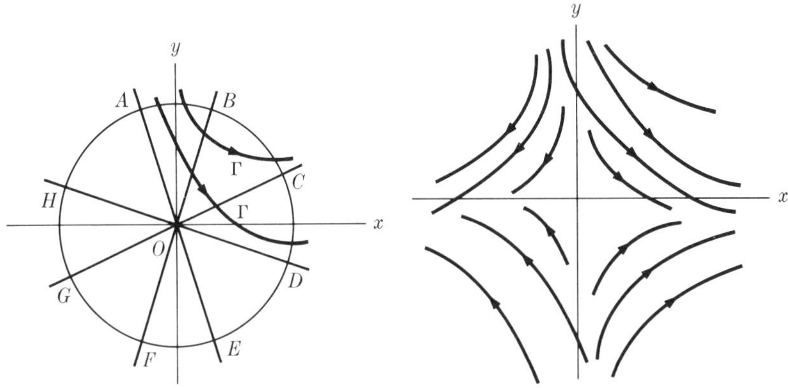

FIG. 31. See proof of
Theorem 3.4.

FIG. 32. See proof of
Theorem 3.4.

$\vec{V} = (\dot{x}, \dot{y})$ in each of the sectors into which the rays OA, OB, etc., divide the disk Ω: $x^2 + y^2 \leqq \delta^2$. The two rays and the circular arc which bound such a sector are understood to belong to the sector, but the vertex 0 is always excepted.

(1) $\dot{x} > 0$ *in the sector* OBE. Everywhere in Ω,

$$\dot{x} = \lambda x + Q_1(x,y) \geqq \lambda x - \frac{\lambda}{2\sqrt{1 + m^2}} \sqrt{x^2 + y^2}.$$

If (x,y) lies in OBE, then $x > 0$ and $y^2 \leqq m^2 x^2$. Hence

$$\sqrt{x^2 + y^2} \leqq \sqrt{x^2 + m^2 x^2} = x\sqrt{1 + m^2},$$

and therefore

$$\dot{x} \geqq \lambda x - \frac{\lambda}{2\sqrt{1 + m^2}} x\sqrt{1 + m^2},$$

or

$$\dot{x} \geqq \frac{\lambda}{2} x > 0.$$

(2) $\dot{y} < 0$ *in the sector OHC.*   Everywhere in Ω,

$$\dot{y} = \mu y + Q_2(x,y) \leqq \mu y - \frac{\mu}{2\sqrt{1 + m^2}}\sqrt{x^2 + y^2}.$$

If (x,y) lies in OHC, then $y > 0$ and $x^2 \leqq m^2 y^2$. Thus $\sqrt{x^2 + y^2} \leqq y\sqrt{1 + m^2}$ and

$$\dot{y} \leqq \mu y - \frac{\mu}{2}y < 0.$$

In the same way, we find that

(3) $\dot{y} > 0$ *in the sector OGD.*
(4) $\dot{x} < 0$ *in the sector OAF.*

Consider now a trajectory $\Gamma: x = u(t)$, $y = v(t)$ which meets the ray OB when $t = 0$. It follows from (1) and (2) that at each point of OB, $\vec{V}$ points into the interior of the sector OBC, so that Γ enters this sector for $t > 0$. Again by (1) and (2), $u(t)$ is increasing and $v(t)$ is decreasing in OBC. Thus Γ must leave the sector, either by crossing the arc BC, in which case it leaves Ω, or by crossing the ray OC. In the latter case, it enters the sector OCD. In this sector, also, $u(t)$ is increasing. Furthermore, on the ray OD, $\vec{V}$ points into the sector OCD, as follows from (1) and (3). Since $\vec{V}$ points into OCD also along the ray OC, Γ must leave the sector by crossing the arc CD. In either case, therefore, Γ leaves Ω with increasing t. This already shows that the origin is unstable.

For $t < 0$, Γ enters the sector OAB. In this sector $\dot{y} < 0$, so that, with decreasing t, $v(t)$ increases. Since $\vec{V}$ points out of the sector on both bounding rays, $-\vec{V}$ points into the sector. Thus, as t decreases, Γ remains in the sector while steadily moving upward, thus eventually crossing the arc AB. Thus Γ leaves Ω also with decreasing t.

Identical reasoning applied to a trajectory which intersects one of the other rays shows that any such trajectory must both enter and leave Ω, and yields the configuration shown in Fig. 32. More refined considerations show that exactly two trajectories approach the origin tangentially to the y-axis as $t \to \infty$, and exactly two approach the origin tangentially to the x-axis as $t \to -\infty$. We thus have the following result.

Theorem 3.4.   *If A has real eigenvalues of opposite sign, the singular point is a saddle point.*

V. *A has purely imaginary eigenvalues $\pm i\beta$.* The canonical form and the corresponding polar equations are obtained by setting $\alpha = 0$ in (3.12) and (3.13):

$$\begin{aligned}
\dot{x} &= -\beta y + Q_1(x,y) \\
\dot{y} &= \beta x + Q_2(x,y)
\end{aligned} \tag{3.26}$$

and

$$\begin{aligned}
\dot{r} &= F_1(r,\theta) \\
\dot{\theta} &= \beta + \frac{1}{r} F_2(r,\theta).
\end{aligned} \tag{3.27}$$

The equation for $\dot{\theta}$ is therefore the same as in (3.13). We shall presently prove that, as in case I, all trajectories near the origin wind around it. However, the equation for $\dot{r}$ no longer allows us to draw any conclusions about the behavior of r. In fact, even if f and g are polynomials— to take the simplest case—the origin may be a stable focus, an unstable focus, or a center, depending on what the polynomials are. The reason is, of course, that an arbitrarily small disturbance (Q_1,Q_2) of the linear vector field $(-\beta y, \beta x)$ can cause a trajectory, after one revolution around the origin, to just miss its starting point, and thereby change the center into a stable or unstable focus. Thus the linear approximation does not decide whether the singular point is stable or unstable. Before proving that, at any rate, the trajectories wind around the origin, we present two illustrative examples.

EXAMPLE 1.
$$\begin{aligned}
\dot{x} &= y + x^3 + xy^2 \\
\dot{y} &= -x + x^2 y + y^3.
\end{aligned} \tag{3.28}$$

The polar equations are

$$\begin{aligned}
\dot{r} &= r^3 \\
\dot{\theta} &= -1,
\end{aligned}$$

and we readily find that

$$r = \frac{1}{\sqrt{c_1 - 2t}}$$

$$\theta = c_2 - t.$$

The origin is therefore an unstable focus. $\square$

EXAMPLE 2.   We pass directly to polar coordinates. Consider the system

$$\dot{r} = r^3 \sin \frac{1}{r^2} \tag{3.29}$$

$$\dot{\theta} = 1.$$

Obviously, $\theta = t + c \to \infty$ as $t \to \infty$. Setting $r^2 = 1/n\pi$ yields a solution of the $\dot{r}$ equation, so that the system has infinitely many orbits $r = 1/\sqrt{n\pi}$, n a positive integer. These orbits form a sequence of circles shrinking onto the origin. In the annulus between two consecutive orbits $r = 1/\sqrt{n\pi}$ and $r = 1/\sqrt{(n+1)\pi}$, $\dot{r}$ is of one sign. If n is even, $\dot{r} > 0$, and if n is odd, $\dot{r} < 0$, in the annulus. It is easy to see that the trajectories in the annulus are spirals which wind onto the even-numbered orbit and unwind from the odd-numbered orbit (Fig. 33). Since every neighborhood of the origin contains orbits as well as spirals, the origin is neither a center nor a focus. A singular point of this kind is usually called a *center-focus*. It is clearly weakly stable.

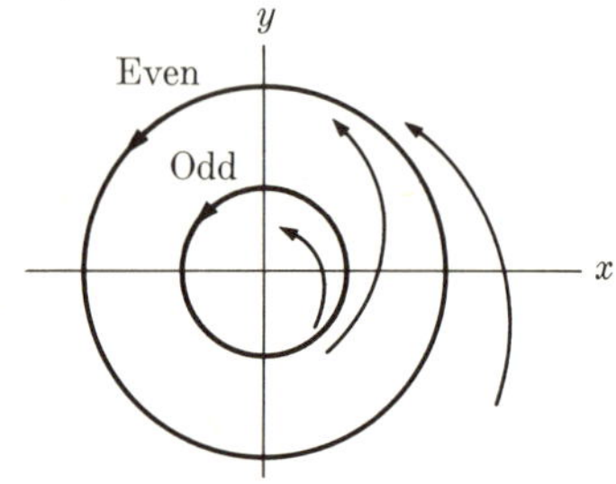

FIG. 33.   See Example 2.

The most significant feature of this example is the existence of isolated orbits. The fact that a non-linear autonomous system can have an isolated orbit—i.e., an orbit such that no nearby trajectory is also an orbit—is essential to the explanation of a host of physical phenomena. Clearly a linear system cannot have such an orbit. The trajectories neighboring an isolated orbit on either side are necessarily spirals winding onto or away from the orbit. If the trajectories on both sides spiral onto the orbit with increasing t, the orbit is said to be asymptotically stable. Isolated orbits are also called *limit cycles*. The operation of an ordinary watch or clock—most obviously, a pendulum clock—is an example of a second-order autonomous equation with a stable limit cycle. A simple example of a system with a unique orbit is given in Exercise 3.   □

We now prove that a trajectory of (3.27) which comes sufficiently close to the origin winds infinitely often around the origin. The first part of the proof consists in showing that such a trajectory winds at least once around the origin. We suppose that $\beta > 0$.

Choose $\delta > 0$ so that $r \leqq \delta$ implies $|F_1(r,\theta)| \leqq (\beta/4\pi)r$ and $|F_2(r,\theta)| \leqq (\beta/2)r$. Then

$$\dot{r} \leqq \frac{\beta}{4\pi}\, r$$

$$\dot{\theta} \geqq \beta - \frac{\beta}{2} = \frac{\beta}{2}$$

for $r \leqq \delta$. Let Γ: $r = r(t)$, $\theta = \theta(t)$ be a trajectory with $r(0) = r_0 < \delta/e$, $\theta(0) = 0$. So long as $r(t) \leqq \delta$, the inequality $\dot\theta(t) \geqq \beta/2$ is valid, and therefore also $\theta(t) \geqq (\beta/2)t$ for $t \geqq 0$ and $\theta(t) \leqq (\beta/2)t$ for $t \leqq 0$. Integrating the inequality.

$$\dot{r}(t) \leqq \frac{\beta}{4\pi}\, r(t)$$

yields

$$r(t) \leqq r_0 e^{\frac{\beta}{4\pi}t}$$

provided $t \geqq 0$ and $r(t) \leqq \delta$. Since

$$r_0 e^{\frac{\beta}{4\pi}t} < \frac{\delta}{e}\, e^{\frac{\beta}{4\pi}t} \leqq \delta$$

if $0 \leqq t \leqq 4\pi/\beta$, the trajectory remains in the disk $r \leqq \delta$ for $0 \leqq t \leqq 4\pi/\beta$. Thus $\theta(t) \geqq (\beta/2)t$ for $0 \leqq t \leqq 4\pi/\beta$, and consequently $\theta(4\pi/\beta) \geqq 2\pi$. The trajectory therefore completes at least one rotation around the origin.

Let p be the smallest positive number such that $\theta(p) = 2\pi$. Obviously $p \leqq 4\pi/\beta$. There are now three cases, according as $r(p)$ is equal to, less than, or greater than, r_0. In the first case, Γ is obviously an orbit. Suppose now that $r(p) < r_0$, and consider the region Ω bounded by the arc $0 \leqq t \leqq p$ of Γ and the segment l connecting its endpoints (Fig. 34). On l, the velocity vector points into Ω, since $\dot\theta > 0$. Since,

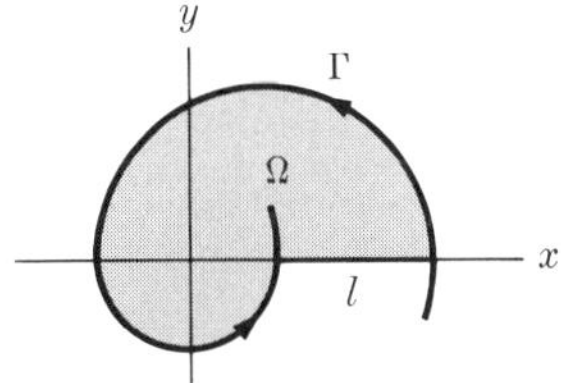

FIG. 34. See proof of Theorem 3.5.

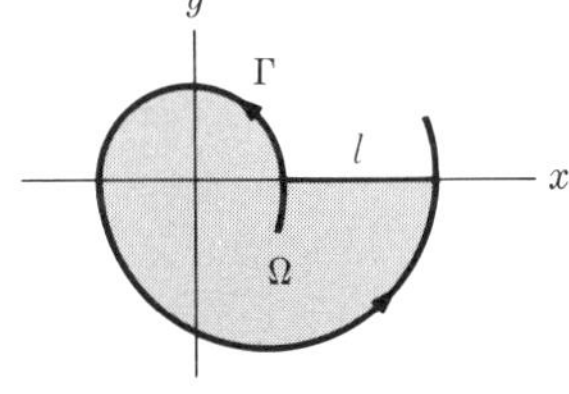

FIG. 35. See proof of Theorem 3.5.

furthermore, Γ cannot intersect itself, it remains in Ω for $t > p$. Hence $\theta(t) \geqq (\beta/2)t$ holds for all $t \geqq 0$. Thus the portion of Γ in Ω is a spiral, with $\theta(t) \to \infty$ as $t \to \infty$.

If $r(p) > r_0$, we have the configuration of Fig. 35. Now Γ is confined to Ω for $t < 0$, and $\theta(t) \leq (\beta/2)t$ holds for $t \leq 0$. Again, the portion of Γ in Ω is a spiral, with $\theta(t) \to -\infty$ as $t \to -\infty$.

We summarize this result.

Theorem 3.5. *If the eigenvalues of A are purely imaginary, the singular point is either a center, a focus, or a center-focus.*

In order to end the discussion of this case with a more conclusive result, we give a sufficient condition for the singular point to be a center.

Theorem 3.6. *If the origin is a singular point of*

$$\begin{aligned} \dot{x} &= f(x,y) \\ \dot{y} &= g(x,y) \end{aligned} \tag{3.30}$$

and is a center for the linear approximation, and if either

(a) $f(x,-y) = -f(x,y),\ g(x,-y) = g(x,y)$

or

(b) $f(-x,y) = f(x,y),\ g(-x,y) = -g(x,y),$

then the origin is a center for (3.30).

Proof. We prove the theorem for case (a). The other case follows by interchanging the roles of x and y.

Suppose that the trajectories near the origin turn counterclockwise with increasing t. Let Γ: $x = u(t)$, $y = v(t)$ be a trajectory with $u(0) > 0$, $v(0) = 0$, and $u(0)$ sufficiently small so that Γ winds around the origin. Define t_1 by the condition that $v(t) > 0$ for $0 < t < t_1$ and $v(t_1) = 0$. Let $u_1(t) = u(t_1 - t)$, $v_1(t) = -v(t_1 - t)$. Differentiating and using (a) shows that u_1, v_1 is a solution. Also $u_1(0) = u(t_1)$, $v_1(0) = v(t_1)$, so that $u_1(t) = u(t + t_1)$, $v_1(t) = v(t + t_1)$. But $u_1(t_1) = u(0)$ and $v_1(t_1) = v_1(0)$. Hence $u(2t_1) = u(0)$ and $v(2t_1) = v(0)$. It follows that Γ is an orbit of period $2t_1$. The proof actually shows that Γ is symmetric with respect ot the x-axis. In case (b), the trajectories are symmetric with respect to the y-axis. In both cases, it is the symmetry of the trajectories which compels them to be orbits. $\square$

Extracting from Theorems 3.1 to 3.4 the results relating to stability, we obtain the following fundamental stability theorem.

Theorem 3.7. *If both eigenvalues of A have negative real part, the singular point (x_0,y_0) of (3.1) is asymptotically stable. If some eigenvalue has positive real part, the singular point is unstable.*

This is the generalization of Theorem 6.7, Chapter 2, to second-order systems. The hypothesis that det $A \neq 0$ is not needed for the instability part of the theorem (see Exercise 5).

EXERCISES

1. To show that an elementary singular point is isolated we suppose that by a translation of coordinates the singular point has been brought to the origin; the system then has the form (3.6). The problem is to show that there is a $\delta > 0$ such that the only singular point of (3.6) in the disk $x^2 + y^2 < \delta^2$ is $(0,0)$.

(a) Show that the singular points of (3.6) are the solutions of the system

$$x = F_1(x,y), \qquad y = F_2(x,y) \tag{1}$$

where

$$F_1(x,y) = \frac{bR_2(x,y) - dR_1(x,y)}{ad - bc}, \qquad F_2(x,y) = \frac{cR_1(x,y) - aR_2(x,y)}{ad - bc}$$

(b) Obviously $F_i(x,y)/\sqrt{x^2 + y^2} \to 0$ as $\sqrt{x^2 + y^2} \to 0$. Choose $\delta > 0$ so that $0 < \sqrt{x^2 + y^2} < \delta$ implies $|F_i(x,y)|/\sqrt{x^2 + y^2} < 1/\sqrt{2}$, $i = 1, 2$. Deduce a contradiction from the assumption that some point (x,y) satisfies both (1) and $0 < \sqrt{x^2 + y^2} < \delta$.

2. To show that $Q_1(x,y)$ and $Q_2(x,y)$ satisfy (3.11), we note that they are linear combinations of $R_1(Pz)$ and $R_2(Pz)$. It therefore suffices to prove that

$$\lim_{r \to 0} \frac{R_i(Pz)}{r} = 0, \qquad i = 1, 2. \tag{2}$$

In this exercise we denote the length of a vector z by $\|z\|$. Setting $P = \begin{bmatrix} \alpha & \beta \\ \gamma & \delta \end{bmatrix}$, we have that $\|Pz\| = \sqrt{(\alpha x + \beta y)^2 + (\gamma x + \delta y)^2}$. Let $k = \sqrt{\alpha^2 + \beta^2 + \gamma^2 + \delta^2}$.

(a) Show that

$$\|Pz\| \leqq k\|z\|$$

for all z.

(b) Let $\epsilon > 0$. By virtue of (3.7) there is a $\Delta > 0$ so that $0 < \|Pz\| < \Delta$ implies $\dfrac{|R_i(Pz)|}{\|Pz\|} < \dfrac{\epsilon}{k}$, $i = 1, 2$. Show that if $0 < \|z\| < \dfrac{\Delta}{k}$, then $\dfrac{|R_i(Pz)|}{\|z\|} < \epsilon$, $i = 1, 2$. This proves (2).

3. Transform the system $\dot{x} = y - x(x^2 + y^2 - 1)$, $\dot{y} = -x - y(x^2 + y^2 - 1)$ to polar coordinates. Show that it has a unique orbit,

and that all trajectories (except the singular point) tend to the orbit
as $t \to \infty$. Note that the origin is an unstable focus. The phase portrait is shown in Fig. 36.

4.   Let $r = r(t)$, $\theta = \theta(t)$ be a non-
constant solution of the polar system
corresponding to (3.1). Show that the
corresponding trajectory is an orbit of
least period p if and only if $r(t + p) =
r(t), \theta(t + p) = \theta(t) + 2\pi n$, where $n = 0$,
1, or -1. (Assume that the origin is a
singular point; this assures that no orbit
passes through the origin.) What properties of the orbit are reflected in the value
of n?

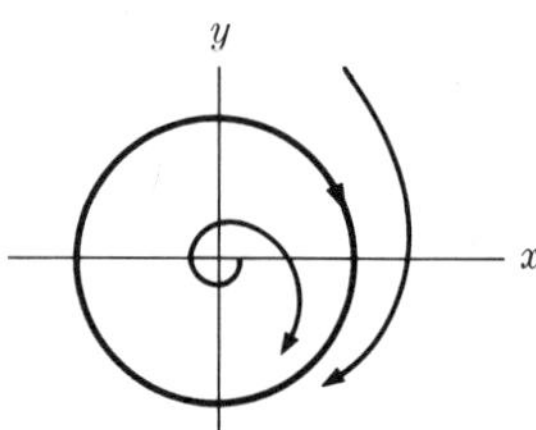

FIG. 36.   See Exercise 3.

5.   Prove: If one eigenvalue of A is
zero, and the other is positive, then the origin is an unstable singular
point of (3.8).

Note that the canonical form is $\dot{x} = \lambda x + Q_1(x,y)$, $\dot{y} = Q_2(x,y)$,
where $\lambda > 0$, and that the corresponding polar equations are $\dot{r} =
\lambda \cos^2 \theta\, r + F_1(r, \theta), \dot{\theta} = -(\lambda/2) \sin 2\theta + (1/r)F_2(r,\theta)$. Choose $\delta > 0$ so
that $r \leq \delta$ implies $|F_i(r,\theta)| \leq (\lambda/4)r$. Now prove the assertion by
showing that a trajectory through any point of the sector $-\pi/4 \leq
\theta \leq \pi/4, 0 < r \leq \delta$, leaves the disk $r \leq \delta$.

6.   Find the singular points of the following systems. Determine
the character of each singular point.

(a) $\dot{x} = xy - 2x - y + 2$, $\dot{y} = x^2 y + y$
(b) $\dot{x} = y$, $\dot{y} = x^2 y - y - x$
(c) $\dot{x} = y$, $\dot{y} = -x + 5y - y^3$
(d) $\dot{x} = -y + xy$, $\dot{y} = -x + y + xy - 1$
(e) $\dot{x} = -x - y + xy + 1$, $\dot{y} = -2x + xy$
(f) $\dot{x} = \sin x + \sin y$, $\dot{y} = -x + y$
(g) $\dot{x} = x^2 + y^2$, $\dot{y} = e^y \sin x$
(h) $\dot{x} = -y + xy^3$, $\dot{y} = x + y^2$
(i) $\dot{x} = y$, $\dot{y} = 1 - x^2$
(j) $\dot{x} = e^{\sin x} - y$, $\dot{y} = e^{\cos x}y - e^{\cos x}$
(k) $\dot{x} = x - x^2$, $\dot{y} = y - y^2$

7.   The equation $\ddot{x} + \mu(x^2 - 1)\dot{x} + x = 0$, $\mu > 0$, was first investigated in 1922 by the Dutch mathematician and engineer B. van
der Pol, who used it to describe an electronic feed-back circuit. (An
essentially equivalent equation was proposed by Lord Rayleigh in
1907 to describe a tuning fork whose oscillations were magnetically

maintained. However, Rayleigh did not attempt an analysis of his equation.) In 1928, the French physicist A. Liénard proved that certain types of non-linear equations, including that of van der Pol, have a phase portrait substantially like Fig. 36. Determine the nature of the singular point of van der Pol's equation for all positive values of μ.

8.   Transform the system

$$\dot{x} = x(x^2 + y^2 - 1)^2 \sqrt{x^2 + y^2} + y$$
$$\dot{y} = -x + y(x^2 + y^2 - 1)^2 \sqrt{x^2 + y^2}$$

to polar coordinates. Describe the phase portrait and make a sketch.

9.   Describe the phase portrait of the system

$$\dot{x} = x^2 - x, \dot{y} = \sin y$$

and make a sketch for $-1 \leq x \leq 2, \quad -\pi \leq y \leq 2\pi$.

4.   Systems with Integrals; Exact Systems

We now consider systems

$$\begin{aligned}\dot{x} &= f(x,y)\\ \dot{y} &= g(x,y)\end{aligned} \tag{4.1}$$

whose trajectories are the level curves of a function $H(x,y)$, called an *integral* of the system. If an integral H is known, the problem of describing the phase portrait reduces to describing the curves $H(x,y) = c$. The orientation of the trajectories cannot, of course, be inferred from H, but is easily found by inspection of the system itself. Strictly speaking, the curves $H(x,y) = c$ are not the trajectories, but the characteristics, of the system.

When we speak of the level curves of a function H, we tacitly assume that the loci $H(x,y) = c$ are, in fact, curves, and that therefore H is not constant on any disk. We shall consider only functions H which have continuous partial derivatives with respect to x and y, and whose critical points are isolated. (The critical points of H are the points at which both partial derivatives vanish.) Then it follows from the implicit function theorem that, apart from the critical points, the loci $H(x,y) = c$ are smooth curves. The implicit function theorem is proved in texts on advanced calculus.

It is a simple matter to decide whether a given function H is an integral of (4.1). The necessary and sufficient condition that the trajectories of (4.1) are level curves of H is clearly that H is constant on any trajectory of (4.1). Let $x = u(t)$, $y = v(t)$ be a trajectory. The condition that $H(u(t),v(t))$ is constant is that

$$\frac{d}{dt} H(u(t),v(t)) = 0 \qquad (4.2)$$

or

$$H_x(u(t),v(t))\dot{u}(t) + H_y(u(t),v(t))\dot{v}(t) = 0$$

for all t for which u, v is defined. Using (4.1), we have

$$H_x(u(t),v(t))f(u(t),v(t)) + H_y(u(t),v(t))g(u(t),v(t)) = 0.$$

If this identity is to hold for all trajectories of (4.1), it must hold for all points (x,y), and we have

$$H_x(x,y)f(x,y) + H_y(x,y)g(x,y) = 0. \qquad (4.3)$$

Conversely, if (4.3) holds for all (x,y), then (4.2) holds for all trajectories. This gives us the following criterion.

Theorem 4.1.   *The necessary and sufficient condition that a function H is an integral of (4.1) is that*

$$H_x f + H_y g = 0. \qquad (4.4)$$

Let Γ: $x = u(t)$, $y = v(t)$ be a trajectory of (4.1) which approaches a singular point (x_0,y_0) as $t \to \infty$. If (4.1) has an integral H, then, since H is continuous, $\lim_{t\to\infty} H(u(t),v(t)) = H(x_0,y_0)$. But H is constant on Γ, and therefore $H(u(t),v(t)) \equiv H(x_0,y_0)$. The same holds if Γ approaches (x_0,y_0) as $t \to -\infty$. Hence the locus $H(x,y) = H(x_0,y_0)$ contains, in addition to (x_0,y_0) itself, all trajectories which approach (x_0,y_0). The converse is obvious: if a level curve of H passes through a singular point of (4.1), it consists of one or more trajectories in addition to the singular point itself.

These considerations show that only systems with rather special phase portraits can have integrals. Suppose that (4.1) has a singular point (x_0,y_0) which is an attractor. Then there is a disk D about (x_0,y_0) such that every trajectory which has a point in D tends to (x_0,y_0) as $t \to \infty$ or $-\infty$. The assumption that (4.1) has an integral H now leads to a contradiction. For if Γ is a trajectory with a point in D, H has the value $H(x_0,y_0)$ on Γ, and consequently $H(x,y) = H(x_0,y_0)$ at every point (x,y) of D. Thus H is constant on D, contrary to the definition of an integral. We thus have the following theorem.

Theorem 4.2.   *If (4.1) has an integral, then none of its singular points is an attractor.*

In particular, a system with an integral has no asymptotically stable singular points.

It follows in the same way that a system with an integral has no limit cycles. For if Γ is a limit cycle, all neighboring trajectories tend toward Γ as $t \to \infty$ or $-\infty$, and by continuity an integral H would have to assume the same value on each of these trajectories as on Γ. Hence there is a band, or annulus, about Γ on which H is constant, contrary to definition. The same reasoning actually shows that it is not even possible for the trajectories on one side of an orbit to approach the orbit. In conjunction with Theorems 3.5 and 4.2, this yields the following result.

Theorem 4.3.   *If* (4.1) *has an integral, and if* (x_0,y_0) *is a singular point which is a center for the linear approximation, then it is a center for* (4.1).

A system which has an integral has infinitely many integrals. Obviously, if H is an integral, so is $c_1 H + c_2$, provided $c_1 \neq 0$. More generally, if $F(s)$ is a function such that $F'(s)$ is continuous and non-zero for all s, then $G(x,y) = F(H(x,y))$ is an integral (Exercise 2).

Exact Systems.   We say that (4.1) is *exact* if

$$f_x + g_y = 0. \tag{4.5}$$

It is a simple matter to compute an integral of such a system. Recall from calculus that if $P(x,y)$ and $Q(x,y)$ have continuous partial derivatives with respect to x and y, and if $P_y = Q_x$, then there exists a function $H(x,y)$ such that $H_x = P$ and $H_y = Q$. A suitable H may be computed by setting $H(x,y) = \int P(x,y)\, dx + R(y)$ and choosing $R(y)$ so that $H_y = Q$.

Theorem 4.4.   *If* $f_x + g_y = 0$, *there exists a function* H *such that* $H_y = f$ *and* $H_x = -g$. *Furthermore, any such* H *is an integral of* (4.1).

Proof.   We use the theorem from calculus cited above, with $P = -g$ and $Q = f$. It remains to show that H is an integral. Because $H_x = -g$ and $H_y = f$, H has continuous partial derivatives, and the critical points of H coincide with the singular points of (4.1). Finally, $H_x f + H_y g = -gf + fg = 0$, so that H is an integral by Theorem 4.1. $\square$

Note that an exact system has the form

$$\dot{x} = H_y$$
$$\dot{y} = -H_x,$$

which readers familiar with classical mechanics will recognize as Hamilton's equations for a dynamical system with one degree of freedom.

EXAMPLE 1.   A linear system

$$\dot{x} = ax + by$$
$$\dot{y} = cx + dy \tag{4.6}$$

is exact if and only if $a + d = 0$. Since $a + d$ is the sum of the eigenvalues of A, the eigenvalues of an exact linear system are negatives of each other. Thus the origin is either a saddle point or a center. Setting $d = -a$, (4.6) becomes

$$\dot{x} = ax + by$$
$$\dot{y} = cx - ay. \tag{4.7}$$

To compute an integral, we need only satisfy $H_y = ax + by$, $H_x = -(cx - ay)$. A simple computation yields $H(x,y) = \frac{1}{2}by^2 + axy - \frac{1}{2}cx^2$. For convenience, we multiply by -2 and replace the integral by $H(x,y) = cx^2 - 2axy - by^2$. The trajectories of (4.7) therefore satisfy

$$cx^2 - 2axy - by^2 = k. \tag{4.8}$$

The analysis of the general case is left to the reader (Exercise 3). We consider two representative cases.

The system

$$\dot{x} = x + 3y$$
$$\dot{y} = x - y$$

has the integral $H(x,y) = x^2 - 2xy - 3y^2$. The level curve through the origin, $H(x,y) = H(0,0)$, is $x^2 - 2xy - 3y^2 = 0$, or $(x - 3y)(x + y) = 0$. It consists of two straight lines, and illustrates our previous remarks about level curves which pass through a singular point. Removing the singular point decomposes these lines into four trajectories. For $k \neq 0$, the curves $H(x,y) = k$ are hyperbolas asymptotic to the two lines.

The system

$$\dot{x} = x - 2y$$
$$\dot{y} = x - y$$

has the integral $H(x,y) = x^2 - 2xy + 2y^2$. The level curve through the origin is $x^2 - 2xy + 2y^2 = 0$, and consists of the origin alone. The remaining level curves are given by $x^2 - 2xy + 2y^2 = k$, where $k > 0$, and are ellipses. Since none of these ellipses passes through the singular point, all of them are trajectories, and consequently orbits. Thus all the solutions are periodic. We know this already, of course, but it is important to be aware that we have here deduced the periodicity by a completely new method. (See Exercise 4 for a basic application to analysis.)   □

EXAMPLE 2.   An example of an exact non-linear system is

$$\dot{x} = 2xy - 2y$$
$$\dot{y} = x - y^2.$$

The singular points are elementary. They are $(0,0)$, a center, and $(1,1)$, $(1,-1)$, both saddle points. An integral is given by $H(x,y) = xy^2 - y^2 - \frac{1}{2}x^2$. The level curve through $(1,1)$ is $2xy^2 - 2y^2 = x^2 - 1$. It consists of the line $x = 1$ and the parabola $2y^2 = x + 1$, both of which also pass through $(1,-1)$. The shape of the remaining curves is found by the usual methods of analytic geometry. The phase portrait is shown in Fig. 37. Note that two trajectories tend to saddle points both as $t \to \infty$ and as $t \to -\infty$. The parabolic sector which they bound is filled with orbits surrounding the center.   □

Knowing an integral of a second-order autonomous system takes us a long way toward constructing the phase portrait, but the task which remains—that of finding the level curves of the integral—may still pose a difficult problem in analytic geometry. At the beginning of Section 5 we shall discuss an important class of systems for which this problem admits of a simple solution.

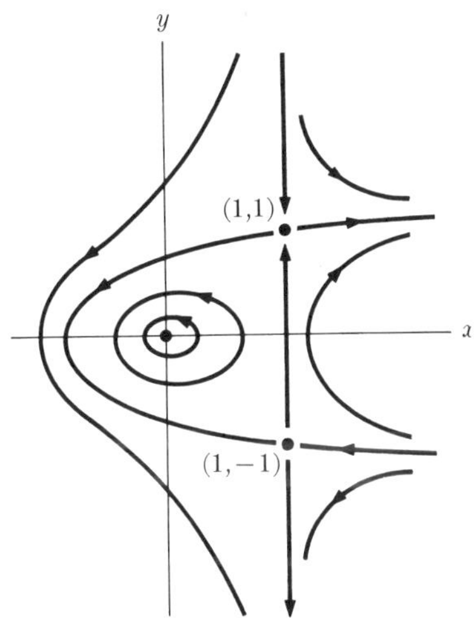

FIG. 37.   See Example 2.

A formal procedure which is frequently used to obtain Theorem 4.4 is the following. We eliminate dt from the system

$$\frac{dx}{dt} = f(x,y)$$
$$\frac{dy}{dt} = g(x,y)$$

(4.9)

and obtain

$$\frac{dy}{dx} = \frac{g(x,y)}{f(x,y)}.$$

Clearing of fractions yields the equation

$$g(x,y)\, dx - f(x,y)\, dy = 0. \tag{4.10}$$

If $g_y = -f_x$, the differential $g\, dx - f\, dy$ is exact. Setting $dH = g\, dx - f\, dy$, (4.10) becomes $dH(x,y) = 0$, whence $H(x,y) = c$. The key to an understanding of this procedure lies in the interpretation of (4.10) as a direction field. To each point (x_0,y_0) which is not a singular point of (4.9) we assign the line parallel to the vector $(f(x_0,y_0),g(x_0,y_0))$. The equation of this line is

$$g(x_0,y_0)(x - x_0) - f(x_0,y_0)(y - y_0) = 0. \tag{4.11}$$

No line is assigned to a singular point. Setting $dx = x - x_0$, $dy = y - y_0$, Eq. (4.11) becomes $g(x_0,y_0)\, dx - f(x_0,y_0)\, dy = 0$. Dropping the subscripts yields (4.10), which is thus seen to be the equation of the line assigned to (x,y) referred to a coordinate system dx, dy whose origin is at the given point. This line is obviously the tangent to the characteristic which passes through the point. The difference between the direction field (4.10) and the kind of direction field defined in Chapter 1 is that we now allow vertical lines, i.e., no direction is excluded. By a solution of (4.10) we mean a curve which is tangent to the direction field defined by (4.10). A solution of (4.10) is therefore the same thing as a characteristic of (4.9). To show that the curves $H(x,y) = c$ are, indeed, the solutions, it is only necessary to compute their tangents.

Integrating Factors.   Let $M(x,y)$ be a function with continuous partial derivatives and such that $M(x,y) > 0$ for all (x,y). Obviously the trajectories of

$$\begin{aligned}
\dot{x} &= f(x,y)\\
\dot{y} &= g(x,y)
\end{aligned} \tag{4.12}$$

are identical with those of

$$\begin{aligned}
\dot{x} &= M(x,y)f(x,y)\\
\dot{y} &= M(x,y)g(x,y),
\end{aligned} \tag{4.13}$$

since the vector assigned to a point (x,y) by (4.13) has the same direction as that assigned by (4.12), differing from the latter only in its length. The factor $M(x,y)$ changes the phase speed, but leaves the phase portrait itself unaltered. [For the connection between the solutions of (4.12) and (4.13), see Exercise 5.] A non-exact system (4.12) can sometimes be made exact by the introduction of a suitable factor M, which is then called an *integrating factor* of the system. Clearly M is an integrating factor if and only if

$$\frac{\partial(Mf)}{\partial x} + \frac{\partial(Mg)}{\partial y} = 0. \tag{4.14}$$

Once an integrating factor has been found, an integral of the exact system can be computed by the usual method. Any such integral is, of course, an integral of the original system.

There is no general rule for deciding whether or not a system has an integrating factor. (It is clear, however, that a system with attractors does not have an integrating factor, since it does not have an integral.) It is sometimes possible to find an integrating factor by inspection. There are also numerous devices which lead to integrating factors in special cases. The most fruitful approach is to work backwards, and to find classes of differential systems having integrating factors of a prescribed form. We illustrate this approach by an example.

By definition, an integrating factor is a solution of the partial differential equation (4.14) which is positive and has continuous partial derivatives. Equation (4.14) may be written

$$M_x f + M_y g + M(f_x + g_y) = 0. \tag{4.15}$$

Suppose that this equation has a solution M which is a function of x alone, so that $M_y = 0$. Then

$$M_x f + M(f_x + g_y) = 0$$

or, writing M' for M_x,

$$\frac{M'}{M} = -\frac{f_x + g_y}{f}.$$

The right side of this equation must therefore be independent of y, and a continuous function of x. These conditions are obviously also sufficient, and yield the integrating factor

$$M = e^{-\int \frac{f_x + g_y}{f} \, dx}. \tag{4.16}$$

We thus have the result that if $(f_x + g_y)/f$ is independent of y and continuous in x, then (4.16) is an integrating factor of (4.12).

EXAMPLE.
$$\dot{x} = x + \cos y$$
$$\dot{y} = xy - y + \sin y.$$

Since $(f_x + g_y)/f = 1$, e^{-x} is an integrating factor, and

$$\dot{x} = e^{-x}(x + \cos y)$$
$$\dot{y} = e^{-x}(xy - y + \sin y)$$

is exact. The usual method yields the integral $H(x,y) = e^{-x}(xy + \sin y)$.

Local Integrals and Integrating Factors. Up to now we have considered only *global* integrals, i.e., functions H defined for *all* (x,y) whose level curves are trajectories of a given system. When such an integral does not exist or cannot be determined, it may still be possible to find an integral defined on some portion of the plane, and such a *local* integral often yields useful information about the given system. For example, if for physical reasons the system is of interest only for positive values of x and y, it is sufficient to find an integral in the first quadrant. It is clear, of course, that no integral exists in any region which contains an attractor. A simple illustration of local integrals is furnished by the system

$$\dot{x} = x, \qquad \dot{y} = y.$$

The origin is a node, and consequently there is no integral in any region which contains the origin. However, every trajectory is a ray $x = x_0 e^t$, $y = y_0 e^t$ and satisfies either $y = cx$ or $x = cy$. Thus y/x is an integral in the region complementary to the y-axis, and x/y is an integral in the region complementary to the x-axis.

Let I_x and I_y be open intervals, not necessarily bounded. The set of points (x,y), where x is in I_x and y is in I_y, will, for lack of a better term, be called a rectangle. It is proved in calculus that if P and Q have continuous partial derivatives and satisfy $P_y = Q_x$ in some rectangle Ω, then there is a function H defined in Ω such that $H_x = P$ and $H_y = Q$. H is computed as before. It follows that if $f_x + g_y = 0$ in some rectangle Ω, then the system $\dot{x} = f$, $\dot{y} = g$ has an integral in Ω. The practical application of this fact is to the computation of local integrating factors. To find an integrating factor in a rectangle Ω, we need only find a function $M(x,y)$ which is positive and has continuous partial derivatives in Ω, and satisfies $\dfrac{\partial(Mf)}{\partial x} + \dfrac{\partial(Mg)}{\partial y} = 0$ in Ω.

For example, the result of the preceding section has the generalization that if $(f_x + g_y)/f$ is independent of y and continuous for all x in some Ω, then (4.16) is an integrating factor in Ω. Applied to the system $\dot{x} = x$, $\dot{y} = y$ considered above, this method yields the integrating factor $1/x^2$ in the left and right half-planes, and thus the integral y/x.

There is no difference between the methods for finding global and local integrating factors. Whether the rectangle Ω may be taken to be the whole plane appears in the course of the computations. We conclude this discussion by deriving another rule for obtaining integrating factors. We ask under what conditions Eq. (4.15) has a solution which is a function of xy alone, say $M(x,y) = \varphi(s)$, where $s = xy$. Let us suppose such a solution exists. Since $M_x(x,y) = \varphi'(xy)y$ and $M_y(x,y) = \varphi'(xy)x$, Eq. (4.15) becomes

$$\varphi'yf + \varphi'xg + \varphi(f_x + g_y) = 0$$

or

$$\frac{\varphi'}{\varphi} = -\frac{f_x + g_y}{yf + xg}.$$

The right side of this equation must therefore be a continuous function ψ of $s = xy$ alone. Conversely, if this condition is satisfied, we obtain the integrating factor

$$M = e^{\int \psi(s)\,ds}. \tag{4.17}$$

EXAMPLE.
$$\dot{x} = ax + bxy$$
$$\dot{y} = cy + dxy. \tag{4.18}$$

Since $\dfrac{f_x + g_y}{yf + xg} = \dfrac{1}{xy}$, we have $\psi(s) = -\dfrac{1}{s}$, so that $M = e^{-\int ds/s} = e^{\log 1/|s|} =$ $1/|s| = 1/|xy|$ is an integrating factor inside each of the four quadrants. Here it is convenient to remove the absolute-value signs and allow M to be negative, which merely reverses the orientation of the trajectories in the second and fourth quadrants. We thus obtain the exact system

$$\dot{x} = \frac{a}{y} + b$$

$$\dot{y} = \frac{c}{x} + d$$

and the integral $H(x,y) = a \log |y| + by - c \log |x| - dx$. In the next section we shall encounter the system (4.18) in connection with a population model.

EXERCISES

1.   Equation (4.8) may be written grad $H \cdot \vec{V} = 0$. Interpret this geometrically.

2.   Let $H(x,y)$ be an integral of (4.1). Show that if $F'(s)$ is continuous and non-zero for all s, then $G(x,y) = F(H(x,y))$ is also an integral.

3.   Show that (4.8) is a family of ellipses if $a^2 + bc < 0$, and a family of hyperbolas together with their asymptotes if $a^2 + bc > 0$. Find the equation of the asymptotes, considering separately the cases $c = 0$ and $c \neq 0$.

4.   A strict analytic definition of the sine and cosine functions is not possible in elementary calculus. In more advanced courses the definitions usually adopted make use of power series. To establish the familiar properties of the sine and cosine functions on the basis of these definitions is awkward and to some extent unsatisfactory, because the definitions would never lead one to discover these properties. The most elegant definition makes use of differential equations.

This definition has the additional advantage that it leads immediately to a geometric interpretation.

Let $x = u(t)$, $y = v(t)$ be the solution of the system

$$\dot{x} = y, \quad \dot{y} = -x \tag{1}$$

satisfying $u(0) = 0$, $v(0) = 1$, and define the sine and cosine functions by $\sin t = u(t)$, $\cos t = v(t)$.

(a)   Show that $\dfrac{d}{dt} \sin t = \cos t$ and $\dfrac{d}{dt} \cos t = -\sin t$.

(b)   By means of an integral for (1) show that (u,v) is periodic. The number π may now be defined as half the period. (In problem 6 of the Miscellaneous Exercises at the end of this chapter the formula

$$\pi = 2 \int_0^1 \frac{ds}{\sqrt{1 - s^2}}$$

is obtained as a special case of a more general result.)

(c)   Show that $\sin^2 t + \cos^2 t = 1$.

(d)   Show that $\sin(-t) = -\sin t$ and $\cos(-t) = \cos t$.

(e)   Show that

$$W(t) = \begin{bmatrix} \cos t & \sin t \\ -\sin t & \cos t \end{bmatrix}$$

is the standard fundamental matrix of (1).

(f)   Derive the identities $\sin(a + b) = \sin a \cos b + \cos a \sin b$ and $\cos(a + b) = \cos a \cos b - \sin a \sin b$ from the formula

$$W(a + b) = W(a)W(b).$$

(See Exercise 5, page 155.)

5.   Let u, v be a solution of (4.12), and let $\varphi(t)$ satisfy $\dot{\varphi}(t) = M[u(\varphi(t)),v(\varphi(t))]$. Show that $\alpha(t) = u(\varphi(t))$, $\beta(t) = v(\varphi(t))$ is a solution of (4.13).

6.   Show that if $\dot{x} = f(x,y)$, $\dot{y} = g(x,y)$ is exact and homogeneous (see Exercise 2, page 201) then it has the integral $H(x,y) = f(x,y)y - g(x,y)x$. (Use the relation $f_x = -g_y$ and Euler's theorem on homogeneous functions to compute H_x and H_y.)

7.   Show that the following systems are exact, compute integrals for them, and sketch the phase portraits. (It is best to begin by locating the singular points and, if they are elementary, to identify them as centers or saddle points.) Classify all singular points as stable or unstable.

(a) $\dot{x} = y^3, \quad \dot{y} = -x^3$

(b) $\dot{x} = x^2 - 2xy, \quad \dot{y} = y^2 - 2xy$

(c) $\dot{x} = x - 2xy, \quad \dot{y} = x - y + y^2$
(d) $\dot{x} = 2xy, \quad \dot{y} = x - y^2$
(e) $\dot{x} = y - 2xy, \quad \dot{y} = -2x + y^2$
(f) $\dot{x} = y, \quad \dot{y} = -4x$
(g) $\dot{x} = -x + y, \quad \dot{y} = y + x^2$
(h) $\dot{x} = y^3, \quad \dot{y} = x^3$

8.   (a)   What is the condition that $\dot{x} = f(x,y)$, $\dot{y} = g(x,y)$ has an integrating factor which is a function of y alone? Compute an integrating factor on the assumption that this condition is satisfied.

(b)   Compute an integral of $\dot{x} = xy - x - \cos x$, $\dot{y} = y + \sin x$.

9.   (a)   What is the condition that $\dot{x} = f(x,y)$, $\dot{y} = g(x,y)$ has an integrating factor which is a function of $x + y$ alone? Compute an integrating factor on the assumption that this condition is satisfied.

(b)   Compute an integral of $\dot{x} = ye^x$, $\dot{y} = xe^y$.

10.   Find local integrating factors by inspection. Specify regions on which the integrating factors are defined.

(a) $\dot{x} = x + x \sin y, \quad \dot{y} = xy + y \sin x$
(b) $\dot{x} = x - xy^2, \quad \dot{y} = x^2 - 1$
(c) $\dot{x} = ye^{x+y}, \quad \dot{y} = -x$

5.   Physical Examples; Cylindrical Phase Space; Bifurcation

Motion on a Line.   We consider a particle of mass m which is constrained to move along a straight line and is acted upon by a force which depends only on the position of the particle. We denote the position of the particle by x and the tangential component of the force by $g(x)$. Then Newton's second law yields the equation of motion $m\ddot{x} = g(x)$. For simplicity we set $m = 1$. Then $\ddot{x} = g(x)$, or equivalently,

$$\dot{x} = y, \qquad \dot{y} = g(x). \tag{5.1}$$

Note that y is the velocity of the particle. The singular points of (5.1) satisfy $y = 0$, $g(x) = 0$, and therefore lie on the x-axis. They correspond to the equilibrium states of the particle.

Our object is to obtain a complete description of the phase portrait, and thus of all possible motions of the particle. To begin with, we note that the vector $\vec{V}(x,y) = (y,g(x))$ has a positive x-component in the upper half-plane, and a negative x-component in the lower half-plane, while on the x-axis itself the x-component vanishes. Thus all trajectories move to the right in the upper half-plane and to the left in the lower half-plane. A trajectory which intersects the x-axis does so at right angles. Furthermore, if a trajectory—or a portion of a trajectory—lying in either half-plane is reflected in the x-axis, its orientation being simultaneously reversed, another trajectory is obtained. For if the given trajectory is $x = u(t)$, $y = v(t)$, then $x = u(-t)$, $y = -v(-t)$ is also a trajectory, as is easily verified. These facts already allow us to conclude that a trajectory is an orbit if and only if it intersects the x-axis in exactly two points. The physical interpretation is obvious: the motion of the particle is periodic when there are exactly two positions in which its velocity is zero. Note that all orbits are symmetric with respect to the x-axis, and that their orientation is clockwise.

The most important property of (5.1) is its exactness. By the usual method, we find the integral $H(x,y) = \frac{1}{2}y^2 - \int g(x)\,dx$. The quantity $\frac{1}{2}y^2$ is the kinetic energy of the particle, while $-\int g(x)\,dx$ is its potential energy. We shall denote the latter by $V(x)$, so that $H(x,y) = \frac{1}{2}y^2 + V(x)$. Note that $V(x)$ is determined only up to an additive constant; it is usually convenient to set $V(0) = 0$. The fact that the sum of the kinetic and potential energies is an integral constitutes the law of conservation of energy: during any motion of the particle its total energy remains constant. The existence of an integral implies that (5.1) has no attractors, and in particular no asymptotically stable singular points. For the same reason, all trajectories near a given orbit are also orbits, so that there are no limit cycles. These facts also follow from the preceding symmetry considerations.

We now explain in some detail how the phase portrait can be obtained from an inspection of the potential function. We first note that the singular points are the points on the x-axis at which $V'(x) = 0$. To simplify matters, we suppose that $V'(x)$ and $V''(x)$ do not vanish simultaneously, i.e., that the singular points are elementary. It is easily checked that a singular point x_0 is a center or a saddle point according as $V''(x_0)$ is positive or negative, so that stable and unstable equilibria correspond, respectively, to minima and maxima of the potential energy. Since the minima and maxima of $V(x)$ alternate, so do the centers and saddle points of (5.1).

The trajectories are given by

$$\tfrac{1}{2}y^2 + V(x) = c. \tag{5.2}$$

If there is a number m such that $V(x) \geqq m$ for all x, then values of c less than m are obviously not admissible: in physical terms, the total energy cannot be less than the potential energy.

We first consider a trajectory which intersects the x-axis, say at $x = \alpha$. Setting $x = \alpha$, $y = 0$ in (5.2) yields $c = V(\alpha)$. The equation of the trajectory is most conveniently written in the form $y =$

$\pm\sqrt{2[V(\alpha) - V(x)]}$. It extends over an x-interval I to the right or left of α according as $V'(\alpha)$ is negative or positive. If I terminates at a point β, so that $V(\beta) = V(\alpha)$, and if β is not a singular point, then the trajectory cuts the x-axis again at β and is therefore an orbit. If β is a saddle point, the trajectory is a loop which approaches β at both ends. If I has no endpoint except α, the trajectory is unbounded, forming a loop which tends to infinity at both ends. Every trajectory which intersects the x-axis belongs to one of these three types.

A trajectory which does not intersect the x-axis is an arc tending at either end to a saddle point or to infinity. Trajectories which do not intersect the x-axis and tend to infinity at both ends arise when there is a number M such that $V(x) \leqq M$ for all x, and correspond to values of c greater than M.

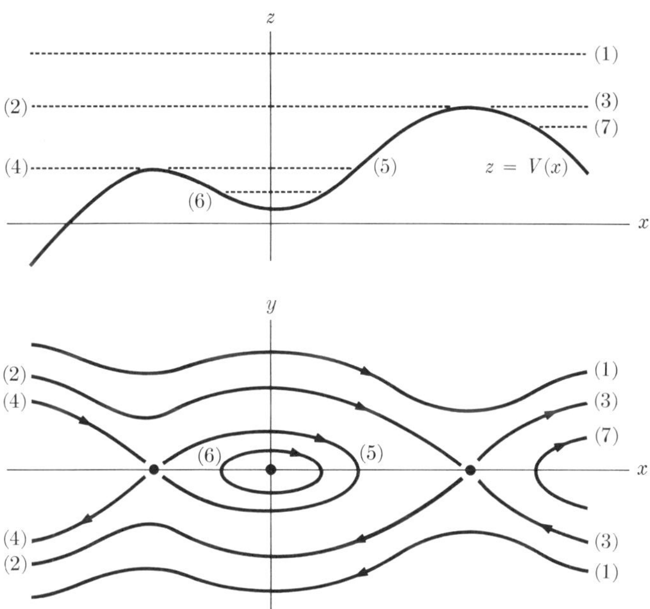

FIG. 38.   Relation between the graph of the potential function and the phase portrait.

The preceding discussion is conveniently summarized in terms of the graph of the potential function, i.e., the curve $z = V(x)$. Each trajectory $H(x,y) = c$ corresponds to a segment of the line $z = c$ which lies above this graph (Fig. 38). The orbits correspond to chordal segments whose endpoints are not maxima. Trajectories which approach saddle points correspond to segments which terminate at one or both ends at a maximum, while trajectories unbounded at both ends correspond to lines which lie wholly above the graph of V or terminate at a single point which is not a maximum. Obviously no line terminates at a minimum.

We mention one further fact of some intrinsic interest. Consider an orbit Γ, and denote the points at which it intersects the x-axis by α and β. Because $V(x)$ assumes the same value c at the endpoints of the interval $[\alpha,\beta]$, and is less than c in the interior of the interval, $V(x)$ has at least one minimum between α and β. It follows that Γ surrounds at least one center. If $V(x)$ has n minima between α and β, then obviously it also has $n - 1$ maxima, one between each pair of successive minima. Thus *the number of centers interior to an orbit of* (5.1) *exceeds by* 1 *the number of saddle points*. An illustration is furnished by the equation $\ddot{x} = x - x^3$, which has two centers and one saddle point. Trajectories corresponding to sufficiently large values of the energy c are orbits surrounding all three singular points. The phase portrait is shown in Fig. 39. In the Appendix we outline a proof of

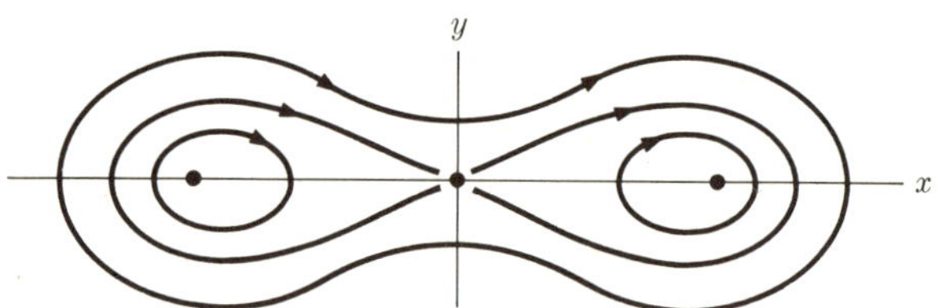

FIG. 39.   Trajectories of $\ddot{x} = x - x^3$.

the following generalization: Let Γ be an orbit of a system $\dot{x} = f(x,y)$, $\dot{y} = g(x,y)$ having only elementary singular points. Let n_1 be the number of saddle points, and n_2 the total number of other singular points, inside Γ. Then $n_2 - n_1 = 1$.

The period of a periodic solution of $\ddot{x} = g(x)$ can be calculated without any knowledge of the solution other than its minimum and maximum values, which we denote as before by α and β. Let the solution be $x = u(t)$, so that the corresponding orbit in the phase plane is $x = u(t)$, $y = \dot{u}(t)$. The period p is the time in which the representative point makes one circuit of the orbit. By symmetry, p is equal to twice the time taken to go from $(\alpha,0)$ to $(\beta,0)$. The upper half of the orbit satisfies $\dot{u}(t) = \sqrt{2[V(\alpha) - V(u(t))]}, \quad 0 \leq t \leq p/2,$

where for convenience we have chosen the phase so that $u(0) = \alpha$. Thus

$$\frac{1}{\sqrt{2}} \frac{\dot{u}(t)}{\sqrt{V(\alpha) - V(u(t))}} = 1$$

for $0 < t < p/2$, and therefore

$$\frac{1}{\sqrt{2}} \int_0^{p/2} \frac{\dot{u}(t)\, dt}{\sqrt{V(\alpha) - V(u(t))}} = \frac{p}{2}.$$

Making the change of variable $x = u(t)$ in the integral and multiplying by 2 yields

$$p = \sqrt{2} \int_\alpha^\beta \frac{dx}{\sqrt{V(\alpha) - V(x)}}.$$

It is not difficult to show that as either α or β approaches a saddle point, p tends to infinity. If α and β approach a center x_0, p approaches the period of the linear approximation $\ddot{x} = g'(x_0)x$, which is

$$\frac{2\pi}{\sqrt{-g'(x_0)}}.$$

It can be shown that p is independent of α and β only in the linear case, $\ddot{x} + \omega^2 x = 0$.

The Pendulum.   A pendulum is the physical system consisting of a particle constrained to move in a circle whose plane is vertical, the only external force acting on the particle being gravity. We may picture a pendulum as a bob of mass m attached to one end of a rigid weightless rod of length l, the rod being hinged at the other end so as to be free to rotate in a vertical plane. We denote the angle between the rod and the vertical by x (Fig. 40). The displacement of the bob from its lowest position is $s = lx$. The only force acting on the bob tangentially to its path is the tangential component of gravity, which is $-mg$ sin x. The equation of motion is therefore $m\ddot{s} = -mg \sin x$, or

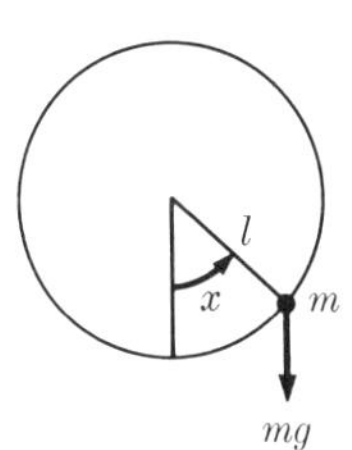

FIG. 40.   Pendulum.

$$\ddot{x} = -\frac{g}{l} \sin x.$$

This equation is of the form (5.1), with $g(x) = -(g/l) \sin x$. Setting $\dot{x} = y$, we obtain the system

$$\dot{x} = y, \qquad \dot{y} = -\frac{g}{l} \sin x.$$

Note that y is the angular velocity of the bob. This system has the integral $H(x,y) = \frac{1}{2}y^2 - (g/l) \cos x$. Since the kinetic energy of the bob is $\frac{1}{2}ml^2y^2$, and its potential energy $-ml \cos x$, the total energy is $ml^2 H(x,y)$, which, of course, is also an integral. As a matter of convenience, we shall suppose that the units are chosen so that m, l, and g are equal to 1. The system then becomes $\dot{x} = y$, $\dot{y} = -\sin x$. The potential energy is now $V(x) = -\cos x$, and the total energy is $H(x,y) = \frac{1}{2}y^2 + V(x)$.

There are infinitely many singular points, all elementary. For every integer n, there is a center at $x = 2n\pi$, and a saddle point at $x = (2n + 1)\pi$. The phase portrait (Fig. 41) is easily obtained from the

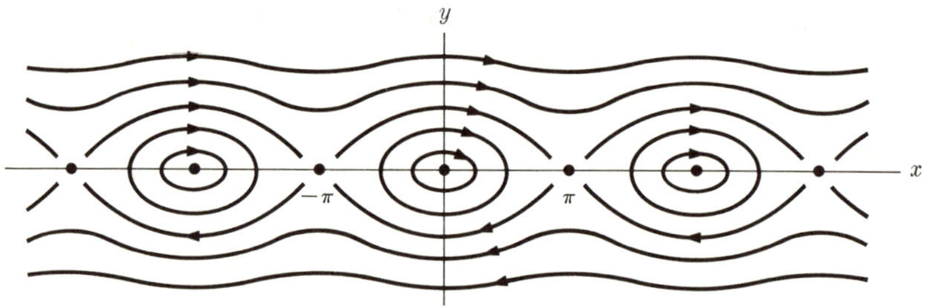

FIG. 41.   Plane phase portrait of the pendulum.

potential function by the method described above. If $-1 < c < 1$, the curve $\frac{1}{2}y^2 - \cos x = c$ consists of infinitely many orbits, one surrounding each center. The centers are given by $c = -1$, and the saddle points by $c = 1$. The trajectories which connect the saddle points are also given by $c = 1$. If $c > 1$, the trajectories are unbounded.

The orbits obviously correspond to oscillations of the pendulum bob about its lowest position. The trajectories connecting the saddle points represent motions in which the total energy of the pendulum is exactly equal to the maximum potential energy, so that the bob creeps toward the unstable equilibrium directly above the point of support. If $c > 1$, the total energy exceeds the maximum potential energy, and x tends to ∞ or $-\infty$ according as y is positive or negative.

The phase portrait as represented in Fig. 41 does not reflect the obvious fact that this "round-and-round" motion is also periodic. Indeed, we have so far ignored a basic fault in our representation of the trajectories: the fact that values of x differing by integer multiples of 2π correspond to identical positions of the pendulum, and that therefore two points (x,y) and $(x + 2\pi n, y)$ correspond to identical states. The way out of this difficulty is to cut the (x,y)-plane along the lines $x = -\pi$ and $x = \pi$ and glue the edges together (Fig. 42). The resulting cylinder is the true phase space of the pendulum, because distinct points of the cylinder correspond to distinct states of the pendulum. On the cylinder there are only two singular points, a center C and a saddle point S. With two exceptions, all the other trajectories are orbits. The exceptions are the two trajectories which issue from and return to the saddle point. The orbits which correspond to the back-and-forth motions of the pendulum are distinguished from the orbits which correspond to the circular motions by their relation to the cylinder. An orbit of the first kind, thought of as a rubber band, can be shrunk to a point (for example, C) without leaving the cylinder, while this is not true of an orbit of the second kind. Analytically, a motion of period p of the second kind is not represented by a periodic solution of the differential equation $\ddot{x} = -\sin x$, but by a solution which has the property that $u(t + p) = u(t) + 2\pi$.

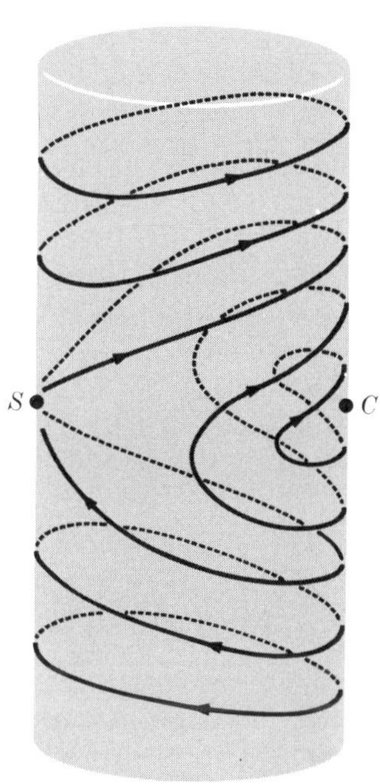

FIG. 42. Cylindrical phase portrait of pendulum.

The Rotating Pendulum. We now consider a particle of mass m moving under gravity on a circle of radius l which rotates with constant angular velocity ω about a vertical diameter (Fig. 43). We may think of the particle as a bead sliding on a spinning circular wire or in a spinning circular tube. As before, we let x be the angle between the vertical and the line from the particle to the center of the circle, and $s = lx$ the displacement of the particle from the lowest point of the circle. The forces on the particle in the direction of increasing s are the tangential components of gravity and of the centrifugal force due to the rotation of the circle. The former is $-mg \sin x$ and the latter $ml\omega^2 \sin x \cos x$. The equation of motion is thus

$$m\ddot{s} = -mg \sin x + ml\omega^2 \sin x \cos x$$

or

$$\ddot{x} = -\frac{g}{l} \sin x + \omega^2 \sin x \cos x,$$

and is again of the form $\ddot{x} = g(x)$. For simplicity, we set $g/l = 1$. The equivalent system is

$$\dot{x} = y, \qquad \dot{y} = -\sin x + \omega^2 \sin x \cos x. \tag{5.3}$$

This example is instructive for a number of reasons. We shall see that the phase portrait is quite different for $\omega \leqq 1$ and $\omega > 1$. (It is understood that $\omega \geqq 0$.) As in the case $\omega = 0$, the phase space is actually a cylinder, and we may therefore restrict ourselves to the strip $-\pi \leqq x \leqq \pi$, the edges of the strip being identified.

The singular points are the points on the x-axis which satisfy

$$\sin x \,(-1 + \omega^2 \cos x) = 0.$$

If $\omega \leqq 1$, the only solutions of this equation are $x = 0$ and $x = \pm\pi$. If $\omega < 1$, both are elementary, 0 being a center and $\pm\pi$ a saddle point, as is easily verified. If $\omega = 1$, the singular point at $x = 0$ is not elementary, but

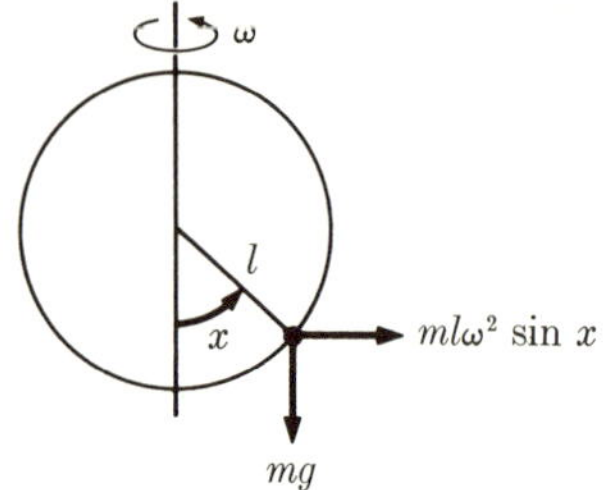

FIG. 43.   Rotating pendulum.

by using an integral of (5.3) it is easy to see that it is still a center. The other singular point remains elementary and a saddle point for $\omega = 1$. The phase portrait for $0 \leq \omega \leq 1$ is therefore qualitatively the same as for $\omega = 0$ (Figs. 41 and 42).

If $\omega > 1$, the solutions of $-1 + \omega^2 \cos x = 0$ furnish two additional singular points, $x = \pm \cos^{-1}(1/\omega^2)$, both of which are centers. Furthermore, $x = 0$ is now a saddle point instead of a center. The saddle point at $x = \pm\pi$ is unaffected by the change in ω. The phase portrait for $\omega > 1$ is shown in Fig. 44. As ω tends to 1 from above, the centers approach the saddle point at the origin. At $\omega = 1$, they coalesce with it to form one center. Or, going in the direction of increasing ω, the center at the origin splits into two centers and a saddle point as ω

passes through 1. Physically, the equilibrium at $x = 0$ becomes unstable, while simultaneously two new equilibria appear near it, both of them stable.

The splitting of a singular point into two or more singular points is called *bifurcation*, and the value of the parameter at which the splitting occurs is called a bifurcation value. The term bifurcation is also applied to more complicated phenomena which can occur in nonconservative systems depending on a parameter (for example, electronic circuits with variable feedback), in which a singular point splits

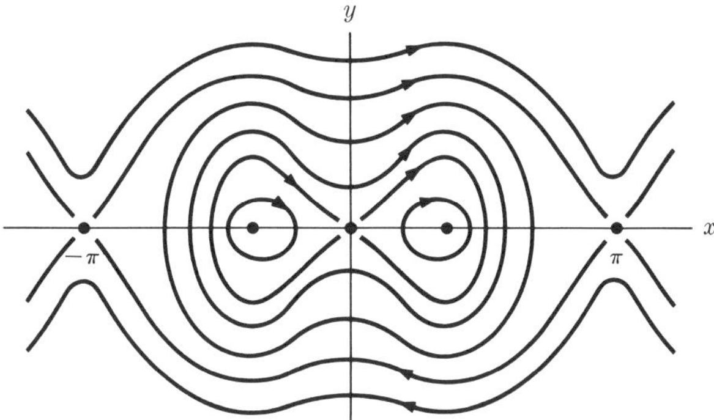

FIG. 44. Plane phase portrait of rotating pendulum for large ω.

into a singular point plus a small limit cycle, or a limit cycle splits into several limit cycles. Lately the term has come to denote any qualitative change in the phase portrait due to small changes in the differential equation.

The appearance and disappearance of singular points and limit cycles, changes in their stability character, and similar but more complicated geometric phenomena in higher dimensions, account for a variety of physical phenomena, known to engineers by such names as hunting, hysteresis, entrainment, synchronization and quenching.

The preceding analysis of the rotating pendulum furnishes an illustration—albeit trivial—of the use of rotating coordinate systems. The singular points at $x = \pm \cos^{-1} (1/\omega^2)$ are equilibria only relative to the rotating circle on which the particle is constrained to move. In reality they correspond to periodic motions in which the particle rotates in a horizontal plane about the vertical diameter of the circle. The fact that these singular points are stable means that the corresponding periodic motions are stable. By the use of suitable coordinates, many non-trivial problems relating to the existence and stability of periodic motions can be reduced to analogous problems involving only singular points.

An Ecological Model.   The interaction of animal populations is an attractive area for mathematical analysis, and one to which Vito Volterra devoted a classic monograph, "Théorie mathématique de la lutte pour la vie." The following is the simplest and best known of the models discussed by Volterra.

A lake contains two species of fish, which we shall simply call big fish and little fish. The big fish subsist primarily by eating the little fish. The sustenance of the little fish, such as vegetation, we suppose to be available in unlimited abundance. Let x denote the number of little fish, and y the number of big fish. We suppose that in isolation, i.e., in the absence of the big fish, the growth of the little fish population is governed by $\dot{x} = \alpha x$, where the growth coefficient α (= birth-rate − death rate) is positive. We also suppose that the little fish are essential to the survival of the big fish, so that the latter's growth coefficient in isolation is negative. Thus if $x = 0$, $\dot{y} = -\gamma y$, where $\gamma > 0$, and the big fish die out.

Now let x and y be positive. We suppose that the fraction of the little fish which die by being eaten is proportional to the number of big fish present, and we denote this fraction by βy, where $\beta > 0$. Thus βy is an addition to the death rate, and $\dot{x} = (\alpha - \beta y)x$. We also suppose that the fraction of the big fish which survive by eating little fish is proportional to the number of little fish present, and denote this fraction by δx, where $\delta > 0$. Since δx constitutes a reduction of the death rate, $\dot{y} = (-\gamma + \delta x)y$. The resulting system

$$\dot{x} = \alpha x - \beta xy$$
$$\dot{y} = -\gamma y + \delta xy \tag{5.4}$$

is of the form (4.18), for which we have already found an integral.

The singular points of (5.4) are the origin and $P = (\gamma/\delta, \alpha/\beta)$. Both are elementary, the origin being a saddle point. P is a center for the linear approximation, and by virtue of the existence of an integral in the first quadrant is also a center for (5.4). We also note the obvious trajectories $x \equiv 0$, $y = y_0 e^{-\gamma t}$ and $x = x_0 e^{\alpha t}$, $y \equiv 0$. It is clear that the interior of each of the four quadrants is an invariant set for (5.4), i.e., a trajectory which has one point inside a given quadrant lies wholly inside that quadrant. Henceforth we consider only the first quadrant.

Since P is a center, solutions with initial values near P are periodic and of small amplitude. Replacing (5.4) by its linear approximation at P shows that the period of these solutions is approximately $2\pi/\sqrt{\alpha\gamma}$. By means of the integral for (5.4), we shall show that *all* the trajectories inside the first quadrant (except P) are orbits surrounding P. The integral found in Section 4 is $\alpha \log y - \beta y + \gamma \log x - \delta x$. It is convenient to multiply this integral by -1 and use instead the integral

$$H(x,y) = \delta x + \beta y - \gamma \log x - \alpha \log y. \qquad (5.5)$$

That the curves $H(x,y) = $ constant are simple closed curves surrounding P follows from the fact that the surface $z = H(x,y)$ is, roughly speaking, a bowl whose lowest point is at P. More precisely, we shall show that *along any ray beginning at P, $H(x,y)$ tends monotonically to infinity as (x,y) tends to infinity or approaches a coordinate axis.* Once this assertion is established, it follows that if $(x_1,y_1) \neq P$, there is exactly one point (x,y) on each ray for which $H(x,y) = H(x_1,y_1)$. Thus the curve $H(x,y) = H(x_1,y_1)$ intersects each ray in exactly one point, and is therefore a simple closed curve about P.

To prove the assertion, we introduce polar coordinates (r,θ) with origin at P:

$$x = x_0 + r \cos \theta, \qquad y = y_0 + r \sin \theta,$$

where $x_0 = \gamma/\delta$ and $y_0 = \alpha/\beta$. A simple computation shows that

$$\frac{\partial H}{\partial r} = \frac{\delta r}{x_0 + r \cos \theta} \cos^2 \theta + \frac{\beta r}{y_0 + r \sin \theta} \sin^2 \theta.$$

Hence $\partial H/\partial r$ is positive at all points inside the first quadrant (except P, where r and θ are not defined), so that H increases with r along each ray $\theta = $ constant. If $0 \leq \theta \leq \pi/2$, then $\partial H/\partial r \to \delta \cos \theta + \beta \sin \theta$ as $r \to \infty$, and since $\delta \cos \theta + \beta \sin \theta > 0, H \to \infty$ as $r \to \infty$. Finally, it is obvious from (5.5) that $H \to \infty$ along the remaining rays $\pi/2 < \theta < 2\pi$, all of which terminate at a coordinate axis. The assertion is therefore proved.

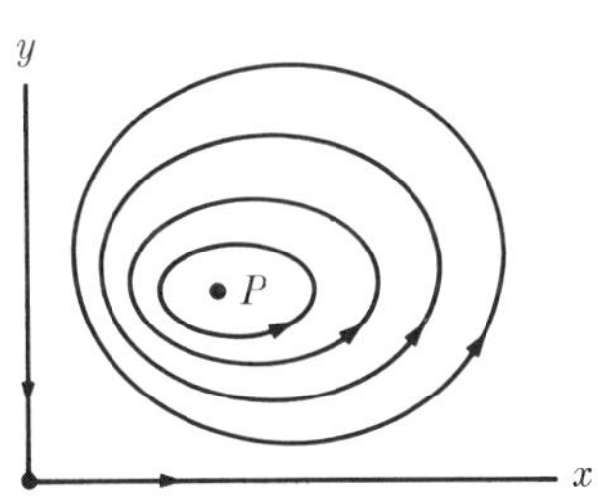

FIG. 45.   Phase portrait of Eq. (5.4).

The phase portrait is shown in Fig. 45. Small changes in the model suffice to change P into a focus, possibly surrounded by a limit cycle, and biological hypotheses which lead to configurations of this sort have been investigated. The failure of certain attempts at biological pest control by the importation of predators is in close agreement with the predictions of our model. The pest population revived after having been virtually exterminated, the predator population declined, and a cycle was created.

Perhaps the reader should be cautioned against the inference that the coexistence of predator and prey must be accounted for by hypotheses as desperate as those underlying this model, which are indeed far from realistic and wholly ignore the existence of instincts which directly control the size of animal populations—most obviously, the instinct of territoriality. Such value as the model has lies primarily in the simplicity of the explanation it offers for the existence of population cycles. Not unlike economic cycles, they puzzle the layman, who sometimes seeks their cause in non-existent external periodic phenomena, in order thereby to explain them as forced oscillations similar to the seasonal fluctuations in the fly population.

EXERCISES

1.   In a well-known textbook of classical mechanics it is erroneously stated that a solution of $\ddot{x} = g(x)$ is either constant, periodic, or unbounded. What kind of behavior is left out of account?

2.   Suppose that at x_0 the graph of the potential function has an inflection point with a horizontal tangent, so that $V'(x_0) = V''(x_0) = 0$ and $V''(x)$ changes sign at x_0. In this case, and only in this case, the singular point $(x_0, 0)$ of $\dot{x} = y$, $\dot{y} = g(x)$ is neither a center nor a saddle point. Describe the phase portrait near such a singular point.

3.   Compute the potential function and sketch the phase portrait for each of the following equations.

$$
\begin{aligned}
&\text{(a)} \quad \ddot{x} + x + x^3 = 0 \\
&\text{(b)} \quad \ddot{x} + x^2 - 1 = 0 \\
&\text{(c)} \quad \ddot{x} - x - x^3 = 0 \\
&\text{(d)} \quad \ddot{x} + xe^{-x} = 0 \\
&\text{(e)} \quad \ddot{x} + x^3 = 0 \\
&\text{(f)} \quad \ddot{x} + x^2 + x^3 = 0 \\
&\text{(g)} \quad \ddot{x} + e^x = 0
\end{aligned}
$$

4.   Suppose that the equation $\ddot{x} = g(x)$ has no constant solution. Show that all the solutions are unbounded.

5.   A pendulum is displaced through an angle α, $0 < \alpha < \pi$, from its stable equilibrium position and released. Find the period of the oscillation in terms of an integral. Assume $g/l = 1$.

6.   Find the linear approximation of Eq. (5.4) at P and verify that its solutions have period $2\pi/\sqrt{\alpha\gamma}$.

MISCELLANEOUS EXERCISES FOR CHAPTER 6

1. The curves shown in Fig. 46 are not the characteristics of any system $\dot{x} = f(x,y)$, $\dot{y} = g(x,y)$, with f and g continuous. Why not?

2. Classify the singular points of the system

$$\dot{x} = x(x^2 + y^2 - 1)(x^2 + y^2 - 9) - y(x^2 + y^2 - 2x - 8)$$
$$\dot{y} = y(x^2 + y^2 - 1)(x^2 + y^2 - 9) + x(x^2 + y^2 - 2x - 8).$$

Transform the system to polar coordinates and sketch the phase portrait. (This example is due to Poincaré. See his *Collected Works*, Vol. 1, page 70.)

3. (a) Let f and g be homogeneous of degree m (see Exercise 2, page 201). Show that for $x \neq 0$ the substitution $y = zx$ transforms the system

$$\dot{x} = f(x,y), \quad \dot{y} = g(x,y) \tag{1}$$

into

$$\dot{x} = x^m f(1,z)$$
$$\dot{z} = x^{m-1}[g(1,z) - zf(1,z)]. \tag{2}$$

(b) Assume that $g(1,z) - zf(1,z)$ does not vanish, and show that $x^m[g(1,z) - zf(1,z)]$ is an integrating factor of (2).

(c) Hence show that

$$\log |x| - \int \frac{f(1,z)\,dz}{g(1,z) - zf(1,z)}$$

is an integral of (2). Substituting y/x for z yields an integral of (1) in the left and right half-planes.

4. Find the phase portrait of $\dot{x} = 2xy$, $\dot{y} = y^2 - x^2$ by the method of Exercise 3. Show that, except for the origin and the two rays on the y-axis, every trajectory is a circle with one point deleted, which issues from, and returns to, the origin. Sketch the phase portrait.

5. (a) Let m and n be positive odd integers. Show that all solutions of $\dot{x} = y^m$, $\dot{y} = -x^n$ are periodic.

(b) Show that the period of the solution satisfying the initial condition $u(0) = 0$, $v(0) = \alpha > 0$ is given by

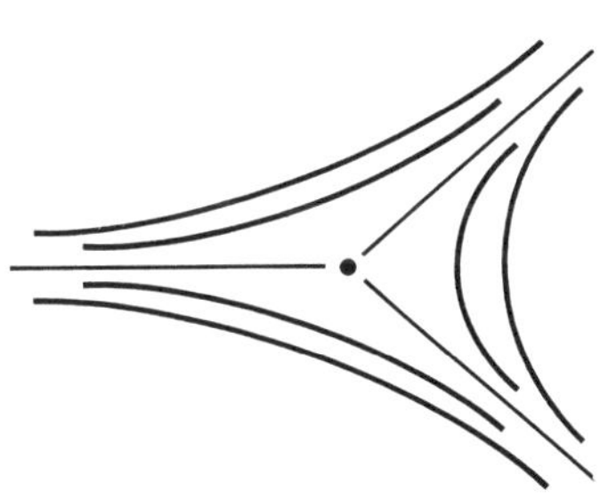

FIG. 46. See Exercise 1.

$$p = 4\left(\frac{m+1}{n+1}\right)^{\frac{n}{n+1}} \alpha^{\frac{1-mn}{n+1}} \int_0^1 \frac{ds}{(1 - s^{m+1})^{\frac{n}{n+1}}}.$$

(c) For which values of m and n is p independent of α? What is the behavior of p as a function of α for other values of m and n?

6. Apply the result of Exercise 5 to the equation $\ddot{x} + x^n = 0$, where n is a positive odd integer. In particular, show that if the number π is defined as half the period of the solutions of $\ddot{x} + x = 0$, then

$$\pi = 2 \int_0^1 \frac{ds}{\sqrt{1 - s^2}}$$

APPENDIX

This appendix is devoted to two topics: (1) the relationship between singular points and the phase portrait as a whole, particularly the orbits; and (2) the geometric significance of integrating factors and of the condition for exactness.

1. The Geometry of Plane Phase Portraits

The main object of this section is to prove Theorem 1.6 below. We shall emphasize the basic geometric features of the proof rather than technical details. The discussion rests on the interpretation of a system

$$\dot{x} = f(x,y), \qquad \dot{y} = g(x,y) \tag{1.1}$$

as a vector field $\vec{V}(x,y) = (f(x,y),g(x,y))$. As always, we assume that the singular points are isolated.

Consider a simple arc C with endpoints α and β which does not pass through any singular point of (1.1). It is not assumed that C is a portion of a trajectory. To each point p of C the vector field assigns a non-zero vector $\vec{V}(p)$. What interests us is the change in direction of this vector as p moves from α to β. Let $\theta(\alpha)$ be the angle between $\vec{V}(\alpha)$ and some fixed direction. To be definite, we shall choose $\theta(\alpha)$ to be the angle between $\vec{V}(\alpha)$ and the positive x-axis, i.e., the unit vector $\vec{i}$, and require that $0 \leqq \theta(\alpha) < 2\pi$. As p moves along C, the angle between $\vec{V}(p)$ and $\vec{i}$ will in general change, and since $\vec{V}$ is a continuous vector field, the angle changes continuously. This statement would be false, however, if we always required the angle to satisfy the condition $0 \leq \theta(p) < 2\pi$. In fact, the angle would make a jump of $\pm 2\pi$ every time that the direction of $\vec{V}(p)$ moved through coincidence with the direction of $\vec{\imath}$, either in the clockwise or in the counterclockwise sense. However, it is possible to choose $\theta(p)$ so that it is a continuous function of p, and the requirement of continuity together with the choice of $\theta(\alpha)$ determines $\theta(p)$ completely. The value $\theta(\beta)$ thus assigned to the direction of $\vec{V}(\beta)$, or, more precisely, the difference $\theta(\beta) - \theta(\alpha)$, is significant. This difference measures not only the angle between the vectors at the two points, but also the net number of complete positive revolutions—counterclockwise minus clockwise—that the vector makes as p moves from α to β. We call $\theta(\beta) - \theta(\alpha)$ the *angular variation of $\vec{V}$ along C from α to β*. If C is traversed in the opposite direction, i.e., from β to α, the angular variation obviously changes sign. Note also that for any point γ of C, the angular variation from α to γ plus the angular variation from γ to β is equal to the angular variation from α to β.

We now define the angular variation of $\vec{V}$ along a simple closed curve C which does not pass through a singular point. We suppose C to be positively oriented, so that as we traverse the curve in the positive sense its interior is to our left. Consider the two arcs into which C is divided by two points α and β. Let v_1 and v_2 be the angular variations as C is traversed in the positive sense from α to β, and from β to α, respectively. The sum $v_1 + v_2$ is defined to be the angular variation of $\vec{V}$ along C. Since $\vec{V}(p)$ returns to its initial value as p goes once around C, this variation is necessarily an integral multiple of 2π, say $2\pi n$. The integer n is called the *index of C with respect to $\vec{V}$*, and is denoted by $\mathcal{I}(C,\vec{V})$. It is obviously equal to the total number of counterclockwise revolutions minus the total number of clockwise revolutions which $\vec{V}(p)$ makes as p traverses C once in the positive sense. This property may be taken as an alternative definition of the index. Note that if C is traversed in the negative sense, the angular variation changes sign. The orientation of C is therefore essential to the definition of the index. We also note that the index remains unchanged if every vector is replaced by its negative, since the rotation of $-\vec{V}$ is the same as that of $\vec{V}$. Thus $\mathcal{I}(C,\vec{V}) = \mathcal{I}(C,-\vec{V})$. Figure 47 shows examples with $\mathcal{I}(C,\vec{V})$ equal to 0, 1, -1, and 2.

We note that the angular variation (and the index) depend only on the behavior of $\vec{V}$ on the given curve itself; our standing hypothesis that $\vec{V}$ is defined in the whole plane is clearly unnecessary if we are only interested in some particular curve. The angular variation is defined as soon as $\vec{V}(p)$ is defined, continuous, and non-zero at every point of the curve. It may also be worth noting that the length of $\vec{V}(p)$ is immaterial; the angular variation remains unchanged if we replace every vector by a unit vector in the same direction.

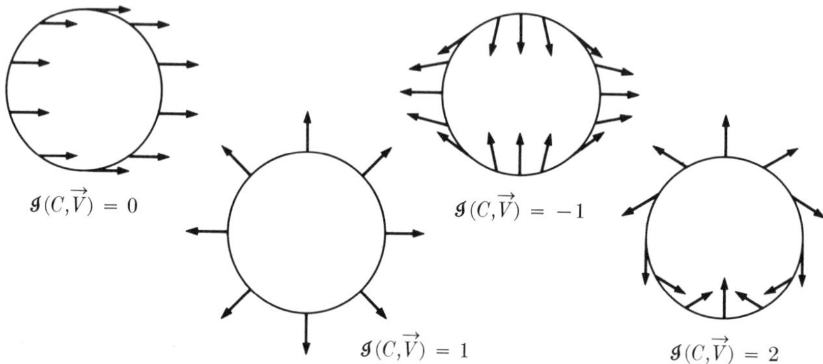

FIG. 47.   Index of a simple closed curve with respect to a vector field.

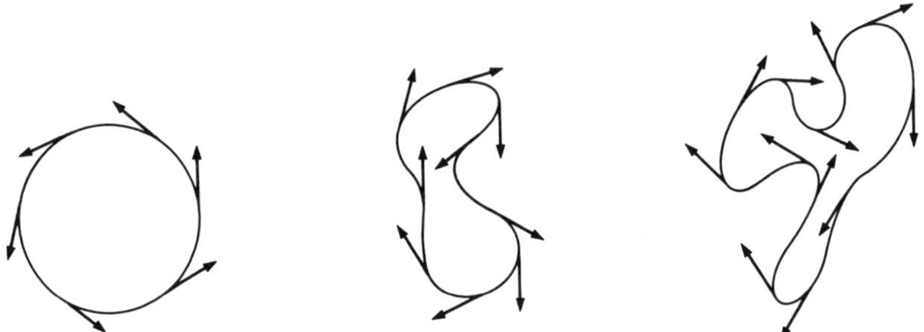

FIG. 48.   Continuous fields of tangent vectors to simple closed curves.

We say that a curve is smooth if it has a parametrization $x = u(t)$, $y = v(t)$, where $\dot{u}$ and $\dot{v}$ are continuous and do not vanish simultaneously. A simple smooth curve has a continuous field of non-zero tangent vectors, for example, $\vec{V} = (\dot{u}(t),\dot{v}(t))$. We shall accept the following fact as intuitively obvious: *The index of a smooth simple closed curve with respect to a continuous non-vanishing field of tangent vectors is equal to* 1. Several examples are shown in Fig. 48.* It

* For a proof of the theorem see Lefschetz, *Differential Equations: Geometric Theory*, page 199.

must be remembered that the orientation of the curve is always taken as positive, so that the orientation induced by a particular parametrization is irrelevant to the index.

Henceforth we suppose $\vec{V}$ to be the vector field defined by (1.1). If C is an orbit of $\vec{V}$, then $\vec{V}$ is tangent to C at every point of C. We note the following consequence.

Theorem 1.1. *Every orbit has index* 1.

We now come to the property of the index on which its utility rests. Let C be a simple closed curve which does not pass through a singular point of $\vec{V}$. Suppose we deform C continuously without allowing it to meet a singular point. Clearly the index varies continuously with C. Since the index assumes only integer values, it follows that it does not change at all.

Theorem 1.2. *Let C_1 and C_2 be simple closed curves which do not pass through a singular point of $\vec{V}$. If C_1 can be deformed into C_2 without crossing a singular point, then $\mathcal{I}(C_1, \vec{V}) = \mathcal{I}(C_2, \vec{V})$.*

This result will now be used to define the index of a *point* with respect to $\vec{V}$. Given a point p, we may choose a disk D with center p which contains no singular point, except possibly p itself. Any two simple closed curves in D which surround p may be deformed into each other without crossing a singular point, and therefore have the same index. This common index is defined to be the index of p with respect to $\vec{V}$, and is denoted by $\mathcal{I}(p, \vec{V})$. Equivalently, the index of p is the index of any sufficiently small circle with center p.

Suppose that p is a regular (non-singular) point. Since $\vec{V}$ is continuous and $\vec{V}(p) \neq 0$, we can choose a disk D about p so that for any point q in D the angle between $\vec{V}(p)$ and $\vec{V}(q)$ is less than, say, $\pi/10$. Thus as q traverses a small circle C with center p, the vector $\vec{V}(q)$ can deviate only very slightly from a fixed direction. It follows that $\mathcal{I}(C, \vec{V}) = 0$, and consequently *the index of a regular point is zero.*

Any simple closed curve which does not surround a singular point may be deformed into a small circle lying in its interior without crossing a singular point, so that the index of such a curve is necessarily zero. Since the index of an orbit is equal to 1, we have reached our first result relating orbits to singular points.

Theorem 1.3.   *Every orbit surrounds at least one singular point.*

We shall presently derive a much stronger theorem. However, even this relatively weak result has obvious applications. If, as often happens, we are looking for orbits, i.e., periodic solutions, of (1.1), it is clear that Theorem 1.3 already narrows the regions of the plane in which orbits might be found. In particular, a system without singular points has no orbits at all. For the case of a single equation $\ddot{x} = f(x,\dot{x})$ this was proved by elementary methods in Chapter 1 (Miscellaneous Exercises, problem 7, page 25).

It follows from the discussion leading up to Theorem 1.3 that if an orbit surrounds only one singular point, then the index of that singular point is equal to 1. It is possible to construct singular points having any prescribed index. Several examples are shown in Fig. 49.

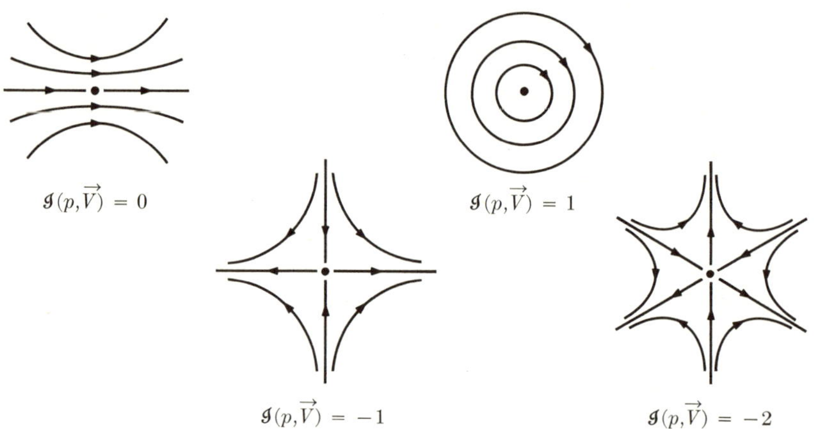

FIG. 49.   Singular points of various indices.

The index of a singular point is easily obtained by inspecting the vector field on a small circle about the singular point, and we shall not pause to discuss analytic methods for computing it. (The index of a curve can be expressed as a line integral. See Exercise 5.) An inspection yields the following result.

Theorem 1.4.   *The index of a center, focus, node, or center-focus is* 1. *The index of a saddle point is* -1.

In case that the singular point is elementary this result may be obtained without much difficulty by analytic means, and is closely related to the fact that the determinant of the linear approximation is negative for a saddle point, and positive in all other cases. Note that if an orbit surrounds only one singular point, that singular point cannot be a saddle. The reader will gain geometric insight by attempting to sketch an orbit surrounding a saddle point but no other singular point.

We now prove the central theorem of the theory of indices.

Theorem 1.5.   *The sum of the indices of all the singular points lying inside any orbit is equal to 1.*

Proof.   Because of our hypothesis that the singular points are isolated, the number of singular points in any bounded region of the plane is finite. Hence there are only a finite number of singular points inside any orbit. Let C be an orbit, and suppose first that only two singular points, p_1 and p_2, lie inside C (Fig. 50). Let C_1 and C_2 be small circles centered on p_1 and p_2. Connect C_1 and C_2 to C and to each other by three auxiliary arcs. These arcs divide the region interior to C and exterior to C_1 and C_2 into two subregions. The boundaries of these subregions are simple closed curves. Since neither subregion contains a singular point, both boundary curves have zero index. Now, the sum of the angular variations of $\vec{V}$ along these curves is equal to the variation along C minus the sum of the variations along C_1 and C_2, since the variations along the auxiliary arcs cancel out, and C_1 and C_2 are traversed in the negative sense. Hence the variation along C equals the sum of the variations along C_1 and C_2, and it follows that $\mathcal{J}(C,\vec{V}) = \mathcal{J}(C_1,\vec{V}) + \mathcal{J}(C_2,\vec{V})$. Since $\mathcal{J}(C,\vec{V}) = 1$ while $\mathcal{J}(C_i,\vec{V}) = \mathcal{J}(p_i,\vec{V})$, $i = 1, 2$, the theorem is proved for the case of two singular points. If there are n singular points, we choose a small circle about each and connect the circles to each other and to C by $n + 1$ auxiliary arcs. The proof then proceeds as before. $\square$

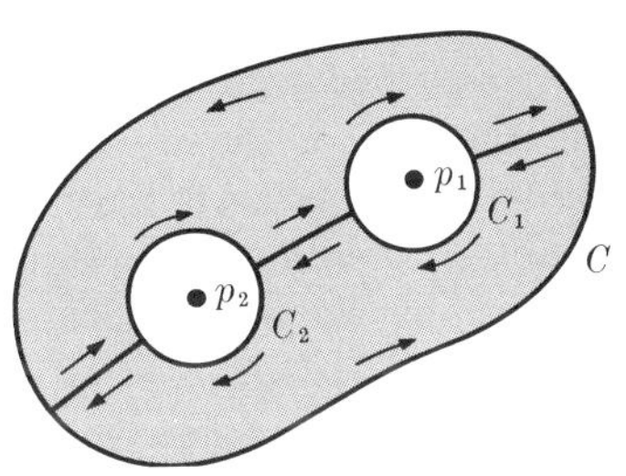

FIG. 50.   Proof of Theorem 1.5.

Combining the two preceding theorems yields our main result.

Theorem 1.6.   *Let Γ be an orbit of a system all of whose singular point are elementary. Let n_1 be the total number of saddle points, and n_2 the total number of other singular points, inside Γ. Then $n_2 - n_1 = 1$.*

As an application to analysis, we consider the distribution of the maxima, minima, and saddle points of a function $H(x,y)$. We suppose that H has continuous partial derivatives of the second order. The critical points of H obviously coincide with the singular points of the system

$$\dot{x} = H_y, \qquad \dot{y} = -H_x. \tag{1.2}$$

We assume that these singular points are elementary—in other words, that the determinant $\Delta = H_{xx}H_{yy} - H_{xy}^2$ does not vanish at a singular point. The critical points of H are then said to be nondegenerate. It is proved in calculus that $\Delta > 0$ at a maximum or minimum, while $\Delta < 0$ at a saddle point (mini-max). Hence the centers of (1.2) correspond to the maxima and minima of H, and the saddle points of (1.2) correspond to the saddle points of H. As we know, the contours, or level curves, of H are the characteristics of (1.2). We therefore conclude that *the total number of maxima and minima inside any simple closed contour of H exceeds by 1 the number of saddle points.*

We proved in Chapter 2, Section 6, that if a first-order autonomous equation $\dot{x} = f(x)$ has no constant solution, then all of its solutions are unbounded both to the right and to the left. The same is true of second-order autonomous systems, but is more difficult to prove in this case. Let Γ: $x = u(t)$, $y = v(t)$ be a trajectory which is neither a singular point nor an orbit. Removing a point $x_0 = u(t_0)$, $y_0 = v(t_0)$ decomposes Γ into two arcs, corresponding to values of t less than, and greater than, t_0. We call these arcs semi-trajectories. The assertion is now that *if there are no singular points, and therefore no orbits, every semi-trajectory is unbounded.* It is an immediate consequence of Theorem 1.3 and of the following theorem, whose proof we omit.

Theorem 1.7 (Poincaré-Bendixson Theorem).   *A bounded semi-trajectory either tends to a singular point or spirals onto a simple closed curve C. The curve C is either an orbit, or consists of one or more singular points together with trajectories tending to these singular points at both ends.*

Now consider a system which has no singular points, and suppose that it has a bounded semi-trajectory γ. Then γ must spiral onto an orbit. But a system without singular points has no orbit. This contradiction yields the desired result.

In conclusion we remark that singular points have considerably less influence on the phase portraits of systems of order $n > 2$. (Such systems define vector fields in n-dimensional Euclidean space.) Consider, for example, the third-order system

$$\dot{x} = y, \qquad \dot{y} = -x, \qquad \dot{z} = 1 - x^2 - y^2.$$

The solutions are $x = a \sin (t + b)$, $y = a \cos (t + b)$, $z = (1 - a^2)t + c$, where a, b, c are arbitrary constants. Although the system has no singular points, it has infinitely many orbits. These orbits, which are obtained by setting $a = 1$, fill out the cylinder $x^2 + y^2 = 1$.

EXERCISES

1.   Let $\vec{V}$ be a continuous non-vanishing vector field on a simple closed curve C. Suppose that $\vec{V}$ always, or never, points into the interior of C. What is $\mathscr{I}(C,\vec{V})$?

2.   (Brouwer Fixed Point Theorem)   Let D be a disk, and let φ be a continuous mapping of D into itself, i.e., a function defined on D such that $\varphi(p)$ belongs to D for all p in D. Show that there is at least one point p in D such that $\varphi(p) = p$. (Hint: If there is no such point, $\vec{V}(p) = \varphi(p) - p$ is a continuous vector field in D without singular points. Compute the index of the boundary of D with respect to $\vec{V}$.)

3.   A simple smooth curve C is said to be without contact (with respect to a vector field $\vec{V}$) if $\vec{V}$ neither vanishes on C nor is tangent to it at any point. Suppose C is closed and without contact. What is $\mathscr{I}(C,\vec{V})$?

4.   Let $\vec{V}$ be the vector field defined by (1.1). Let C be a smooth simple closed curve such that (1) $\vec{V}$ does not vanish on C; (2) there are at most a finite number of points on C at which $\vec{V}$ is tangent to C; and (3) no trajectory cuts C at a point of tangency. In the neighborhood of a point q at which a trajectory is tangent to C, the trajectory therefore lies on one side of C, either inside or outside. In the first case q is called an interior contact, and in the second case, an exterior contact, of C. Let i and e be the number of interior and exterior contacts of C. Then $\mathscr{I}(C,\vec{V}) = (i - e)/2 + 1$. We divide the proof into steps.

(a)   Show that, as p moves along C in the positive sense through an exterior (interior) contact, $\vec{V}(p)$ rotates clockwise (counterclockwise).

(b)   Label the contacts $q_1, q_2, \ldots , q_n = q_1$ in the order in which they are encountered as C is traversed in the positive sense, starting at an arbitrary contact q_1. Let $\vec{T}$ be a continuous field of non-zero tangent vectors to C. Show that the angular variation of $\vec{V}$ minus the angular variation of $\vec{T}$ along an arc $q_k q_{k+1}$ is π, $-\pi$, or 0, according as the endpoints are both interior contacts, both exterior contacts, or contacts of different kinds.

(c)   Let $\sigma(q_k) = 1$ or -1 according as q_k is an interior or exterior contact. Show that the angular variation of $\vec{V}$ minus the angular variation of $\vec{T}$ along C is given by

$$\pi \sum_{k=1}^{n-1} \frac{\sigma(q_k) + \sigma(q_{k+1})}{2},$$

and therefore by $\pi(i - e)$.

(d)   Conclude that $\mathscr{I}(C,\vec{V}) = (i - e)/2 + 1$.

(e)   Illustrate this formula by taking C to be a small circle about (1) a regular point, (2) a saddle point, and (3) a center surrounded by elliptical orbits.

5.   Let C be a smooth simple closed curve with parametrization $x = u(t)$, $y = v(t)$, where $\dot{u}$ and $\dot{v}$ are continuous. Assume C passes through no singular point of (1.1). On C, the angle between $\vec{V}$ and $\vec{i}$ is a function of t and satisfies

$$\cos \theta(t) = \frac{f(u(t),v(t))}{\sqrt{f^2(u(t),v(t)) + g^2(u(t),v(t))}}$$

$$\sin \theta(t) = \frac{g(u(t),v(t))}{\sqrt{f^2(u(t),v(t)) + g^2(u(t),v(t))}}.$$

(a)   Accepting the fact that $\theta(t)$ can be chosen so as to be differentiable, show that

$$\dot{\theta} = \frac{(g_x\dot{u} + g_y\dot{v})f - (f_x\dot{u} + f_y\dot{v})g}{f^2 + g^2}$$

and therefore that

$$d\theta = \frac{fdg - gdf}{f^2 + g^2}.$$

(b)   Conclude that

$$\mathscr{I}(C,\vec{V}) = \frac{1}{2\pi} \int_C \frac{fdg - gdf}{f^2 + g^2}.$$

2.   Integrating Factors as Density Functions

In discussing exact systems and integrating factors we did not raise the issue of the possible geometric significance of the condition of exactness $f_x + g_y = 0$, or, more generally, of the condition $\partial(Mf)/\partial x + \partial(Mg)/\partial y = 0$ satisfied by an integrating factor. The definitions of exactness and of integrating factors were motivated by purely computational considerations. The purpose of this section is to show that integrating factors and systems which have integrating factors are characterized by a simple qualitative property. This property, which is easily generalized to systems of higher order, is basic to statistical mechanics and ergodic theory, and is therefore important in its own right.

Let us imagine the vector field $\vec{V}(x,y) = (f(x,y),g(x,y))$ as describing the stationary flow of a fluid over the (x,y)-plane, the velocity of the fluid at any point (x,y) being given by $\vec{V}(x,y)$. The term "stationary" reflects the fact that this velocity is independent of the time. The trajectories of the system $\dot{x} = f(x,y)$, $\dot{y} = g(x,y)$ are the stream lines of the flow. Of course, since the solutions of the system may not be defined for all t, such a flow may represent a considerable abstraction from reality. This however, is a relatively minor point, and nothing essential will be lost if we assume the solutions to be defined for all t. Flows defined by arbitrary vector fields depart in a much more significant respect from flows of real fluids. In fact, it is precisely by asking which flows deserve to be called stationary fluid flows that we arrive at a geometrical interpretation of integrating factors. In a stationary fluid flow the total mass of fluid contained in a given bounded region D of the plane, such as a disk, is constant in time, the mass of fluid entering the region in any interval of time being equal to the mass of fluid leaving the region. The flow therefore has a stationary mass distribution, or, what amounts to the same thing, a stationary density $\rho(x,y)$. The density is defined by the condition that the mass of fluid in any region D is given by $\iint_D \rho(x,y)\,dx\,dy$. It is easy to see that this density function is closely related to the flow of the fluid. Consider the fluid particles occupying some region D_0 at a given instant of time, which we may take as $t = 0$. At a later time t the flow will have carried the particles to new positions, and they now occupy a region D_t. Since their total mass remains unchanged, we have

$$\iint_{D_0} \rho(x,y)\,dx\,dy = \iint_{D_t} \rho(x,y)\,dx\,dy. \tag{2.1}$$

This relation is clearly also satisfied for negative values of t. The fact that $\rho(x,y)$ must satisfy (2.1) for all regions D_0 and for all t constitutes the connection between the density and the flow. The relation (2.1) does not by itself determine $\rho(x,y)$, since we may, for example, multiply ρ by a constant without disturbing this relation, but it is nevertheless a significant restriction which the flow imposes on the possible density functions. An important case arises if $\rho(x,y) = $ constant. The density of the fluid is then constant throughout the plane, and we say the flow is incompressible. Equation (2.1) now implies that the area of D_0 is equal to the area of D_t, and for this reason we also say that the flow is area-preserving. If the flow is compressible, i.e., if $\rho(x,y)$ is not constant, the flow still preserves something, namely, the mass-distribution. Areas and mass-distributions are examples of *measures*, i.e., of functions which assign numbers to sets rather than to points, subject to such general requirements as that the measure of the union of two disjoint sets is equal to the sum of their measures. In mathematical terminology, therefore, stationary fluid flows are said to be measure-preserving, or to have an invariant measure.

Our object now is to establish the remarkable result that $\rho(x,y)$ is a density function for the flow defined by a system

$$\dot{x} = f(x,y), \qquad \dot{y} = g(x,y) \tag{2.2}$$

if and only if $\rho(x,y)$ is an integrating factor of the system. We require that $\rho(x,y)$ be positive for all (x,y) and have continuous partial derivatives.

We shall need the so-called rule for change of variables in multiple integrals. This rule, which is proved in texts on advanced calculus, may be stated as follows. Let φ be a mapping of the plane into itself i.e., a function which assigns to each point (x,y) a point $\varphi(x,y) = (u(x,y),v(x,y))$. We suppose that φ is one-to-one, i.e., $\varphi(p) = \varphi(q)$ only if $p = q$. We suppose also that u and v have continuous partial derivatives, and that the determinant

$$\mathcal{J}(x,y) = \begin{vmatrix} u_x(x,y) & u_y(x,y) \\ v_x(x,y) & v_y(x,y) \end{vmatrix}$$

is positive for all (x,y). This determinant is called the Jacobian of φ. Let D be a bounded region and $\varphi(D)$ the region consisting of all points $\varphi(p)$, where p is a point of D. Then for any continuous function $\rho(x,y)$,

$$\iint\limits_{\varphi(D)} \rho(x,y)\, dx\, dy = \iint\limits_{D} \rho(u(x,y),v(x,y))\mathcal{J}(x,y)\, dx\, dy. \tag{2.3}$$

In particular, $\iint\limits_{D} \mathcal{J}(x,y)\, dx\, dy$ is the area of $\varphi(D)$. It follows that φ preserves areas if and only if $\mathcal{J}(x,y) \equiv 1$.

We shall also make use of the fact that if $x = u(t,\xi,\eta)$, $y = v(t,\xi,\eta)$ is the solution of (2.2) satisfying $u(0,\xi,\eta) = \xi$, $v(0,\xi,\eta) = \eta$, then the mixed partial derivatives $u_{t,\xi}$, $u_{\xi,t}$, $u_{t,\eta}$, $u_{\eta,t}$ exist and are continuous, and similarly for the derivatives of v. The proof of this theorem may be found in advanced texts on differential equations.

To simplify the formulas, we write (2.2) in the form

$$\dot{x}_1 = f_1(x_1,x_2), \qquad \dot{x}_2 = f_2(x_1,x_2) \tag{2.4}$$

or, in vector notation,

$$\dot{x} = f(x), \tag{2.5}$$

and let $x = u(t,\eta)$ be the solution of (2.5) satisfying the initial condition $u(0,\eta) = \eta$. Thus $x_i = u_i(t,\eta_1,\eta_2)$ is the solution of (2.4) which satisfies $u_i(0,\eta_1,\eta_2) = \eta_i$, $i = 1$, 2. To avoid a minor technical complication we assume that u is defined for all (t,η).

The function u is the flow defined by (2.5): a particle, or representative point, which is initially at the point η is at the point $u(t,\eta)$ at time t. (We leave it as an exercise for the reader to show that a particle which is at the point η at time τ is at the point $u(t - \tau, \eta)$ at time t.) For each value of t, u defines a mapping φ_t of the plane into itself, namely $\varphi_t(\eta) = u(t,\eta)$. By the uniqueness theorem, φ_t is one-to-one for every t, and by the theorem mentioned above, its coordinate functions have continuous partial derivatives with respect to η_1 and η_2. We now compute $\mathcal{J}(t,\eta)$, the Jacobian of φ_t.

Lemma 2.1. $\quad \mathcal{J}(t,\eta) = e^{\int_0^t \left(\frac{\partial f_1}{\partial x_1} + \frac{\partial f_2}{\partial x_2}\right) ds}$, *the partial derivatives being evaluated at* $u(s,\eta)$.

Proof. Let (a_{ij}) denote the matrix whose (i,j)th entry is a_{ij}. Setting $W(t,\eta) = \left(\dfrac{\partial u_i(t,\eta)}{\partial \eta_j}\right)$, we have $\mathcal{J}(t,\eta) = \det W(t,\eta)$. Since $u_i(0,\eta) = \eta_i$, it follows that $\dfrac{\partial u_i(0,\eta)}{\partial \eta_j} = \dfrac{\partial \eta_i}{\partial \eta_j} = 1$ or 0 according as $i = j$ or $i \neq j$. Hence $W(0,\eta) = I$ for all η. Now

$$\begin{aligned}
\dot{W}(t,\eta) &= \left(\frac{\partial^2 u_i(t,\eta)}{\partial t \partial \eta_j}\right) \\
&= \left(\frac{\partial^2 u_i(t,\eta)}{\partial \eta_j \partial t}\right) \\
&= \left(\frac{\partial \dot{u}_i(t,\eta)}{\partial \eta_j}\right) \\
&= \left(\frac{\partial f_i(u(t,\eta))}{\partial \eta_j}\right) \\
&= \left(\sum_{k=1}^{2} \frac{\partial f_i(u(t,\eta))}{\partial x_k} \frac{\partial u_k(t,\eta)}{\partial \eta_j}\right),
\end{aligned}$$

where the change in the order of differentiation is justified by the continuity of the mixed partial derivatives. Setting $A(t,\eta) = \dfrac{\partial f_i(u(t,\eta))}{\partial x_j}$, we thus have $\dot{W}(t,\eta) = A(t,\eta)W(t,\eta)$. Hence W is a solution matrix of the system $\dot{x} = A(t,\eta)x$. Note that η occurs merely as a parameter in this system. Since $W(0,\eta) = I$, W is the standard fundamental matrix. Liouville's formula for $\det W(t,\eta)$ now yields the desired formula for $\mathcal{J}(t,\eta)$. $\square$

We note that $\mathcal{J}(t,\eta) > 0$, so that formula (2.3) applies to the mappings φ_t.

Theorem 2.2. $\rho(x_1,x_2)$ *is a density function for the flow defined by the system* $\dot{x} = f(x)$ *if and only if*

$$\frac{\partial(\rho f_1)}{\partial x_1} + \frac{\partial(\rho f_2)}{\partial x_2} = 0 \tag{2.6}$$

Proof. By definition, ρ is a density function if

$$\iint\limits_{D} \rho(\eta)\, d\eta = \iint\limits_{\varphi_t(D)} \rho(\eta)\, d\eta \tag{2.7}$$

for all bounded regions D and all t. (For the sake of brevity we have written $d\eta$ instead of $d\eta_1\, d\eta_2$.) By (2.3),

$$\iint\limits_{\varphi_t(D)} \rho(\eta)\, d\eta = \iint\limits_{D} \rho(\varphi_t(\eta))\mathcal{J}(t,\eta)\, d\eta,$$

so that (2.7) becomes

$$\iint\limits_{D} \rho(\eta)\, d\eta = \iint\limits_{D} \rho(\varphi_t(\eta))\mathcal{J}(t,\eta)\, d\eta.$$

The necessary and sufficient condition that this relation holds for all D and t is that

$$\rho(\eta) = \rho(\varphi_t(\eta))\mathcal{J}(t,\eta)$$

for all (t,η), in other words, that the right side is independent of t. The condition on ρ thus becomes

$$\frac{\partial}{\partial t}[\rho(\varphi_t(\eta))\mathcal{J}(t,\eta)] = 0$$

for all (t,η). Using the definition of φ_t and differentiating yields

$$\left(\frac{\partial\rho}{\partial x_1}\frac{\partial u_1}{\partial t} + \frac{\partial\rho}{\partial x_2}\frac{\partial u_2}{\partial t}\right)\mathcal{J} + \rho\frac{\partial\mathcal{J}}{\partial t} = 0, \tag{2.8}$$

where for brevity we have omitted the arguments of the functions. Now $\dfrac{\partial u_i}{\partial t} = f_i(u)$, $i = 1, 2$, and by Lemma 2.1, $\dfrac{\partial \mathcal{J}}{\partial t} = \left(\dfrac{\partial f_1}{\partial x_1} + \dfrac{\partial f_2}{\partial x_2}\right)\mathcal{J}$. Thus (2.8) becomes

$$\left(\frac{\partial \rho}{\partial x_1}f_1 + \frac{\partial \rho}{\partial x_2}f_2\right)\mathcal{J} + \left(\frac{\partial f_1}{\partial x_1} + \frac{\partial f_2}{\partial x_2}\right)\rho\mathcal{J} = 0.$$

Dividing by $\mathcal{J}$ yields

$$\frac{\partial \rho}{\partial x_1}f_1 + \frac{\partial \rho}{\partial x_2}f_2 + \rho\frac{\partial f_1}{\partial x_1} + \rho\frac{\partial f_2}{\partial x_2} = 0,$$

or

$$\frac{\partial(\rho f_1)}{\partial x_1} + \frac{\partial(\rho f_2)}{\partial x_2} = 0. \tag{2.9}$$

In this formula, ρ, f_1, and f_2 are evaluated at $u(t,\eta)$. Since (2.9) holds for all (t,η) if and only if it holds for all x, the theorem is proved. $\square$

A density function is by definition positive. We have therefore reached our main result.

Theorem 2.3. *ρ is a density function if and only if it is an integrating factor.*

Thus a system has an integrating factor if and only if it has an invariant measure. More concretely put, a system with an integrating factor defines a flow like a steady-state fluid flow. This gives us another proof of the fact that a system with attractors cannot have an integrating factor. Consider the case of an asymptotically stable singular point p. We can choose a disk D of radius r and center p such that every trajectory in D tends to p as $t \to \infty$. For t sufficiently large, $\varphi_t(D)$ is therefore contained in a disk of radius $r/2$ about p. Now the assumption that the system has an integrating factor, and therefore a density function, leads to the absurd conclusion that the mass of $\varphi_t(D)$ is equal to the mass of D.

We remark that in hydrodynamics, (2.6) is called the equation of continuity for a steady-state fluid flow.

The condition that the flow is incompressible is that (2.6) is satisfied by $\rho = $ constant, and therefore by $\rho = 1$. Hence the necessary and sufficient condition that a system $\dot{x}_1 = f_1(x_1,x_2)$, $\dot{x}_2 = f_2(x_1,x_2)$ defines an incompressible flow is that $\partial f_1/\partial x_1 + \partial f_2/\partial x_2 = 0$. But this is precisely the condition that the system is exact. We therefore have another significant result.

Theorem 2.4.   *A system is exact if and only if its flow preserves areas.*

In vector analysis the function $\partial f_1/\partial x_1 + \partial f_2/\partial x_2$ is called the divergence of the vector field $\vec{V} = (f_1, f_2)$ and is denoted by div $\vec{V}$. Short intuitive proofs of the fact that stationary and incompressible flows are characterized by div $(\rho\vec{V}) = 0$ and div $\vec{V} = 0$, respectively, may be found in most texts on engineering mathematics.

Except for the obvious notational changes, Theorem 2.2 and its proof remain valid for nth-order systems $\dot{x}_1 = f_i(x_1, \ldots, x_n)$, $i = 1, \ldots, n$. Such a system defines a flow in n-dimensional space, and the condition that $\rho(x_1, \ldots, x_n)$ is a density function is that

$$\frac{\partial(\rho f_1)}{\partial x_1} + \frac{\partial(\rho f_2)}{\partial x_2} + \cdots + \frac{\partial(\rho f_n)}{\partial x_n} = 0.$$

Thus the flow preserves n-dimensional volume if

$$\frac{\partial f_1}{\partial x_1} + \frac{\partial f_2}{\partial x_2} + \cdots + \frac{\partial f_n}{\partial x_n} = 0. \tag{2.10}$$

A case of the greatest importance is that of a Hamiltonian system

$$\dot{x}_i = \frac{\partial H}{\partial y_i}, \qquad \dot{y}_i = -\frac{\partial H}{\partial x_i} \qquad i = 1, \ldots, n$$

where H is a function of the $2n$ variables $x_1, \ldots, x_n, y_1, \ldots, y_n$. The reader will readily verify that such a system satisfies (2.10). With a suitable choice of coordinates the equations of motion of conservative dynamical systems—e.g., the solar system—assume this form. The fact that Hamiltonian flows preserve volume was first pointed out by Liouville.

EXERCISES

1.   (a)   Let $u(t,\eta)$ be the solution of (2.5) which satisfies $u(0,\eta) = \eta$, and suppose u is defined for all (t,η). Show that $u(t,u(\tau,\eta)) = u(t + \tau,\eta)$, and consequently that $\varphi_{t_1}(\varphi_{t_2}(\eta)) = \varphi_{t_1+t_2}(\eta)$.

(b)   Let $v(t,\tau,\eta)$ be the solution of (2.5) which satisfies $v(\tau,\tau,\eta) = \eta$, so that $v(t,0,\eta) = u(t,\eta)$. Show that $v(t,\tau,\eta) = u(t - \tau,\eta)$.

2.   Let D be a region bounded by a smooth simple closed curve. Gauss's theorem states that

$$\iint_D (f_x + g_y)\, dx\, dy = \int_C f\, dy - g\, dx,$$

where C is the positively oriented boundary of D. Prove:

(a)   If D is bounded by an orbit of $\dot{x} = f$, $\dot{y} = g$, then

$$\iint_D (f_x + g_y)\, dx\, dy = 0.$$

(b)   If $f_x + g_y$ does not vanish at any point of D, then D contains no orbit.

3.   Let D be the square $0 \leq x \leq 1$, $0 \leq y \leq 1$, and let φ_t be the flow defined by $\dot{x} = \lambda x$, $\dot{y} = -\mu y$, where λ, $\mu > 0$.

(a)   Find $\varphi_t(D)$ and compute its area.

(b)   What is the condition that the area of $\varphi_t(D)$ (1) increases with t; (2) decreases with t; (3) is constant?

4.   Can compressible and incompressible steady state flows be distinguished by their phase portraits?

7

EXISTENCE AND UNIQUENESS THEOREMS

In this chapter we prove the fundamental existence and uniqueness theorems which we have used throughout the book. The case of complex linear systems is discussed at the end of Section 3. With this exception, all numbers occurring in this chapter are real. We assume that the reader has some familiarity with the basic notions of analysis, particularly that of uniform convergence.

1. First-Order Equations

The Local Existence and Uniqueness Theorem.   We wish to show that a first-order equation $\dot{x} = f(t,x)$ has a unique solution satisfying a given initial condition, assuming that f and $\partial f/\partial x$ are continuous. More precisely, we shall show that given a point (t_0,x_0) there is an interval about t_0 on which a solution u satisfying $u(t_0) = x_0$ is defined, and that no other solution defined on the same interval satisfies the same initial condition. The size of the interval on which a solution is shown to exist is dictated by the method of proof, and has no intrinsic significance. In particular, the method of proof does not lead to the construction of a maximal solution. The various questions connected with maximal solutions will be taken up later. All that concerns us at the moment is to show that there is *some* interval about t_0 on which a solution exists. For this reason, there is no advantage in assuming that f is defined in the whole plane. In fact, the only assumption that it is profitable for us to make is that f is defined on some rectangle containing the initial point (t_0,x_0).

Hypothesis. We assume that f and $\partial f/\partial x$ are defined and continuous on a rectangle

$$Q: \quad |t - t_0| \leqq a, \qquad |x - x_0| \leqq b, \qquad a, b > 0. \tag{1.1}$$

The first step in the proof is to show that the differential equation together with the initial condition may be replaced by a single integral equation. The remainder of the proof then consists in showing that the integral equation has a unique solution.

Lemma 1.1. *u is a solution of $\dot{x} = f(t,x)$ satisfying the initial condition $u(t_0) = x_0$ if and only if*

$$u(t) = x_0 + \int_{t_0}^{t} f(s,u(s))\, ds.$$

Proof. Suppose that u is a solution of $\dot{x} = f(t,x)$ defined on an interval I and satisfying $u(t_0) = x_0$. We integrate both sides of the equation $\dot{u}(t) = f(t,u(t))$ from t_0 to t, where t is in I:

$$\int_{t_0}^{t} \dot{u}(s)\, ds = \int_{t_0}^{t} f(s,u(s))\, ds,$$

or

$$u(t) - u(t_0) = \int_{t_0}^{t} f(s,u(s))\, ds.$$

Since $u(t_0) = x_0$, we have

$$u(t) = x_0 + \int_{t_0}^{t} f(s,u(s))\, ds, \qquad t \text{ in } I. \tag{1.2}$$

We shall now show that, conversely, any function which satisfies this integral equation satisfies both the differential equation and the initial condition. Suppose that u is a function defined on an interval I and satisfies (1.2). Setting $t = t_0$ yields $u(t_0) = x_0$, so that u satisfies the initial condition. Next, we note that an integral is always a continuous function, so that a solution of (1.2) is automatically continuous. Since both u and f are continuous, it follows that the integrand $f(s,u(s))$ is continuous. We may therefore apply the fundamental theorem of calculus to (1.2) and conclude that u is differentiable, and that $\dot{u}(t) = f(t,u(t))$. $\square$

We must now say something about the hypothesis that f has a continuous partial derivative with respect to x. We shall use this hypothesis to prove both the existence of a solution and its uniqueness. Strictly speaking, however, it is not this hypothesis itself, but a simple consequence of it, which actually enters into the proof. This consequence is the following.

Lemma 1.2.   *There is a positive number K such that*

$$|f(t,x_1) - f(t,x_2)| \leq K|x_1 - x_2| \qquad (1.3)$$

for any two points (t,x_1) and (t,x_2) in the rectangle Q.

Proof.   Choose two points (t,x_1), (t,x_2) in Q having the same abscissa. By the mean value theorem of differential calculus applied to f as a function of x only, there is a number x_3 between x_1 and x_2 such that

$$f(t,x_1) - f(t,x_2) = \frac{\partial f(t,x_3)}{\partial x} (x_1 - x_2).$$

Taking absolute values yields

$$|f(t,x_1) - f(t,x_2)| = \left| \frac{\partial f(t,x_3)}{\partial x} \right| |x_1 - x_2|.$$

Since Q is a closed bounded subset of the plane, and since $\partial f/\partial x$ is continuous on Q, there is a positive number K such that $|\partial f(t,x)/\partial x| \leq K$ for all (t,x) in Q. It follows that

$$|f(t,x_1) - f(t,x_2)| \leq K|x_1 - x_2|. \qquad \square$$

Any number K for which (1.3) holds is called a *Lipschitz constant* for f (with respect to x). In the following we shall not make any explicit use of the fact that $\partial f/\partial x$ is continuous, but merely of the fact that f has a Lipschitz constant. Since any number larger than a Lipschitz constant is itself a Lipschitz constant, we may always assume that the Lipschitz constant is positive. (The case that f has a zero Lipschitz constant is actually trivial. See Exercise 1.)

We are now ready to give a proof of the uniqueness theorem. It is much simpler than the existence proof, and benefits by being isolated from it. In the proof we shall use a notation which will also be of great service in the existence proof. Let I be a closed bounded interval, and u a function which is defined and continuous on I, so that $|u(t)|$ assumes a maximum. This maximum is called the *norm* of u, and is denoted by $\|u\|$:

$$\|u\| = \max_{t \text{ in } I} |u(t)|.$$

If u and v are two functions defined and continuous on I, so is their difference. We may therefore take the norm of this difference:

$$\|u - v\| = \max_{t \text{ in } I} |u(t) - v(t)|.$$

Note that the norm of a function is zero if and only if the function is identically zero, and is positive otherwise. In particular, $\|u - v\| = 0$ if and only if $u = v$.

Theorem 1.3 (Uniqueness Theorem). *Let d be a number which satisfies $0 < d < 1/K$, where K is a positive Lipschitz constant for f. If u and v are solutions of $\dot{x} = f(t,x)$ defined on the interval $I: |t - t_0| \leqq d$, and if $u(t_0) = v(t_0)$, then $u = v$.*

Proof. Let $u(t_0) = v(t_0) = x_0$. By Lemma 1.1,

$$u(t) = x_0 + \int_{t_0}^{t} f(s,u(s))\, ds$$

and

$$v(t) = x_0 + \int_{t_0}^{t} f(s,v(s))\, dx$$

for t in I. Hence, subtracting,

$$\begin{aligned}
u(t) - v(t) &= \int_{t_0}^{t} f(s,u(s))\, ds - \int_{t_0}^{t} f(s,v(s))\, ds \\
&= \int_{t_0}^{t} [f(s,u(s)) - f(s,v(s))]\, ds.
\end{aligned}$$

Taking absolute values yields

$$|u(t) - v(t)| \leqq \left| \int_{t_0}^{t} |f(s,u(s)) - f(s,v(s))|\, ds \right|$$

By hypothesis, the points $(s,u(s))$ and $(s,v(s))$ belong to Q, for each s in I. Hence

$$|f(s,u(s)) - f(s,v(s))| \leqq K|u(s) - v(s)|,$$

and it follows that

$$\begin{aligned}
|u(t) - v(t)| &\leqq \left| \int_{t_0}^{t} K|u(s) - v(s)|\, ds \right| \\
&\leqq \left| \int_{t_0}^{t} K\|u - v\|\, ds \right| \\
&= K\|u - v\|\, |t - t_0| \\
&\leqq K\|u - v\|d.
\end{aligned}$$

Since this inequality is valid for all t in I, the maximum of $|u(t) - v(t)|$ cannot exceed the right side, that is,

$$\|u - v\| \leqq K\|u - v\|d. \tag{1.4}$$

We now obtain a contradiction by supposing that $\|u - v\| > 0$. For, dividing (1.4) by $\|u - v\|$ yields $1 \leqq Kd$, or $d \geqq 1/K$, contrary to the choice of d. It follows that $\|u - v\| = 0$, and therefore $u = v$. $\qquad\square$

We have established that if there is a solution of $\dot{x} = f(t,x)$ defined on a sufficiently small interval about t_0 and assuming the value x_0 at t_0, then there is only one such. We now turn to the more difficult problem of showing that such a solution exists. Before embarking on the proof, we recall a theorem from analysis. A sequence of functions $u_0, u_1, u_2, \ldots$ will be denoted by (u_n). Let I be a closed bounded interval, and (u_n) a sequence of continuous functions defined on I. Suppose that for every $\epsilon > 0$ there is an integer m such that $n > m$ implies that $\|u_n - u_{n+k}\| < \epsilon$ for all k. Then there is a continuous function u on I such that $\lim_{n \to \infty} \|u - u_n\| = 0$. We say that the sequence (u_n) converges uniformly to u.

An outline of the existence proof, which is due to Emile Picard, is easily sketched. We choose a small interval I about t_0, and a function u_0 which is defined and continuous on I and whose graph lies in Q. We may, for example, take $u_0(t) \equiv x_0$. We then define another function u_1 on I by setting $u_1(t) = x_0 + \int_{t_0}^{t} f(s, u_0(s))\, ds$. Continuing in this fashion we define, for every positive integer n, a function u_n on I by $u_n(t) = x_0 + \int_{t_0}^{t} f(s, u_{n-1}(s))\, ds$. We then prove that the sequence (u_n) thus obtained converges uniformly to a function u, and that this function satisfies $u(t) = x_0 + \int_{t_0}^{t} f(s, u(s))\, ds$.

The first thing we must do is to assure that the process by which we construct each u_n from the preceding one does, in fact, work, i.e., that all the functions defined by the process are continuous and have graphs which lie in Q. Obviously, if at some stage we obtained a function whose graph did not lie in Q, the process would stop. We shall see that this does not happen if we choose I to be sufficiently small. In the following, M denotes a positive number such that $|f(t,x)| \leq M$ for all (t,x) in Q. Since f is continuous, such a number certainly exists.

Lemma 1.4.   *Let d be a positive number such that $d \leq a$ and $d \leq b/M$. Let u be a continuous function defined on the interval I: $|t - t_0| \leq d$ and such that the graph of u lies in Q, i.e., $|u(t) - x_0| \leq b$. Then the function $\bar{u}$ defined on I by*

$$\bar{u}(t) = x_0 + \int_{t_0}^{t} f(s, u(s))\, ds$$

is continuous, and its graph is contained in Q.

Proof.   Obviously $\bar{u}$ is not merely continuous, but actually has a continuous derivative, since $\dot{\bar{u}}(t) = f(t, u(t))$. Note, incidentally, that $\bar{u}(t_0) = x_0$. Now

$$\bar{u}(t) - x_0 = \int_{t_0}^{t} f(s, u(s))\, ds,$$

so

$$
\begin{aligned}
|\bar{u}(t) - x_0| &\leq \left| \int_{t_0}^{t} |f(s, u(s))|\, ds \right| \\
&\leq \left| \int_{t_0}^{t} M\, ds \right| = M|t - t_0| \\
&\leq Md \leq b.
\end{aligned}
$$

But this is precisely the condition that the graph of $\bar{u}$ lies in Q.

The interval I constructed in the lemma therefore has the property that we need in order to construct the sequence (u_n). However, in order to prove that the sequence converges uniformly, we shall impose a further restriction on the interval, namely, that $d < 1/K$. Note that this requirement already occurred in the uniqueness proof.

Theorem 1.5 (Existence Theorem). *Let d be a positive number such that $d \leqq a$, $d \leqq b/M$ and $d < 1/K$. Let u_0 be a continuous function defined on the interval I: $|t - t_0| \leqq d$, and such that $|u_0(t) - x_0| \leqq b$. Then the sequence (u_n) defined on I by*

$$u_{n+1}(t) = x_0 + \int_{t_0}^{t} f(s,u_n(s)) \; ds, \qquad n = 0, 1, 2, \ldots \qquad (1.5)$$

converges uniformly to a function u, and

$$u(t) = x_0 + \int_{t_0}^{t} f(s,u(s)) \; ds. \qquad (1.6)$$

Proof. We begin by finding an estimate for $\|u_n - u_{n+1}\|$. Since

$$u_n(t) - u_{n+1}(t) = \int_{t_0}^{t} [f(s,u_{n-1}(s)) - f(s,u_n(s))] \; ds,$$

we have, using the Lipschitz condition,

$$\begin{aligned}
|u_n(t) - u_{n+1}(t)| &\leqq \left| \int_{t_0}^{t} K|u_{n-1}(s) - u_n(s)| \; ds \right| \\
&\leqq \left| \int_{t_0}^{t} K\|u_{n-1} - u_n\| \; ds \right| \\
&\leqq Kd\|u_{n-1} - u_n\|.
\end{aligned}$$

Hence

$$\|u_n - u_{n+1}\| \leqq Kd\|u_{n-1} - u_n\|. \qquad (1.7)$$

Let $c = Kd$. Because $d < 1/K$, $c < 1$. This fact is basic to the rest of the proof. Set $n = 1$ in (1.7):

$$\|u_1 - u_2\| \leqq c\|u_0 - u_1\|. \qquad (1.8)$$

Next, set $n = 2$ and use (1.8):

$$\|u_2 - u_3\| \leqq c\|u_1 - u_2\| \leqq c^2\|u_0 - u_1\|.$$

Setting $n = 3$ yields in the same way

$$\|u_3 - u_4\| \leqq c^3\|u_0 - u_1\|,$$

and by an obvious induction we have

$$\|u_n - u_{n+1}\| \leqq c^n\|u_0 - u_1\|$$

for all n.

Now (see Exercise 2)

$$
\begin{aligned}
\|u_n - u_{n+k}\| &\leq \|u_n - u_{n+1}\| + \|u_{n+1} - u_{n+2}\| \\
&\qquad\qquad + \cdots + \|u_{n+k-1} - u_{n+k}\| \\
&\leq c^n\|u_0 - u_1\| + c^{n+1}\|u_0 - u_1\| \\
&\qquad\qquad + \cdots + c^{n+k-1}\|u_0 - u_1\| \\
&= c^n(1 + c + \cdots + c^{k-1})\|u_0 - u_1\| \\
&\leq \frac{c^n}{1 - c}\|u_0 - u_1\|.
\end{aligned}
\tag{1.9}
$$

Because $c < 1$, $\lim_{n \to \infty} c^n = 0$. Hence the right side of (1.9) tends to zero as $n \to \infty$, and therefore, given $\epsilon > 0$, there is an m such that $n > m$ implies that $\|u_n - u_{n+k}\| < \epsilon$ for every k. We conclude that (u_n) converges uniformly to a function u. Note that the graph of u lies in Q.

We now show that for any t in I,

$$
\lim_{n \to \infty} \int_{t_0}^{t} f(s, u_n(s))\, ds = \int_{t_0}^{t} f(s, u(s))\, ds.
\tag{1.10}
$$

In fact,

$$
\begin{aligned}
\left| \int_{t_0}^{t} f(s, u(s))\, ds - \int_{t_0}^{t} f(s, u_n(s))\, ds \right| &= \left| \int_{t_0}^{t} [f(s, u(s)) - f(s, u_n(s))]\, ds \right| \\
&\leq \left| \int_{t_0}^{t} K|u(s) - u_n(s)|\, ds \right| \\
&\leq \left| \int_{t_0}^{t} K\|u - u_n\|\, ds \right| \\
&\leq Kd\|u - u_n\|,
\end{aligned}
$$

and since $\|u - u_n\| \to 0$, (1.10) is established.

Finally, it is obvious that $\lim_{n \to \infty} u_{n+1}(t) = u(t)$, so that equating the limits of the left and right sides of (1.5) yields (1.6). The theorem is therefore proved. $\square$

Combining Theorems 1.3 and 1.5, we obtain the following result.

Theorem 1.6 (Local Existence and Uniqueness Theorem).
Let f and $\partial x/\partial x$ be continuous on a rectangle $|t - t_0| \leq a$, $|x - x_0| \leq b$. If d is a sufficiently small positive number, then the differential equation $\dot{x} = f(t,x)$ has a solution u which is defined on the interval $|t - t_0| \leq d$ and satisfies the initial condition $u(t_0) = x_0$, and no other solution defined on this interval satisfies the same initial condition.

As we have already pointed out, the hypothesis that f has a continuous partial derivative with respect to x is not essential to the proof, and may be replaced by the less stringent assumption that f satisfies a Lipschitz condition, i.e., has a Lipschitz constant. However, the functions usually encountered do in fact have continuous partial derivatives. An example of a function which satisfies a Lipschitz condition but is not differentiable is $f(t,x) = |x|$. Clearly $\partial f/\partial x$ does not exist at $x = 0$, but $K = 1$ is a Lipschitz constant in the whole plane.

Uniqueness in the Large; Maximal Solutions. We now suppose that f and $\partial f/\partial x$ are defined and continuous on an open set Ω. The reader may assume that Ω is the whole plane, which is certainly the most important case. It follows immediately from Theorem 1.6 that given a point (t_0,x_0) in Ω there is a positive number d such that the equation $\dot{x} = f(t,x)$ has a solution u defined on $|t - t_0| \leqq d$ satisfying $u(t_0) = x_0$, and that no other solution defined on this interval satisfies the same initial condition. For the proof, we need only choose positive numbers a and b such that the rectangle $|t - t_0| \leqq a$, $|x - x_0| \leqq b$ is contained in Ω, and apply Theorem 1.6. We remark parenthetically that instead of assuming f to have a continuous partial derivative with respect to x, it suffices to assume that f is locally Lipschitzian in x, i.e., that corresponding to every closed bounded rectangle Q in Ω there is a number K such that $|f(t,x_1) - f(t,x_2)| \leqq K|x_1 - x_2|$ for all points (t,x_1), (t,x_2) in Q.

Our first object is to prove that through each point of Ω there passes a unique maximal solution. The idea of a maximal solution was already introduced in Chapter 1, but for the convenience of the reader we restate it here. Let u and v be solutions defined on intervals I_u and I_v, and let I_u be properly contained in I_v. If $u(t) = v(t)$ for all t in I_u, we say that v is a *continuation* of u. A maximal solution is one which has no continuation. It is easy to see that a solution u defined on an interval of the form $|t - t_0| \leqq d$ is not maximal. Consider the right-hand endpoint $t_1 = t_0 + d$ and let $x_1 = u(t_1)$. By the result just obtained, there exists a positive number d_1 and a solution u_1 defined on the interval $|t - t_1| \leqq d_1$ which satisfies $u_1(t_1) = x_1$. Define the function v on the interval $[t_0 - d, t_1 + d_1]$ by $v(t) = u(t)$ if $|t - t_0| \leqq d$ and $v(t) = u_1(t)$ if $t_1 \leqq t \leqq t_1 + d_1$. Obviously v is a continuation of u "to the right." By considering the left-hand endpoint $t_0 - d$, we find in the same way that there is a continuation of u to the left. It follows that if a solution is defined on an interval with an endpoint which belongs to the interval, then the solution can be continued. Thus *the interval of definition of a maximal solution is open.*

Lemma 1.7.   *Let u and v be solutions of $\dot{x} = f(t,x)$, and suppose that for some t_0, $u(t_0) = v(t_0)$. Then $u(t) = v(t)$ for all t for which both u and v are defined.*

Proof.   We consider first the case that u and v are defined on open intervals I_u and I_v. The intersection of I_u and I_v is itself an open interval I. We must show that u and v agree on I. Suppose that for some $t < t_0$, $u(t) \neq v(t)$, and let t_1 be the least upper bound of all such numbers t. Then $t_1 < t_0$, since by Theorem 1.6 u and v agree on a sufficiently small interval about t_0. We shall show that neither $u(t_1) = v(t_1)$ nor $u(t_1) \neq v(t_1)$ is possible. If $u(t_1) = v(t_1)$, Theorem 1.6 tells us that u and v agree on some interval about t_1, whereas the least upper bound of a set has the property that every interval about it contains a member of the set. If $u(t_1) \neq v(t_1)$, it follows from the mere continuity of u and v that there is an interval about t_1 on which $u(t) \neq v(t)$. Hence there are numbers t between t_1 and t_0 for which $u(t) \neq v(t)$, contradicting the fact that t_1 is an upper bound for the numbers less than t_0 at which u and v disagree. We conclude that u and v agree at all points of I less than t_0. The assumption that $u(t) \neq v(t)$ for some number $t > t_0$ leads to a like contradiction when we consider the greatest lower bound of all such numbers. It follows that $u(t) = v(t)$ for all t in I.

If u and v are not defined on open intervals, they obviously have continuations which are. Since these continuations agree on the intersection of their intervals of definition, so do u and v.   □

We are now able to prove an existence and uniqueness theorem in terms of maximal solutions.

Theorem 1.8.   *For every (t_0,x_0) in Ω there is a maximal solution u such that $u(t_0) = x_0$. If v is any other solution such that $v(t_0) = x_0$, then u is a continuation of v.*

Proof.   Consider the family $\mathfrak{F}$ of all solutions through (t_0,x_0). The union of the intervals on which these solutions are defined is an interval I. By Lemma 1.7, all solutions of the family which are defined at a given t in I have the same value there, say $u(t)$. The function u thus defined on I is itself obviously a member of $\mathfrak{F}$. It follows from the definition of I that u is maximal, and it is obvious that u is a continuation of every other member of $\mathfrak{F}$.   □

This result may also be expressed by saying that (1) through each point of Ω there passes a maximal solution, and (2) the graphs of any two maximal solutions are either identical or disjoint.

Theorem 1.9.   *Let f and $\partial f/\partial x$ be continuous in the whole plane. If u is a maximal solution whose interval of definition has an endpoint α, then $|u(t)| \to \infty$ as $t \to \alpha$.*

Proof.   We shall treat explicitly only the case that α is a right endpoint, the proof for a left endpoint being essentially the same. The proof is by contradiction. The assertion of the theorem is that for every $L > 0$ there is an $\epsilon > 0$ such that $\alpha - \epsilon < t < \alpha$ implies $|u(t)| \geqq L$. Suppose the theorem is false. Then there exists an $L > 0$ such that for every $\epsilon > 0$ the interval $(\alpha - \epsilon, \alpha)$ contains at least one t for which $|u(t)| < L$. Choose a positive number δ, and let Q be the rectangle

$$|t - \alpha| \leqq \delta, \qquad |x| \leqq 2L.$$

Let M be a positive number such that $|f(t,x)| \leqq M$ for (t,x) in Q, and let K be a positive Lipschitz constant for f on Q. Let ϵ be the smallest of the numbers $\delta/2$, $1/K$, and L/M. Choose t_0 so that $\alpha - \epsilon < t_0 < \alpha$ and $|u(t_0)| < L$. Let $x_0 = u(t_0)$, and let Q' be the rectangle

$$|t - t_0| \leqq \frac{\delta}{2}, \qquad |x - x_0| \leqq L.$$

We show that Q' is contained in Q. Suppose that (t,x) belongs to Q'. Then

$$|t - \alpha| \leqq |t - t_0| + |t_0 - \alpha|$$
$$< \frac{\delta}{2} + \epsilon \leqq \frac{\delta}{2} + \frac{\delta}{2} = \delta.$$

Likewise

$$|x| \leqq |x - x_0| + |x_0| < L + L = 2L.$$

Hence (t,x) belongs to Q. It follows that M is a bound, and K a Lipschitz constant, for f on Q'. By Theorem 1.5, there is a solution v satisfying $v(t_0) = x_0$ defined on the interval $|t - t_0| \leqq d$, where d is any positive number less than ϵ. Since $\alpha - t_0 < \epsilon$, we may choose d so that $\alpha - t_0 < d < \epsilon$. Then v is defined for $t_0 \leqq t \leqq t_0 + d$, where $t_0 + d > t_0 + \alpha - t_0 = \alpha$. By uniqueness, $u(t) = v(t)$ for $t_0 \leqq t < \alpha$. Hence v continues u through α. This contradiction proves the theorem. $\square$

EXERCISES

1.   Show that if f has a zero Lipschitz constant on a rectangle $|t - t_0| \leqq a$, $|x - x_0| \leqq b$, the existence and uniqueness theorem follows trivially from the fundamental theorem of calculus.

2.   (a)   Let u and v be functions defined and continuous on a closed bounded interval I. Show that $\|u + v\| \leqq \|u\| + \|v\|$.

(b)   Let $u_1, u_2, \ldots, u_n$ be defined and continuous on I. Show that

$$\|u_1 + u_2 + \cdots + u_n\| \leqq \|u_1\| + \|u_2\| + \cdots + \|u_n\|$$

and that

$$\|u_1 - u_n\| \leqq \|u_1 - u_2\| + \|u_2 - u_3\| + \cdots + \|u_{n-1} - u_n\|.$$

3.   Theorem 1.5 can be used, at least in principle, to approximate a solution to any desired degree of accuracy in a neighborhood of t_0 by means of the functions u_n. As an example we take the differential equation $dx/dt = x$ and the solution u satisfying $u(0) = 1$. (Obviously $u(t) = e^t$.) Choose $u_0(t) \equiv 1$ and compute $u_n(t)$ for all n. Show that $\lim_{n \to \infty} u_n(t) = e^t$ for all t.

4.   Solutions which are not maximal arise almost invariably when we try to "solve" a differential equation by means of a power series. To illustrate the solution of a differential equation by a power series we take the equation $dx/dt = x^2$. Let u be the solution satisfying $u(0) = 1$ and assume that u can be represented by a power series in t, that is, $u(t) = \displaystyle\sum_{n=0}^{\infty} \frac{u^{(n)}(0)}{n!} t^n$. Our object is to determine the coefficients of this series, and we must therefore calculate the derivatives $u^{(n)}(0)$. We are given that $u(0) = 1$, and since $u'(t) = u^2(t)$ we have that $u'(0) = u^2(0) = 1$. Now $u''(t) = 2u(t)u'(t)$, so $u''(0) = 2$. Show that $u^{(n)}(0) = n!$ for all n, whence $u(t) = \displaystyle\sum_{n=0}^{\infty} t^n$. Note that $u(t) = 1/(1 - t)$ for $-1 < t < 1$, but that u is not defined outside this interval. Obviously u is not maximal.

2.   Systems

Preliminaries.   We now consider a system of N first-order equations

$$\dot{x}_i = f_i(t, x_1, x_2, \ldots, x_N), \qquad i = 1, 2, \ldots, N \qquad (2.1)$$

where f_i and $\partial f_i/\partial x_j$, $i, j = 1, \ldots, N$, are continuous. The theorems and proofs of the preceding section can be extended without difficulty to this general case. Whenever possible, we use vector notation. We shall adhere, but with some additions, to the notations and terminology of Chapter 5. By suitably generalizing the concepts of absolute value and norm as they entered into the proofs of the preceding section, we shall be able to give proofs for arbitrary N which are virtually identical with those for the case $N = 1$.

We shall denote the set of all real N-tuples by R^N, and write $x = (x_1, \ldots ,x_N)$, $(t,x) = (t,x_1, \ldots ,x_N)$. The function which assigns to a point (t,x) in R^{N+1} the point $(f_1(t,x), \ldots ,f_N(t,x))$ in R^N will be denoted by f, so that

$$f(t,x) = (f_1(t,x), \ldots ,f_N(t,x)).$$

The system (2.1) therefore becomes

$$\dot{x} = f(t,x). \tag{2.2}$$

A solution of (2.1) is an N-tuple of real-valued functions $u_1, \ldots , u_N$ defined on an interval I such that $\dot{u}_i(t) = f_i(t,u_1(t), \ldots ,u_N(t))$, $i = 1, \ldots , N$. These N equations can be condensed into the single vector equation $\dot{u}(t) = f(t,u(t))$, where u and $\dot{u}$ are the functions from I to R^N defined by $u(t) = (u_1(t), \ldots ,u_N(t))$, $\dot{u}(t) = (\dot{u}_1(t), \ldots ,\dot{u}_N(t))$.

Since subscripts will henceforth be needed to designate cordinates, we shall use superscripts to label vectors and vector-valued functions.

Let x be an N-vector. The number $|x_1| + |x_2| + \cdots + |x_N|$ will be called the absolute value of x, and denoted by $|x|$. If α is a number, obviously $|\alpha x| = |\alpha||x|$. We leave it to the reader to show that

$$|x^1 - x^n| \leqq |x^1 - x^2| + |x^2 - x^3| + \cdots + |x^{n-1} - x^n|$$

for any $x^1, x^2, \ldots , x^n$.

By definition, a function u is continuous if its coordinate functions $u_1, \ldots , u_N$ are continuous. Let u be defined and continuous on a closed bounded interval I. We define the norm of u to be the sum of the norms of its coordinate functions: $\|u\| = \|u_1\| + \cdots + \|u_N\|$. Obviously, $\|u\| = 0$ if and only if $u(t) \equiv 0$. Note that for each t, $|u_i(t)| \leqq \|u_i\|$, $i = 1, \ldots , N$. Adding these N inequalities yields $\sum_{i=1}^{N} |u_i(t)| \leqq \sum_{i=1}^{N} \|u_i\| = \|u\|$. The left side of this inequality is just $|u(t)|$. Hence

$$|u(t)| \leqq \|u\| \tag{2.3}$$

for all t in I. It is obvious that, given functions $u^1, \ldots , u^n$ defined and continuous on I,

$$\|u^1 - u^n\| \leqq \|u^1 - u^2\| + \|u^2 - u^3\| + \cdots + \|u^{n-1} - u^n\|. \tag{2.4}$$

Note, incidentally, that if $u(t)$ is continuous, so is $|u(t)|$.

Integrals, like derivatives, are defined coordinate-wise, as in Chapter 5. The reader should show (Exercise 1) that

$$\left| \int_{t_0}^{t} u(s) \, ds \right| \leqq \left| \int_{t_0}^{t} |u(s)| \, ds \right|. \tag{2.5}$$

With the definitions of absolute value and norm that we have adopted, it is not true that $|u(t)| \leqq c$ implies $\|u\| \leqq c$. However, if $|u(t)| \leqq c$, then $|u_i(t)| \leqq c$, and therefore $\|u_i\| \leqq c$, $i = 1, \ldots, N$. Adding these N inequalities yields $\|u\| \leqq Nc$. Thus:

$$|u(t)| \leqq c \; implies \; \|u\| \leqq Nc. \tag{2.6}$$

We say that a sequence of continuous functions (u^n) defined on a closed bounded interval I converges uniformly to a function u if $\lim_{n \to \infty} \|u^n - u\| = 0$. Evidently, (u^n) converges uniformly to u if and only if (u_i^n) converges uniformly to u_i, $i = 1, \ldots, N$. In other words, uniform convergence for a sequence of vector-valued functions means uniform convergence of each of the N coordinate sequences (u_1^n), $(u_2^n), \ldots, (u_N^n)$. It is easy to see that (u^n) converges uniformly if for every $\epsilon > 0$ there is an m such that $n > m$ implies that

$$\|u^n - u^{n+k}\| < \epsilon$$

for all k.

By a rectangle in R^{N+1} we shall mean a set of points (t,x) which satisfy inequalities of the form

$$|t - t_0| \leqq a, \qquad |x - x^0| \leqq b,$$

a and b being given positive numbers, and (t_0,x^0) a given point in R^{N+1}. Note that a rectangle is a closed and bounded set.

The Local Existence and Uniqueness Theorem.   As in the case of a single first-order equation, we begin by considering the system $\dot{x} = f(t,x)$ only on some rectangle about a given point (t_0,x^0). We shall prove that there is an interval about t_0 on which the system has a unique solution u satisfying $u(t_0) = x^0$.

Hypothesis.   We assume that f_i and $\dfrac{\partial f_i}{\partial x_j}$, $i, j = 1, \ldots, N$, are defined and continuous on a rectangle

$$Q: \quad |t - t_0| \leqq a, \qquad |x - x^0| \leqq b.$$

The theorems and proofs of the preceding section may now be adopted verbatim, or with only minor changes. We shall call attention to any modifications needed in the proofs, but otherwise refer the reader to the preceding section. We shall thus avoid tedious repetitions and emphasize the new features of the proofs.

Lemma 2.1. *u is a solution of $\dot{x} = f(t,x)$ satisfying the initial condition $u(t_0) = x^0$ if and only if*

$$u(t) = x^0 + \int_{t_0}^{t} f(s,u(s))\ ds.$$

Proof. The proof is identical to the proof of Lemma 1.1. Note that the fundamental theorem of calculus remains valid for vector-valued functions, since differentiation and integration are defined coordinate-wise. $\square$

Lemma 2.2. *There is a positive number K such that*

$$|f(t,\xi) - f(t,\eta)| \leq K|\xi - \eta| \tag{2.7}$$

for any two points (t,ξ), (t,η) in the rectangle Q.

Proof. We consider the coordinate functions $f_1, \ldots, f_N$ separately, applying the mean value theorem for functions of several variables to each of them in turn. This theorem follows easily from the one-variable case (Exercise 2). Let f_i be one of the coordinate functions. It follows from the mean value theorem that there is a point (t,θ^i) on the line segment joining (t,ξ) and (t,η) such that

$$f_i(t,\xi) - f_i(t,\eta) = \sum_{j=1}^{N} \frac{\partial f_i}{\partial x_j}(t,\theta^i)(\xi_j - \eta_j).$$

Since Q is closed and bounded, and since the partial derivatives are continuous on Q, they are bounded. Hence there are positive numbers $c_1^i, c_2^i, \ldots, c_N^i$ such that $\left|\dfrac{\partial f_i}{\partial x_j}(t,x)\right| \leq c_j^i$ for all (t,x) in Q. Thus

$$|f_i(t,\xi) - f_i(t,\eta)| \leq \sum_{j=1}^{N} c_j^i|\xi_j - \eta_j|.$$

There are N of these inequalities, one for each coordinate function. Adding them yields

$$|f(t,\xi) - f(t,\eta)| \leq \sum_{i=1}^{N}\sum_{j=1}^{N} c_j^i|\xi_j - \eta_j|.$$

Let c be the largest of the numbers c_j^i. Then

$$|f(t,\xi) - f(t,\eta)| \leq \sum_{i=1}^{N} \sum_{j=1}^{N} c|\xi_j - \eta_j|$$

$$= \sum_{i=1}^{N} c|\xi - \eta| = Nc|\xi - \eta|.$$

It only remains to set $K = Nc$, and the lemma is proved.   □

As before, any number K which satisfies (2.7) is called a Lipschitz constant for f with respect to x.

Theorem 2.3 (Uniqueness Theorem).   *Let d be a number which satisfies $0 < d < 1/NK$ where K is a positive Lipschitz constant for f. If u and v are solutions of $\dot{x} = f(t,x)$ defined on the interval I:*

$$|t - t_0| \leq d,$$

and if $u(t_0) = v(t_0)$, then $u = v$.

Proof.   The proof differs from that of Theorem 1.3 in only one step. As before, we find that

$$|u(t) - v(t)| \leq Kd\|u - v\|.$$

Using (2.6), we therefore conclude that

$$\|u - v\| \leq NKd\|u - v\|.$$

The assumption that $\|u - v\| > 0$ now leads to the contradiction that $d \geq 1/NK$. Hence $\|u - v\| = 0$, or $u = v$.   □

The proof of the next lemma is the same as that of Lemma 1.4. M is a positive number such that $|f(t,x)| \leq M$ for (t,x) in Q.

Lemma 2.4.   *Let d be a positive number such that $d \leq a$ and $d \leq b/M$. Let u be a continuous function defined on the interval I:*

$$|t - t_0| \leq d$$

and such that the graph of u lies in Q, i.e., $|u(t) - x^0| \leq b$. Then the function $\bar{u}$ defined on I by $\bar{u}(t) = x^0 + \int_{t_0}^{t} f(s,u(s))\, ds$ is continuous, and its graph lies in Q.

Theorem 2.5.   *Let d be a positive number such that $d \leq a$, $d \leq b/M$ and $d < 1/NK$. Let u^0 be a continuous function defined on the interval*

I: $|t - t_0| \leq d$ whose graph lies in Q. Then the sequence (u^n) defined on I by

$$u^{n+1}(t) = x^0 + \int_{t_0}^{t} f(s, u^n(s))\, ds, \qquad n = 0, 1, \ldots$$

converges uniformly to a function u, and

$$u(t) = x^0 + \int_{t_0}^{t} f(s, u(s))\, ds.$$

Proof.   Exactly as in Theorem 1.5, we obtain the inequality

$$|u^{n+1}(t) - u^n(t)| \leq Kd\|u^n - u^{n-1}\|.$$

Using (2.6), we thus have that

$$\|u^{n+1} - u^n\| \leq NKd\|u^n - u^{n-1}\|.$$

We set $c = NKd$ and note that $c < 1$. The remainder of the proof is identical with that of Theorem 1.5. $\square$

Theorem 2.6 (Local Existence and Uniqueness Theorem).
Let f_i and $\partial f_i / \partial x_j$, $i, j = 1, \ldots, N$, be continuous on a rectangle

$$|t - t_0| \leq a, \qquad |x - x^0| \leq b.$$

If d is a sufficiently small positive number, then the equation $\dot{x} = f(t,x)$ has a solution u which is defined on the interval $|t - t_0| \leq d$ and satisfies the initial condition $u(t_0) = x^0$, and no other solution defined on this interval satisfies the same initial condition.

This theorem follows directly from Theorems 2.3 and 2.5. As before, the differentiability hypotheses can be replaced by a Lipschitz condition.

To obtain the generalizations of Theorems 1.7–1.9, we suppose f_i and $\partial f_i / \partial x_j$ to be continuous on an open set Ω of R^{N+1}. The definition of a maximal solution remains unchanged, as does the proof that the interval of definition of a maximal solution is open. The theorems remain valid without change for arbitrary N, as do the proofs, except that in generalizing the proof of Theorem 1.9 the number $1/K$ must be replaced by $1/NK$.

Lemma 2.7.   *Let u and v be solutions of $\dot{x} = f(t,x)$, and suppose that for some t_0, $u(t_0) = v(t_0)$. Then $u(t) = v(t)$ for all t for which both u and v are defined.*

Lemma 2.8.   *For every (t_0, x^0) in Ω there is a maximal solution u such that $u(t_0) = x^0$. If v is any other solution such that $v(t_0) = x^0$, then u is a continuation of v.*

Theorem 2.9.   *Let f_i and $\partial f_j/\partial x_j$, $i, j = 1, \ldots, N$ be continuous in R^{N+1}. If u is a maximal solution whose interval of definition has an endpoint α, then $|u(t)| \to \infty$ as $t \to \alpha$.*

We conclude this discussion by noting that in the case $N = 1$, $|u(t)| \to \infty$ implies that either $u(t) \to \infty$ or $u(t) \to -\infty$. For if $|u(t)| \geq L > 0$ for $\alpha - \epsilon < t < \alpha$ (or $\alpha < t < \alpha + \epsilon$), $u(t)$ cannot assume both positive and negative values in this interval. If it did, then by the intermediate value theorem it would also have to vanish somewhere in the interval, contrary to the assumption that $|u(t)| \geq L$. If $N > 1$, the situation is quite different, because $|u(t)| \to \infty$ does not imply that $|u_i(t)| \to \infty$ for some i. As an example, consider the function $u(t) = (u_1(t), u_2(t))$, where

$$u_1(t) = \frac{1}{\sqrt{2 - 2t}} \cos \log \frac{1}{\sqrt{2 - 2t}}$$

$$u_2(t) = \frac{1}{\sqrt{2 - 2t}} \sin \log \frac{1}{\sqrt{2 - 2t}}.$$

u is defined only for $t < 1$, and $|u(t)| \geq 1/\sqrt{2 - 2t} \to \infty$ as $t \to 1$. Both $|u_1(t)|$ and $|u_2(t)|$ are unbounded, but neither tends to infinity as $t \to 1$. (We remark that u is a solution of $\dot{x}_1 = (x_1 - x_2)(x_1^2 + x_2^2)$, $\dot{x}_2 = (x_1 + x_2)(x_1^2 + x_2^2)$. The solutions of this system can be computed by changing to polar coordinates.)

EXERCISES

1.   Prove inequality (2.5).
2.   (a)   Let (t, ξ) and (t, η) be two points of a rectangle Q:

$$|t - t_0| \leq a, \qquad |x - x_0| \leq b.$$

Show that for any s in the interval $[0,1]$, the point $(t, (1 - s)\eta + s\xi)$ also belongs to Q. [The set of all such points is the line segment connecting (t, ξ) and (t, η).]

(b)   Let g and $\partial g/\partial x_j$, $j = 1, \ldots, N$, be real-valued continuous functions on the rectangle Q. Define the function h on the interval $[0,1]$ by $h(s) = g(t, (1 - s)\eta + s\xi)$. Show that

$$h'(s) = \sum_{j=1}^{n} \frac{\partial g}{\partial x_j}(t, (1 - s)\eta + s\xi)(\xi_j - \eta_j).$$

(c) Apply the mean value theorem to h to show the existence of a number s_0 for which $h(1) - h(0) = h'(s_0)$. Let $\theta = (1 - s_0)\eta + s_0\xi$ and conclude that

$$g(t,\xi) - g(t,\eta) = \sum_{j=1}^{N} \frac{\partial g}{\partial x_j}(t,\theta)(\xi_j - \eta_j).$$

3. Gronwall's Lemma; Linear Systems

Theorem 2.9 may be paraphrased by saying that a solution can be continued, either forwards or backwards, so long as it does not escape to infinity. We used this fact many times in Chapter 6, where the simple geometric situation made it appear obvious. We now want to use Theorem 2.9 to prove that all solutions of certain kinds of systems are defined for all t, and it is necessary to be clear about the reasoning involved. Suppose we are given (1) a real-valued function g continuous for all t; (2) a maximal solution u defined on an interval I; and (3) that $|u(t)| \leq g(t)$ for all t in I. We conclude that $I = R$. For if I had an endpoint α, then $|u(t)| \to \infty$ as $t \to \alpha$, and this implies that $g(t) \to \infty$ as $t \to \alpha$, contrary to the assumption that g is continuous at α. What is curious about this argument is that we use bounds on the values of a function to draw a conclusion about the domain of definition of the function. As a last step we may, of course, conclude that $|u(t)| \leq g(t)$ for all t.

The problem of finding a suitable g is solved in certain simple but important cases by a celebrated lemma due to Thomas H. Gronwall, which allows us to deduce bounds for the solutions of a system $\dot{x} = f(t,x)$ from bounds for f. The importance of Gronwall's lemma extends considerably beyond this particular application, however. The lemma can be used to prove the uniqueness theorem, the continuous dependence of solutions on parameters and on the initial conditions, as well as numerous results concerning boundedness and stability.

Lemma 3.1 (Gronwall's Lemma). *Let $\varphi(t)$, $\alpha(t)$ and $\beta(t)$ be continuous and non-negative in an interval I. Let $c \geq 0$, and let t_0 and t be numbers in I. If*

$$\varphi(t) \leq c + \left| \int_{t_0}^{t} (\alpha(s)\varphi(s) + \beta(s))\, ds \right| \tag{3.1}$$

then

$$\varphi(t) \leq \left[c + \left| \int_{t_0}^{t} \beta(s)\, ds \right| \right] e^{\left| \int_{t_0}^{t} \alpha(s)\, ds \right|}. \tag{3.2}$$

Proof.   We consider first the case that $c > 0$. It is also necessary to distinguish between the cases $t \geqq t_0$ and $t < t_0$. For the sake of brevity we write $\psi(t) = \alpha(t)\varphi(t) + \beta(t)$.

If $t \geqq t_0$, (3.1) becomes

$$\varphi(t) \leqq c + \int_{t_0}^{t} \psi(s) \, ds.$$

Since the right side of this inequality is positive, we may divide by it and obtain

$$\frac{\varphi(t)}{c + \int_{t_0}^{t} \psi(s) \, ds} \leqq 1.$$

Multiplying by $\alpha(t)$ yields

$$\frac{\alpha(t)\varphi(t)}{c + \int_{t_0}^{t} \psi(s) \, ds} \leqq \alpha(t),$$

or

$$\frac{\psi(t)}{c + \int_{t_0}^{t} \psi(s) \, ds} - \frac{\beta(t)}{c + \int_{t_0}^{t} \psi(s) \, ds} \leqq \alpha(t).$$

Replacing the denominator of the second fraction by $c + \int_{t_0}^{t} \beta(s) \, ds$ can only increase the fraction, and thus diminish the left side of the inequality. Hence

$$\frac{\psi(t)}{c + \int_{t_0}^{t} \psi(s) \, ds} - \frac{\beta(t)}{c + \int_{t_0}^{t} \beta(s) \, ds} \leqq \alpha(t).$$

Because $t \geqq t_0$, integrating from t_0 to t preserves the inequality. Integration yields

$$\log \frac{c + \int_{t_0}^{t} \psi(s) \, ds}{c + \int_{t_0}^{t} \beta(s) \, ds} \leqq \int_{t_0}^{t} \alpha(s) \, ds.$$

We take exponentials and multiply by $c + \int_{t_0}^{t} \beta(s) \, ds$ to obtain

$$c + \int_{t_0}^{t} \psi(s) \, ds \leqq \left[c + \int_{t_0}^{t} \beta(s) \, ds \right] e^{\int_{t_0}^{t} \alpha(s) \, ds}.$$

By hypothesis, the left side of this inequality is $\geqq \varphi(t)$. Hence

$$\varphi(t) \leqq \left[c + \int_{t_0}^{t} \beta(s) \, ds \right] e^{\int_{t_0}^{t} \alpha(s) \, ds},$$

which is (3.2) for the case $t \geqq t_0$.

If $t < t_0$, (3.1) becomes

$$\varphi(t) \leqq c - \int_{t_0}^{t} \psi(s)\, ds$$

and the proof proceeds as before, except that we must now integrate from t to t_0 in order to preserve the inequality. We find that

$$\varphi(t) \leqq \left[c - \int_{t_0}^{t} \beta(s)\, ds \right] e^{-\int_{t_0}^{t} \alpha(s)\, ds}.$$

This concludes the proof for the case $c > 0$.

Now suppose that

$$\varphi(t) \leqq \left| \int_{t_0}^{t} \psi(s)\, ds \right|.$$

Then (3.1) holds for every positive c, and therefore (3.2) does also. It follows that (3.2) holds for $c = 0$. $\square$

The case that $c = 0$ and $\beta(t) = 0$ is worthy of special note.

Corollary 3.2. *If $\varphi(t)$ and $\alpha(t)$ are continuous and non-negative, and if*

$$\varphi(t) \leqq \left| \int_{t_0}^{t} \alpha(s)\varphi(s)\, ds \right|,$$

then $\varphi(t) \equiv 0$.

This special case of Gronwall's lemma yields a very simple proof of the uniqueness theorem (Exercise 1).

Theorem 3.3. *Let f and $\partial f_i/\partial x_j$, $i, j = 1, \ldots, N$ be continuous for all (t,x) in R^{N+1}. Suppose that there exist continuous and non-negative real-valued functions $\alpha(t)$ and $\beta(t)$ such that $|f(t,x)| \leqq \alpha(t)|x| + \beta(t)$. Then every maximal solution of $\dot{x} = f(t,x)$ is defined for all t.*

Proof. Let u be a maximal solution satisfying $u(t_0) = x^0$ and having the interval of definition I. Then for t in I,

$$u(t) = x^0 + \int_{t_0}^{t} f(s,u(s))\, ds.$$

Hence

$$|u(t)| \leqq |x^0| + \left| \int_{t_0}^{t} |f(s,u(s))|\, ds \right|$$

$$\leqq |x^0| + \left| \int_{t_0}^{t} [\alpha(s)|u(s)| + \beta(s)]\, ds \right|.$$

By Gronwall's lemma,

$$|u(t)| \leqq \left[|x^0| + \left| \int_{t_0}^{t} \beta(s)\, ds \right| \right] e^{\left| \int_{t_0}^{t} \alpha(s)\, ds \right|}.$$

The right side of this inequality is continuous for all t. It follows that $I = R$. $\quad\square$

Theorem 3.4. *Consider the linear system $\dot{x} = A(t)x + b(t)$, where A and b are continuous for all t. For every (t_0, x^0) the system has a unique solution u defined for all t satisfying $u(t_0) = x^0$.*

Proof. Obviously the functions $f(t,x) = A(t)x + b(t)$ and

$$\frac{\partial f_i}{\partial x_j}(t,x) = a_{ij}(t)$$

are continuous for all (t,x), so that the system has a unique maximal solution u satisfying $u(t_0) = x^0$. Now

$$f_i(t,x) = \sum_{j=1}^{N} a_{ij}(t)x_j + b_i(t),$$

so

$$|f_i(t,x) \leq \sum_{j=1}^{N} |a_{ij}(t)||x_j| + |b_i(t)|.$$

Adding these N equalities yields

$$|f(t,x)| \leq \sum_{i,j=1}^{N} |a_{ij}(t)||x_j| + |b(t)|$$

$$\leq \left[\sum_{i,j=1}^{N} |a_{ij}(t)| \right] |x| + |b(t)|.$$

Hence $|f(t,x)| \leq \alpha(t)|x| + \beta(t)$, where $\alpha(t) = \sum_{i,j=1}^{N} |a_{ij}(t)|$ and $\beta(t) = |b(t)|$. It follows from Theorem 3.3 that u is defined for all t. $\quad\square$

Complex Linear Systems. So far we have considered only real systems and real solutions. We complete the discussion of existence theorems by showing that Theorem 3.4 remains valid without change for systems

$$\dot{x} = A(t)x + b(t) \tag{3.3}$$

when $a_{ij}(t)$ and $b_i(t)$ are complex-valued functions continuous for all t, and for complex initial vectors x^0. In fact, a complex system of order N is easily seen to be equivalent to a real system of order $2N$.

Let $A(t) = P(t) + iQ(t)$, $b(t) = r(t) + is(t)$, and $x = \xi + i\eta$. The system then becomes

$$\dot{\xi} + i\dot{\eta} = [P(t) + iQ(t)](\xi + i\eta) + r(t) + is(t),$$

or, equating real and imaginary parts,

$$\dot{\xi} = P(t)\xi - Q(t)\eta + r(t)$$
$$\dot{\eta} = Q(t)\xi + P(t)\eta + s(t).$$

This is a system of $2N$ real linear equations, which we can write in the form

$$\dot{y} = C(t)y + d(t) \tag{3.4}$$

by simply setting $y = (\xi,\eta)$, $d(t) = (r(t),s(t))$, and

$$C(t) = \begin{bmatrix} P(t) & -Q(t) \\ Q(t) & P(t) \end{bmatrix}.$$

Thus $u(t) = \varphi(t) + i\psi(t)$ is a solution of (3.3) if and only if $v(t) = (\varphi(t),\psi(t))$ is a solution of (3.4). Since the solutions of (3.4) are defined for all t, so are the solutions of (3.3), and since (3.4) has a unique solution satisfying the initial condition $v(t_0) = y^0 = (\xi^0,\eta^0)$, (3.3) has a unique solution satisfying the initial condition $u(t_0) = x^0 = \xi^0 + i\eta^0$. Thus Theorem 3.4 has been proved for the complex case also.

Throughout this chapter we have dealt with systems of first-order equations. Since every equation or system can be written in this form, the theorems we have obtained apply to any systems whatever. For the case of a single Nth-order linear equation, we obtain the following special case of Theorem 3.4.

Corollary 3.5.   *If $a_1(t)$, ... , $a_N(t)$ and $b(t)$ are continuous for all t, the equation*

$$x^{(N)} + a_1(t)x^{(N-1)} + \cdots + a_N(t)x = b(t)$$

has a unique solution defined for all t satisfying a given initial condition $u^{(k)}(t_0) = x_k$, $k = 0, 1, . . . , N - 1$.

The proof follows by writing the equation as a system of first-order equations.

EXERCISES

1. Use Corollary 3.2 to prove the following uniqueness theorem. Let $f(t,x)$ be continuous and satisfy a Lipschitz condition (with respect

to x) on a rectangle Q, let u and v be solutions of $\dot{x} = f(t,x)$ defined on a common interval and having graphs in Q, and let $u(t_0) = v(t_0)$ for some t_0. Then $u = v$.

2.   The uniqueness theorem is a special case of a result concerning the continuous dependence of solutions on the initial values. Let f satisfy the conditions of Exercise 1, let (t_0,x^0) and (t_0,y^0) be points of Q, let u and v be solutions of $\dot{x} = f(t,x)$ defined on a common interval I and having graphs in Q, and let $u(t_0) = x^0$, $v(t_0) = y^0$. Show that

$$|u(t) - v(t)| \leq |x^0 - y^0|e^{K|t-t_0|}$$

for t in I, where K is a Lipschitz constant. (The uniqueness theorem follows by setting $x^0 = y^0$.)

3.   The result obtained in Exercise 2 is incomplete, because it involves a hypothesis about the interval of definition of both solutions. Prove: If u is a solution of $\dot{x} = f(t,x)$ defined for $a \leq t \leq b$ then there exists a positive number ϵ such that any solution v which satisfies $|u(t_0) - v(t_0)| < \epsilon$ for some t_0 in $[a,b]$ can also be defined for $a \leq t \leq b$. Assume that f is continuous and satisfies a Lipschitz condition on every rectangle.

4.   A system with parameters can always be written in a form which involves no parameters. Consider the system

$$\dot{x} = f(t,x,\mu) \tag{1}$$

where $x = (x_1, \ldots ,x_n)$ and $\mu = (\mu_1, \ldots ,\mu_m)$. Let $y_i = x_i$ for $1 \leq i \leq n$ and let $y_{n+i} = \mu_i$ for $1 \leq i \leq m$, so $y = (x,\mu)$. Similarly, let $g_i(t,y) = f_i(t,y)$ for $1 \leq i \leq n$ and let $g_{n+i}(t,y) = 0$ for $1 \leq i \leq m$. Then (1) is equivalent to

$$\dot{y} = g(t,y), \tag{2}$$

but contains no parameters. The continuous dependence of the solutions of (1) on the parameters is now seen to reduce to the continuous dependence of the solutions of (2) on the initial conditions. We suppose for simplicity that f and $\partial f_i/\partial x_j$ are continuous for all (t,x,μ), $i, j = 1, \ldots , n$. Let $u(t,x^0,\eta)$ be the solution of (1) whose value at $t = t_0$ is x^0. Use Exercises 2 and 3 to show that u is continuous.

5.   Suppose that: (1) for every (t_0,x^0) the system $\dot{x} = f(t,x)$ has a solution satisfying $u(t_0) = x^0$; (2) all solutions have a common interval of definition; and (3) any linear combination of solutions is itself a solution. Show that there is a matrix $A(t)$ such that $f(t,x) = A(t)x$. [Hint: Generalize the proof of Theorem 3.2, page 40, by considering the solutions satisfying $u^j(t) = e^j$, $j = 1, \ldots , n$.]

ANSWERS TO SELECTED EXERCISES

Chapter 1, Section 2, page 9

1. 1, NA, L, NH

2. 4, A, L, H

3. 2, NA, L, H

4. 2, A, NL

5. 1, A, NL

6. 2, NA, NL

7. 4, NA, L, NH

8. 2, A, NL

Chapter 1, Section 3, page 12

5. (a) $u(t) \equiv 1$

   (b) $u(t) \equiv c$ for any c

   (c) $u(t) \equiv 1$ or -1

   (d) none

   (e) $u(t) \equiv 1$ or -2

Chapter 1, Section 4, page 14

1. $$\frac{dx}{dt} = r_1 c - r_2 \frac{x}{v}, \quad \frac{dv}{dt} = r_1 - r_2$$

2. $$\frac{dx_1}{dt} = -\frac{r}{v_1} x_1, \quad \frac{dx_2}{dt} = \frac{r}{v_1} x_1 - \frac{r}{v_2} x_2$$

3. $$\frac{d^2 x}{dt^2} = g - \frac{\pi r^2}{m} x$$

4. $$\frac{d^2 x}{dt^2} = g - \frac{\pi r^2}{m} x - \frac{\pi r^2}{m} \sin \omega t$$

5. $$\frac{d^2 x}{dt^2} = \frac{r^2}{R^2 - r^2} \left(g - \frac{\pi R^2}{m} x \right)$$

6. $$\frac{d^2 x}{dt^2} = g - \frac{k}{m} \left(1 - \frac{s}{\sqrt{s^2 + x^2}} \right) x$$

Chapter 1, Miscellaneous Exercises, page 24

1. For example, $\dfrac{dx}{dt} = (1 - \mu) f(t,x) + \mu g(t,x)$

2. If $\mu = 0$ or 1, there is one constant solution, $x \equiv \mu$. If $\mu > 1$, or $\mu < 0$, there are two constant solutions, $x \equiv \mu \pm \sqrt{\mu^2 - \mu}$. If $0 < \mu < 1$, there is no constant solution.

3. (a) $x \equiv 0,\ y \equiv 0;\ x \equiv -1,\ y \equiv 1$
   (b) $x \equiv 0,\ y \equiv 0$
4. (a) The constant solutions are $x \equiv c,\ y \equiv c$, where c is arbitrary if $\mu = 1$, and $c = 0$ if $\mu \neq 1$.
   (b) $x \equiv \dfrac{\mu^3}{1 - \mu^3},\ y \equiv \dfrac{\mu}{1 - \mu^3}$; there is no constant solution if $\mu = 1$.

5. $v_i = \displaystyle\sum_{j=1}^{n} a_{ij} u_j$. Because each term on the right is differentiable, v_i is also. Differentiate:

$$v_i' = \sum_{j=1}^{n} a_{ij} u_j' = \sum_{j=1}^{n} a_{ij} v_j.$$

6. (a) $x \equiv c,\ c$ arbitrary
   (b) $x \equiv 0$
   (c) $x \equiv n\pi,\ n$ an arbitrary integer
7. $f(x,0)$ never vanishes [if we had $f(c,0) = 0$, then $u(t) \equiv c$ would be a constant solution] and, because it is continuous, it is of one sign for all x. If $f(x,0) > 0$, then no solution has a relative maximum [because $u'(t_0) = 0$ implies $u''(t_0) = f(u(t_0),u'(t_0)) = f(u(t_0),0) > 0$] and if $f(x,0) < 0$, then no solution has a relative minimum. But a continuous periodic function has both.

Chapter 2, Section 1, page 31

1. (a) $u(t) = \pi + \arctan t$
   (b) $u(t) = \arctan (t - 1)$
   (c) same as (b)
   (d) $u(t) \equiv \dfrac{\pi}{2}$

2. (a) $x \equiv 0$ and $x = \dfrac{1}{c - t};\qquad u(t) = \dfrac{1}{1 - t},\ t < 1$
   (b) $x = 1 - ce^{-t};\qquad u(t) = 1 - e^{-t}$, all t
   (c) $x = ce^{-\frac{1}{2}t^2};\qquad u(t) \equiv 0$
   (d) $x = \arcsin (ce^{\frac{1}{2}t^2})$, where $|c| < 1$
       $u(t) = \arcsin (\tfrac{1}{2}e^{\frac{1}{2}t^2}),\ |t| < \sqrt{\log 4}$
   (e) $x = ce^{\sin t};\qquad u(t) = e^{\sin t}$, all t
   (f) $x = \log (e^t + c);$
       $u(t) = \log (e^t + 1 - e),\quad t > \log (e - 1)$

(g) $x \equiv 0$ and $x = \dfrac{1}{1 - ce^t}$;

$$u(t) = \frac{1}{1 - \frac{1}{2}e^t}, \quad t < \log 2$$

(h) $x \equiv 0$ and $x = \dfrac{1}{\pm \sqrt{c - 2t}}$;

$$u(t) = \frac{-1}{\sqrt{11 - 2t}}, \quad t < \frac{11}{2}$$

(i) $x = n\pi + 2 \arctan (ce^{t^2})$; $u(t) = 2 \arctan e^{t^2}$, all t
(j) $x = \log (t + c)$; $u(t) = \log (t + 1), \, t > -1$

4.   (a) $x = ce^t - t - 1$
   (b) $x = ce^{-2t} + \frac{1}{2}t - \frac{1}{4}$
   (c) $x = t - (n + \frac{1}{2})\pi$ and $x = t - n\pi - \arctan (t + c)$
   (d) $x = \tan (t + c) - t$

Chapter 2, Section 2, page 37

2.   (a) $u(t) = e^t$, all t

(b) $u(t) = \dfrac{1}{1 - t}, \, t < 1$

(c) $u(t) = \tan t, \, -\dfrac{\pi}{2} < t < \dfrac{\pi}{2}$

(d) $u(t) = -\log (1 - \frac{1}{2}t^2), \, -\sqrt{2} < t < \sqrt{2}$

3.   This is the case if and only if some (and hence every) indefinite integral of f on I assumes all real values. It is true for $f(t) = t^2$ but not for $f(t) = t$, to take two simple examples with $I = R$.

4.   No. F is not an indefinite integral of $1/x$ since it is not defined on an interval.

5.   (a)   Separation of variables produces the constant solution $x \equiv 0$ and the half-parabolas $x = (c + \frac{1}{2}t)^2, \, c + \frac{1}{2}t > 0$, and $x = -(c - \frac{1}{2}t)^2, \, c - \frac{1}{2}t > 0$. The maximal continuations of the solution $-\frac{1}{4}t^2 (t < 0)$ are given by the function $u(t) = -\frac{1}{4}t^2$ for $t < 0$, $u(t) = 0$ for $t \geq 0$, and the functions u defined for any $c \leq 0$ by $u(t) = -\frac{1}{4}t^2$ for $t < 0$, $u(t) = 0$ for $0 \leq t \leq -2c$, $u(t) = (c + \frac{1}{2}t)^2$ for $t > -2c$. The same procedure applies to the other half-parabolas.

(b)   Separation of variables produces the constant solution $x \equiv 0$ and the curves $x = \pm (t + c)^{3/2}, \, t + c > 0$. Each of the latter has a unique maximal continuation via the constant solution.

Chapter 2, Section 3, page 42

3. (a) $x = \frac{1}{2}(\sin t - \cos t) + ce^t$
    (b) $x = \frac{1}{2}t^3 + ct$

    (c) $x = \frac{1}{2}t \log t - \frac{1}{4}t + c\,\dfrac{1}{t}$

    (d) $x = t \log t - 1 + ct$
    (e) $x = (1 + t)(t - \log(1 + t) + c)$
    (f) $x = e^{-t} + ce^{-2t}$
    (g) $x = te^{-2t} + ce^{-2t}$

7. (a) only $x \equiv 0$
    (b) all
    (c) only $x = -\frac{1}{2}(\sin t + \cos t)$
    (d) none
    (e) only $x \equiv 0$

14. (a) $x = ce^{-\cos t}$;   (b) $u_M - u_m = |u(0)|(e^2 - 1)$
15. (a) The periodic solutions are those for which $u(0) < \frac{1}{2}$.
    (b) All solutions are periodic.
    (c) No solution is periodic.

Chapter 2, Section 4, page 47

2. (b) $\dfrac{\pi}{2|a|}$

4. (a) $\lim\limits_{a \to 0} u(t,a) = u(t,0)$

    (b) The limit does not exist.

    (c) $\lim\limits_{a \to 0} u(t,a) = 1 - \dfrac{1}{(t + 1)^2}$, $t < -1$. This is a solution of
    the equation for $a = 0$.

Chapter 2, Section 5, page 54

3. (a) $u(t) = c_0 + (x_0 - c_0 + \frac{1}{2})e^{-t} + \frac{1}{2}(\sin t - \cos t)$
    (b) $x_0 = c_0 - \frac{1}{2}$
    (d) The concentration in the vessel attains its maxima $\pi/4$
    seconds later than in the inflow. The maximum value of the

    former is $c_0 + \dfrac{1}{\sqrt{2}}$, that of the latter $c_0 + 1$.

4. (b) $v = (r - R)t + v_0$

    $$m = m_0\left[\left(\frac{r - R}{v_0}\right)t + 1\right]^{\frac{R}{R-r}}$$

    $$t > \frac{v_0}{R - r}$$

(c)   same formulas, $t < \dfrac{v_0}{R - r}$

5.   (b) $u(t) = c + (x_0 - c)e^{-t}$
      $v(t) = c + (x_0 - c)te^{-t} + (y_0 - c)e^{-t}$

   (d) $t = \dfrac{x_0 - y_0}{x_0}$

6.   (a) $u(t) = x_0 e^{-\lambda t}$
      $v(t) = (y_0 + \lambda x_0 t)e^{-\lambda t}$

Chapter 2, Section 6, page 63

5.   (a) $x = n\pi$ is asymptotically stable if n is odd and unstable if n is even.

   (b) $x = n\pi$ is unstable for all n.

   (c) $x = 0$ is unstable.

   (d) $x = 0$ is weakly stable; $x = \dfrac{1}{n\pi}$ is asymptotically stable if n is positive and even or negative and odd, unstable if n is positive and odd or negative and even.

   (e) $x = c < 0$ is weakly stable; $x = 0$ is unstable; $x = 1$ is asymptotically stable.

Chapter 3, Section 1, page 77

6.   (a) $1, -\dfrac{1}{2} \pm \dfrac{\sqrt{3}}{2} i$

   (b) $\pm 1, \pm i$

   (c) $\pm 1, \pm \dfrac{1}{2} \pm \dfrac{\sqrt{3}}{2} i$

7. $-1, \dfrac{1}{2} \pm \dfrac{\sqrt{3}}{2} i$

8. $\pm \dfrac{1}{\sqrt{2}} (1 + i)$

10.   (a) ce^{it}

   (b) $(c + t)e^{it}$

   (c) $(c + it)e^{t}$

Chapter 3, Section 2, page 82

3.   (a)   dependent
   (b)   independent
   (c)   independent

    (d) dependent

    (e) dependent

Chapter 3, Section 3, page 89

1. (a) If $a_s \neq 0$, let $f(t) = t^s$; then $Lf \neq 0$.

   (b) This follows from the corresponding fact about polynomials, since by (a) all the coefficients of $L_1 L_2$ vanish.

4. (a) $c_1 e^t + c_2 e^{-2t}$

   (b) $c_1 e^{-t} + c_2 t e^{-t}$

   (c) $c_1 e^{it} + c_2 t e^{it}$

   (d) $c_1 + c_2 e^{-t}$

   (e) $c_1 + c_2 t$

   (f) $c_1 e^t + c_2 e^{-t}$

   (g) $c_1 e^{(1+i)t} + c_2 e^{-(1+i)t}$

   (h) $c_1 e^{-2t} + c_2 e^{-t}$

   (i) $c_1 e^t + c_2 t e^t$

5. (a) $c_1 \cos t + c_2 \sin t$

   (b) $(c_1 \cos t + c_2 \sin t)e^{-t}$

   (c) $\left(c_1 \cos \dfrac{\sqrt{3}}{2} t + c_2 \sin \dfrac{\sqrt{3}}{2} t \right) e^{-\frac{1}{2}t}$

   (d) $(c_1 \cos \sqrt{2}\, t + c_2 \sin \sqrt{2}\, t)e^{-t}$

7. (a) $c_1 e^{-t} + c_2 e^t - \frac{1}{2} t e^{-t}$

   (b) $c_1 \cos 2t + c_2 \sin 2t + \frac{1}{3} e^{it}$

   (c) $c_1 e^{2t} + c_2 e^t + \frac{1}{10} \cos t - \frac{3}{10} \sin t$

   (d) $(c_1 + c_2 t)e^t + t + 2$

11. (a) $c_1 e^{-t} \int e^{\frac{3}{2}t - \frac{1}{4}\sin 2t}\, dt + c_2 e^{-t}$

   (b) $c_1 e^{-t} \int e^{\frac{1}{2}t^2 + t}\, dt + c_2 e^{-t}$

   (c) $c_1(t - 1) + c_2 e^{-t}$

Chapter 3, Section 4, page 96

2. (a) $c_1 e^t + c_2 e^{-2t} + (-\frac{1}{2}t^2 + \frac{1}{2}t - \frac{5}{4})e^{-t}$

   (b) $(c_1 + c_2 t)e^{-t} + \frac{1}{2}t^2 e^{-t} + \frac{1}{6}t^3 e^{-t}$

   (c) $c_1 e^{-t} + c_2 e^{-2t} + (-\frac{1}{3}t^3 - t^2 - 2t)e^{-2t}$

   (d) $c_1 + c_2 e^{-t} + \frac{1}{2}t^2 - t$

   (e) $c_1 \cos t + c_2 \sin t - \frac{1}{3} \sin 2t$

   (f) $\left(c_1 \cos \dfrac{\sqrt{3}}{2} t + c_2 \sin \dfrac{\sqrt{3}}{2} t \right) e^{\frac{1}{2}t} - \cos t - t \sin t - 2t \sin t$

   (g) $c_1 e^t + c_2 e^{-t} - (\frac{1}{5} \sin t + \frac{2}{5} \cos t)e^t$

Chapter 3, Section 5, page 108

1.   (b) $\bar{w} = \sqrt{\omega_0^2 - 2k^2} \to \omega_0$ as $k \to 0$

2.   (a) $\frac{1}{2}e^{-t} + \frac{1}{2}te^{-t} - \frac{1}{2}\cos t$

     (b) $u(0) = -\frac{1}{2},\ u'(0) = 0,\ \phi = \dfrac{3\pi}{2}$

4.   (a) $p = 2;\ u(t) = \dfrac{1}{1 - \pi^2}\sin \pi t$, period 2

     (b) $p = 2\pi;\ u(t) = c_1 \cos t + c_2 \sin t + \frac{4}{3}\sin 2t + \frac{9}{8}\sin 3t$, period 2π

     (c) $p = 2\pi$; no periodic solutions

     (d) $p = \pi;\ u(t) = c_1 \cos t + c_2 \sin t + 1 + \frac{1}{3}\cos 2t$, period 2π; $u(t) = 1 + \frac{1}{3}\cos 2t$, period π

Chapter 3, Section 6, page 118

1.   (a) $c_1 + (c_2 + c_3 t)e^{-t}$

     (b) $(c_1 + c_2 t + c_3 t^2)e^{-t}$

     (c) $c_1 e^t + \left(c_2 \cos \dfrac{\sqrt{3}}{2}t + c_3 \sin \dfrac{\sqrt{3}}{2}t\right)e^{-\frac{1}{2}t}$

     (d) $c_1 e^t + c_2 e^{-t} + c_3 \cos t + c_4 \sin t$

     (e) $c_1 e^t + (c_2 + c_3 t)\cos t + (c_4 + c_5 t)\sin t$

     (f) $(c_1 + c_2 t + c_3 t^2 + c_4 t^3 + c_5 t^4)e^{-t}$

2.   (a) $c_1 e^t + \left(c_2 \cos \dfrac{\sqrt{3}}{2}t + c_3 \sin \dfrac{\sqrt{3}}{2}t\right)e^{-\frac{1}{2}t} - \frac{1}{2}e^{-t}$

     (b) $c_1 e^{-t} + \left(c_2 \cos \dfrac{\sqrt{3}}{2}t + c_3 \sin \dfrac{\sqrt{3}}{2}t\right)e^{\frac{1}{2}t} + \frac{1}{2}\cos t + \frac{1}{2}\sin t$

     (c) $c_1 e^{-t} + \left(c_2 \cos \dfrac{\sqrt{3}}{2}t + c_3 \sin \dfrac{\sqrt{3}}{2}t\right)e^{\frac{1}{2}t} + \frac{1}{3}te^{-t}$

     (d) $c_1 e^t + c_2 e^{-t} + c_3 \cos t + c_4 \sin t + \frac{1}{4}t \cos t$

Chapter 4, Section 5, page 137

1.   (a) $\lambda_1 = i,\ \lambda_2 = -i;\ x^1 = (i,1),\ x^2 = (1,i)$

     (b) $\lambda = 1$ is a double eigenvalue; $x = (1,0)$

     (c) $\lambda = 1$ is a triple eigenvalue; $(1,0,0)$ and $(0,1,-1)$ are linearly independent eigenvectors belonging to 1

     (d) $\lambda_1 = 0,\ \lambda_2 = 2;\ x^1 = (1,i),\ x^2 = (i,1)$

2. (a) $P = \begin{bmatrix} i & 1 \\ 1 & i \end{bmatrix}$

(b) $P = \begin{bmatrix} 1 & 1 \\ -1 & 2 \end{bmatrix}$

(c) $P = \begin{bmatrix} 1 & 1 \\ -1 & 1 \end{bmatrix}$

(d) $P = \begin{bmatrix} 1 & 1 & 0 & 0 \\ i & -i & 0 & 0 \\ 0 & 0 & 1 & 1 \\ 0 & 0 & i & -i \end{bmatrix}$

(e) $P = \begin{bmatrix} 1 & 1 & 1 \\ 0 & 1 & 0 \\ 0 & 0 & 2 \end{bmatrix}$

Chapter 5, Section 1, page 141

2. Given a real number t_0 and two n-vectors η and μ, there exists a unique solution $x = u(t)$ defined for all t such that $u(t_0) = \eta$ and $u'(t_0) = \mu$.

Chapter 5, Section 5, page 155

1. (a) $\frac{1}{4} \begin{bmatrix} 3e^{2t} + e^{-2t} & e^{2t} - e^{-2t} \\ 3e^{2t} - 3e^{-2t} & e^{2t} + 3e^{-2t} \end{bmatrix}$

(b) $u_1(t) = \frac{1}{4}e^{2t} + \frac{1}{12}e^{-2t} - \frac{1}{3}e^{t}$
$u_2(t) = \frac{1}{4}e^{2t} - \frac{1}{4}e^{-2t}$

2. (a) $x_1 = c_1 e^{3t} - c_2 e^{-t} + \frac{1}{2}t e^{3t} - \frac{1}{2}e^{t} + \frac{1}{8}e^{3t}$
$x_2 = c_1 e^{3t} + c_2 e^{-t} + \frac{1}{2}t e^{3t} - \frac{1}{8}e^{3t}$

(b) $x_1 = c_1 + c_2 e^{2t}$
$x_2 = -c_1 + c_2 e^{2t}$

(c) $x_1 = c_1 e^{t} + c_2 e^{2t} + c_3 e^{-t}$
$x_2 = c_1 e^{t} - 3c_3 e^{-t}$
$x_3 = c_1 e^{t} + c_2 e^{2t} - 5c_3 e^{-t}$

(d) $x_1 = c_1 + i c_2 e^{2t}$
$x_2 = i c_1 + c_2 e^{2t}$

(e) $x_1 = c_1 e^{t} + c_2 e^{5t}, \quad x_2 = -c_1 e^{t} + 3c_2 e^{5t}$

(f) $x_1 = c_1 e^{t} + c_2 e^{5t} - \frac{1}{4}t e^{t} - \frac{1}{16}e^{t}$
$x_2 = -c_1 e^{t} + 3c_2 e^{5t} + \frac{1}{4}t e^{t} - \frac{3}{16}e^{t}$

(g) $x_1 = c_1 e^{t} + c_2 i e^{it} - c_3 i e^{-it}$
$x_2 = c_1 e^{t} - c_2 e^{it} - c_3 e^{-it}$
$x_3 = c_1 e^{t} + c_2 e^{it} + c_3 e^{-it}$

(h) $x_1 = 3c_1 e^{2t} - c_2 e^{-2t}$
$x_2 = c_1 e^{2t} + c_2 e^{-2t}$

3. $\begin{bmatrix} e^{\alpha t} \cos \beta t & e^{\alpha t} \sin \beta t \\ -e^{\alpha t} \sin \beta t & e^{\alpha t} \cos \beta t \end{bmatrix}$

4. (a) $x_1 = c_1 e^{-t} + c_2 e^{10t}$

   $x_2 = -c_1 e^{-t} + 10 c_2 e^{10t}$

   (b) $u_1(t) = e^{-t}, \quad u_2(t) = -e^{-t}$

Chapter 6, Section 2, page 201

6. (a) saddle; $m = \pm \sqrt{3}$

   (b) saddle; $m = \frac{1}{2} \pm \dfrac{\sqrt{5}}{2}$

   (c) saddle; $m = \pm 1$

   (d) center

   (e) unstable two-tangent node; $m = 1, x \equiv 0$

   (f) stable focus

   (g) stable focus

   (h) stable one-tangent node; $m = 1$

   (i) center

   (j) unstable two-tangent node; $m = -1, 3$

   (k) unstable focus

7. (a) unstable focus

   (b) stable focus

   (c) center

   (d) saddle

   (e) stable one-tangent node

   (f) unstable two-tangent node

Chapter 6, Section 3, page 217

6. (a) (1,0) is a saddle

   (b) (0,0) is a stable focus

   (c) (0,0) is an unstable (two-tangent) node

   (d) (1,1) is an unstable (two-tangent) node; $(-1,0)$ is a saddle

   (e) (0,1) is a saddle; (1,2) is an unstable (stellar) node

   (f) $(n\pi, n\pi)$ is an unstable focus if n is even, and a saddle if n is odd

   (g) (0,0) is a non-elementary singular point; it is obviously unstable

   (h) (0,0) is a center (by Theorem 3.6)

   (i) $(-1,0)$ is a saddle; (1,0) is a center (by Theorem 3.6)

    (j) $(n\pi,1)$ is an unstable (two-tangent) node if n is even, and a saddle if n is odd

    (k) $(1,0)$ and $(0,1)$ are saddles; $(1,1)$ is a stable, and $(0,0)$ an unstable (stellar) node

7.   Unstable focus if $0 < \mu < 2$, unstable node if $\mu \geqq 2$

Chapter 6, Section 4, page 227

7.   (a) $(0,0)$ is a center

     (b) $(0,0)$ is called a monkey saddle: two depressions for the legs and one for the tail

     (c) $(\tfrac{1}{4},\tfrac{1}{2})$ is a center; $(0,0)$ and $(0,1)$ are saddles

     (d) $(0,0)$ is unstable

     (e) $(0,0)$ is a center; $(\tfrac{1}{2},1)$ is a saddle

     (h) $(0,0)$ is a saddle

8.   (a) $\dfrac{f_x + g_y}{g}$ is independent of x and continuous in y;

$$M = e^{-\int \frac{f_x+g_y}{g}\,dy}$$

     (b) $H = e^{-y}(\cos x - xy)$

9.   (a) $\dfrac{f_x + g_y}{f + g}$ depends only on $s = x + y$ and is continuous;

$$M = e^{-\int \frac{f_x+g_y}{f+g}\,ds}$$

10.   (a) $M = \dfrac{1}{|xy|}$ in each quadrant

     (b) $M = \dfrac{1}{|x|}$ in the left and right half-planes

     (c) $M = e^{-x}$ in the whole plane

Chapter 6, Section 5, page 240

5. $p = \sqrt{2} \displaystyle\int_{-\alpha}^{\alpha} \frac{dx}{\sqrt{\cos x - \cos \alpha}}$

Chapter 6, Miscellaneous Exercises, page 241

2. $(0,0)$ is an unstable focus, $\left(\dfrac{1}{2}, \dfrac{\sqrt{35}}{2}\right)$ is an unstable node, $\left(\dfrac{1}{2}, -\dfrac{\sqrt{35}}{2}\right)$ is a saddle point; $r = 1$ is a stable limit cycle.

SUGGESTIONS FOR FURTHER READING

Bieberbach, L., *Einführung in die Theorie der Differentialgleichungen im reellen Gebiet*, Springer-Verlag, Berlin, 1956.

Cesari, L., *Asymptotic Behavior and Stability Problems in Ordinary Differential Equations*, 2nd ed., Springer-Verlag, Berlin, 1963.

Coddington, E. A., and N. Levinson, *Theory of Ordinary Differential Equations*, McGraw-Hill, New York, 1955.

Hurewicz, W., *Lectures on Ordinary Differential Equations*, M.I.T., Cambridge, Mass., 1958.

Lefschetz, S., *Differential Equations: Geometric Theory*, 2nd ed., Wiley, New York, 1963.

Minorsky, N., *Non-Linear Oscillations*, Van Nostrand-Reinhold, Princeton, N.J., 1962.

Nemytskii, V. V., and V. V. Stepanov, *Qualitative Theory of Differential Equations*, Princeton University Press, Princeton, N.J., 1960.

Pontryagin, L. S., *Ordinary Differential Equations*, Addison-Wesley, Reading, Mass., 1962.

Sansone, G., and R. Conti, *Non-Linear. Differential Equations*, Pergamon, New York, 1964.

Stoker, J. J., *Nonlinear Vibrations in Mechanical and Electrical Systems*, Interscience-Wiley, New York, 1950. ·

INDEX